AF559861

TEXTBOOK OF MODERN PHYSICS

TEXTBOOK OF MODERN PHYSICS

By

Dr. Ajit Kumar Sharma

Deptt. of Physics

D.A.V. (P.G.) College

Bulandshahr

(U.P.)

DISCOVERY PUBLISHING HOUSE PVT. LTD.

NEW DELHI-110 002

Published by:
DISCOVERY PUBLISHING HOUSE PVT. LTD.
4383/4B, Ansari Road, Darya Ganj
New Delhi-110 002 (India)
Phone: +91-11-23279245; 23253475; 43596065
+91 9811179893 / +91 9871656464
E-mail: discoverybooksindia@gmail.com
orderdphbooks@gmail.com
namitwasan9@gmail.com
web: www.discoverypublishinggroup.com

First Edition: **2011**
Reprinted: **2023**

ISBN: 978-81-8356-844-9

Text book of Modern Physics

Printed at:
Infinity Imaging Systems
Delhi

Preface

The book "Text Book of Modern Physics" has been written to meet the requirements of graduate students of all Indian Universities. The subject matter of this book has been presented in this book has complete theory and large number of solved examples. We have been selected sufficient problems from various Universities examination papers.

I have spared no pains in applying my long experience of degree class teaching to make the book useful to the students. We have tried to our best to keep the book free from the misprint. The author shall be grateful to the readers who point out errors and omissions which inspite of all care might have been there.

The author in general hope. That the present book will be warmly received by the students and teachers. We shall indeed be very thankful to our colleague for their recommendations this book of for their students.

---*Author*

Contents

1

Series Relationship in Atomic Spectra

INTRODUCTION

A quantitative experimental study of these spectral lines was made in the second half of the 19th century when several regularities were observed in the spacings of the lines. The atom spectra consist of a large number of lines. For example, in 1870. Lveing and Dewar noticed that the spectral lines of various elements could be grouped into distinct 'series'. *In each series the spacing and intensity of lines decrease regularly towards shorter wavelengths,* until it becomes impossible to distinguish the individual lines. The point at which the lines of the series finally converge is called the 'series limit'. The various series in complicated spectra overlap. When a beam of white light is passed through a prism (or grating) it breaks up into the beams of constituent colours. The different coloured beams, when focussed on a screen by a converging lens, from an array of colours on the screen. This is called the 'spectrum' of white light.

The are two main types of spectra :

(i) 'emission spectra' and

(ii) 'absorption spectra'.

On the other hand, when light from a source showing a continuous emission spectrum is passed through an absorbing material and then into the spectroscope, and 'absorption spectrum' of the material is observed. Emission and absorption spectra are further classified according to their appearance. When light coming directly from a source is allowed to enter a prism or grating spectroscope, and 'emission spectrum' of the source is observed in the spectroscope.

Band Spectra

Such spectra are obtained by the radiation from gas molecules such as oxygen (O_2), nitrogen (N_2), cyanogen (CN), etc. They consist of illuminated regions, called 'bands', separated by dark spaces. With a high-resolving instrument each band is seen to consist of very fine lines which become closer and closer on one side of the band until they coincide. This side thus has a sharp and bright edge called the 'head' of the band.

Continuous Spectra

When the emitting source is an incandescent solid, or liquid, such as a lamp filament or a gas at high pressure, the spectrum is continuous containing all colours (wavelengths) from red to violet. Its appearance is like an unbroken luminous band of light in which the colour changes gradually from point to point but without any sharp boundary. In this spectrum, the intensity is maximum at a certain point and decreases on both sides of it. The point of maximum intensity shifts towards the violet end of the spectrum as the temperature of the source increases.

Line Spectra

If the light source is a low-pressure gas (as in a discharge tube), flame, arc or spark, the spectrum is discontinuous, showing a number of sharp bright coloured lines. These lines are the images of the slit of the spectroscope formed by lights of different colours. The entire series of images is called a 'line spectrum'. The different lines differ in intensity and nature. Some are sharp, some are sharp on one side and diffused on the other, and others are diffused on both sides.

The line spectrum is the characteristic of the atom or the ion. It means that a particular atom or ion always gives the characteristic set of spectral lines, and no two atoms or ions can give the same spectral line. For example, sodium atom gives two intense yellow lines called D_1 and D_2 lines.

Absorption Spectra

The absorption spectra are obtained when the absorbing substance is placed between a source emitting a continuous spectrum and the slit of the spectroscope. In such cases, certain colours (wavelengths) are absorbed by the substance. Hence the spectrum is found to consist of dark lines or bands against a bright background. An example of such spectra is sun's spectrum. It is a line absorption spectrum. It was studied in detail by Fraunhoffer offer who named the absorption lines as A, B, C, D... These lines inform us about the elements which are present around the sun.

Similarly, when an intense beam of continuous white light is passed through sodium vapour and then sent into a spectroscope, we obtain two dark lines on a continuous background in the same positions as the yellow D_1 and D_2 lines in the sodium, emission spectrum. If the sodium vapour is replaced by Iodine vapour (I2), an absorption band spectrum of I2 molecule is obtained.

BOHR'S THEORY OF HYDROGEN SPECTRUM

The existence of sharp spectral lines cannot be explained from the classical electro-magnetic theory. Bohr explained it by applying Planck's quantum hypothesis to the Rutherford's atomic model. The Rutherfod's atom consists of a central massive nucleus containing the positive charge of the atom, and the electrons move round the nucleus in circular planetary orbits. The centripetal force required for the orbital motion is provided by the electrostatic attraction between the positively-charge nucleus and the negatively-charged electron.

Bohr proposed two postulates.

(i) An electron can move only in those orbits for which the angular momentum L of the electron is an integral multiple of $h/2\pi$ where h is Planck's constant. (Thus, Bohr quantised toe angular momentum of the electron.). The electron moving in any of the permitted orbits does not radiated energy in spite of its acceleration towards the centre of the orbit. The atom, therefore, is said to exist in a *stationary* state.

(ii) The emission (or absorption) of radiation by the atom takes place when an electron jumps for one permitted orbit to another. The radiation is emitted (or absorbed) as a single quantum (photon) whose energy hv is equal to the difference in energies of the electron in the two orbits involved. Thus, if E_i be the energy of the initial orbit of the electron and Z_f that of the final orbit then we have

$$h\nu = E_i - E_f,$$

where ν is the frequency of the emitted (or absorbed) radiation.

Let e, m and v be the charge, mass and velocity of the electron (measured in column, kg and meter/sec respectively) and r the radius of the orbit measured in meter. The positive charge on the nucleus is Z e, where Z is the atomic number Fig. 1. In case of hydrogen Z = 1, so that positive charge on the nucleus is e. As the centripetal force is provided by the electrostatic attraction, we have

$$\frac{mv^2}{r} = \frac{1}{4\pi\varepsilon_0}\frac{e^2}{r^2}$$

or $$m\,v^2 = \frac{e^2}{4\pi\varepsilon_0 r}. \qquad ...(1)$$

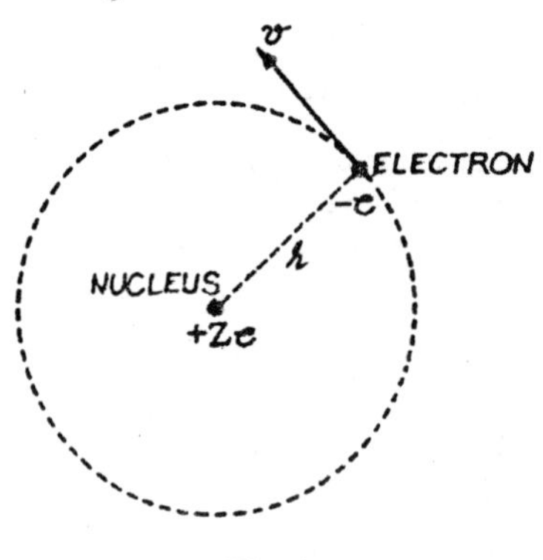

Fig. 1

From the first postulate, the angular momentum of the electron is given by

$$L = m\,v\,r = n\frac{h}{2\pi}, \qquad ...(2)$$

where n is called as 'quantum number' having values 1, 2, 3, ...

Squaring eq. (2) and dividing by eq. (1) we get

$$r = n^2\frac{h^2\varepsilon_0}{\pi m e^2},\ n = 1, 2, 3, ... \qquad ...(3)$$

This is the expression for the radius of the permitted orbits.

Now, the energy E of the electron in an orbit is the sum of kinetic and potential energies. The kinetic energy of the electron is

$$K = \frac{1}{2}mv^2 = \frac{e^2}{8\pi\varepsilon_0 r}. \qquad \text{[from eq. (i)]}$$

The potential energy at a distance r to infinity against the electrostatic attraction $\left(-\frac{e^2}{4\pi\varepsilon_0 r^2}\right)$, and is given by

$$U = \int_r^{\infty} -\frac{e^2}{4\pi\varepsilon_0 r^2}\,dr = \frac{1}{4\pi\varepsilon_0}\left[\frac{e^2}{r}\right]_r^{\infty} = -\frac{e^2}{4\pi\varepsilon_0 r}.$$

Hence the total energy of the electron is

$$E = K + U = \frac{e^2}{8\pi\varepsilon_0 r} - \frac{e^2}{4\pi\varepsilon_0 r} = -\frac{e^2}{8\pi\varepsilon_0 r}.$$

Substituting for r from eq. (3), we get

$$E = -\frac{me^4}{8\varepsilon_0^2h^2}\left(\frac{1}{n^2}\right) \quad n = 1, 2, 3... \qquad ...(4)$$

This is the expression for the energy of the electron in the n th orbit. We see that it in negative.

Let E_i and E_f be the energies of the electron corresponding to the initial (higher) and final (lower) orbits of the excited atom. Then, we have

$$E_i = \frac{me^4}{8\varepsilon_0^2h^2}\left(\frac{1}{n_i^2}\right)$$

and

$$E_f = -\frac{me^4}{8\varepsilon_0^2h^2}\left(\frac{1}{n_f^2}\right),$$

where n_i and n_f are the corresponding quantum numbers. The energy difference between these states is

$$E_i - E_f = \frac{me^4}{8\varepsilon_0^2h^2}\left(\frac{1}{n_f^2} - \frac{1}{n_i^2}\right).$$

Hence, from Bohr's second postulate, the frequency v of the emitted photon is

$$v = \frac{E_i - E_f}{h}.$$

$$= \frac{me^4}{8\varepsilon_0^2h^2}\left(\frac{1}{n_f^2} - \frac{1}{n_i^2}\right)$$

The corresponding wavelength l is given by

$$\frac{1}{\lambda} = \frac{v}{c} = \frac{me^4}{8\varepsilon_0^2 ch^3}\left(\frac{1}{n_f^2} - \frac{1}{n_i^2}\right).$$

This equation indicates that, *Since ni and nf can take only integral values, the radiation emitted by excited hydrogen atoms should contain certain discarte wavelengths only.*

The value of the constant *term* $\frac{me^4}{8\varepsilon_0^2 ch^3}$ comes out to be the same as the Rydberg constant R in the Balmer's empirical formula. Thus we have

$$\frac{1}{\lambda} = R\left(\frac{1}{n_f^2} - \frac{1}{n_i^2}\right).$$

EMISSION OF SPECTRUM

When the hydrogen atom gets sufficient energy from outside by some means, the electron from an inner orbit of lower energy goes up to an outer orbit of higher energy. This excited state of the atom lasts for jumping down, it emits the difference in energy between the two orbits as electromagentic radiation. If the electron jumps from an orbit ni to an orbit nf, the wavelength of the emitted radiation will be

$$\frac{1}{\lambda} = R\left(\frac{1}{n_f^2} - \frac{1}{n_i^2}\right).$$

It is found that for

$n_f = 1$,	$n_i = 2, 3, 4, \ldots$	we obtain	Lyman series,
$n_f = 2$,	$n_i = 3, 4, 5, \ldots$	we obtain	Balmer series,
$n_f = 3$,	$n_i = 4, 5, 6$	we obtain	Paschen series,
$n_f = 4$,	$n_i = 5, 6, 7, \ldots$	we obtain	Brackett series,
$n_f = 5$,	$n_i = 6, 7, 8, \ldots$	we obtain	Pfund series,

The corresponding energy level diagram is shown in Fig. 2. The top horizontal line represents zero energy, that is, energy of the electron outside the atom ($n = \infty$). The other horizontal lines represent energies of different orbits given by the formula

$$E = -\frac{me^4}{8\varepsilon_0^2 h^2}\left(\frac{1}{n^2}\right).$$

The arrows ending at the lines n = 1, 2, 3, 4, and 5 represent the transitions responsible for the Lyman, Balmer, Paschen, Brackett and Pfund series respectively.

Shortcomings of Bohr's Theory

Bohr's theory although very successful l in explaining the hydrogen spectrum and giving valuable information about atomic structure, has the following shortcomings.

(i) An individual line of hydrogen spectrum, when examined under a high resolving spectroscope, is found to be accompanied by a number of theory as such. It can, however, be explained when the relativistic variation in the mass of the electron and the electron 'spin' are taken into account.

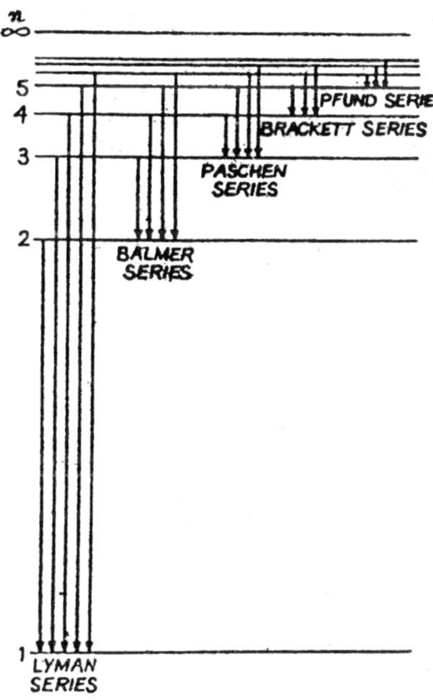

Fig. 2

(ii) Bohr's theory cannot explain the variation in intensity of the spectral lines of an element. The intensity can be explained by quantum mechanics.

(iii) The theory is only applicable to one-electron atoms such as hydrogen isotopes, singly-ionized helium, doubly-ionised lithium, etc. It does not explain the spectra of complex atoms.

(iv) The success of Bohr's theory in explaining the effect of magnetic field on spectral lines is only partial. The theory cannot explain the 'anomalous' Zeeman effect.

(v) The theory does not satisfactorily explain the distribution of electrons in atoms.

No Balmer Lines in Absorption Spectrum of Hydrogen

The Bohr theory also explains the absorption line spectrum of hydrogen. When a beam of - continuous light (containing all wavelengths) is passed through hydrogen and then sent into a spectrograph, a set of dark lines is obtained. In terms of quantum theory the incident light is a beam of quanta (photons) of all sorts of energies. Now according to Bohr theory the hydrogen atoms absorb only those quanta whose energies correspond to transitions between its discrete energy levels.

The resulting excited hydrogen atoms re-radiate the absorbed energy almost atones but these photons come off in random directions with only a few in the same direction as the original beam of continuous light. The dark lines in the absorption spectrum are therefore never completely black. Obviously the absorption lines will have exactly the same frequencies as the emission lines.

Now it is found that all the emission lines of hydrogen spectrum do not appear in the absorption spectrum. The reason is that normally the atom is always in the ground state $n = 1$. Therefore, absorption transitions can only occur from $n = 1$ to $n > 1$. Hence lines of only the Lyman series can appear in absorption spectrum.

To obtain Balmer series in absorption, the atom must initially be in the state $n = 2$, because Balmer lines require transitions from $n = 2$ to $n > 2$. Since atoms are usually in the ground state. Balmer lines are not obtained in absorption.

The simplest atomic spectrum is that of hydrogen. Its visible part consists of a single series which was first observed by Balmer in 1885. This is called the 'Balmer series' of hydrogen. Its first line having longest wavelength of 6563 Å (Fig. 3), is named Ha, the next Hb, and so on, the series limit reaching at 3646 Å (Fig. 3) Besides this, there is a series of lines in the ultraviolet part of the hydrogen spectrum which is known as 'Lyman series, and three series in the infra-red part which are known as 'Paschan series', 'brackett series' and 'Pfund series'.

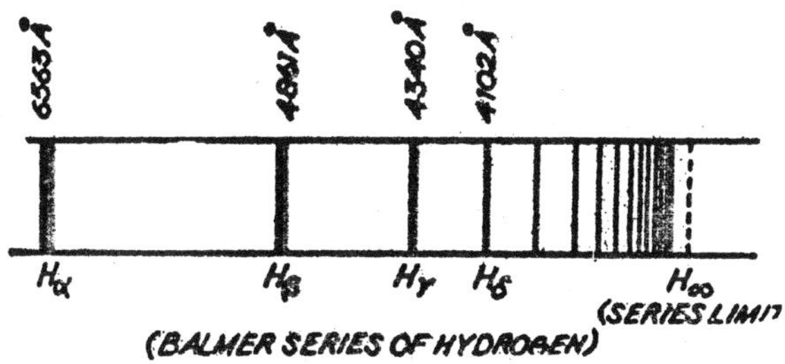

Fig. 3

Balmer discovered a formula for the wavelengths of all the lines of the Balmer series. His formula is

$$\lambda = 3646 \frac{n^2}{n^2 - 4} \text{ Å}, \; n = 3, 4, 5, \ldots$$

n = 3 gives the wavelength of H_α line, n = 4 gives H_β line, ... and n = ¥ gives the series limit.

Subsequently, Rydberg found that Balmer's formula was a special case of a more general formula which is as follows.

$$\bar{v} = \frac{1}{\lambda} = R_H \left(\frac{1}{m^2} - \frac{1}{n^2} \right),$$

where $\bar{v}$ is wave number (reciprocal of wavelength), m and n are positive integers (n > m) and RH is the Rydberg constant for hydrogen. Its value is 1.097 × 107 m^{-1}. To obtain formula for Lyman series we set m = 1 and n = 2, 3, 4, ..., for Balmer series m = 2 and m = 3, 4, 5, ... and so on. Thus.

$$\bar{v} = R_H \left(\frac{1}{1^2} - \frac{1}{n^2} \right), \; n = 2, 3, 4, \ldots \qquad \text{(Lyman)}$$

$$\bar{v} = R_H \left(\frac{1}{2^2} - \frac{1}{n^2} \right), \; n = 3, 4, 5, \ldots \qquad \text{(Balmer)}$$

$$\bar{v} = R_H \left(\frac{1}{3^2} - \frac{1}{n^2} \right), \; n = 4, 5, 6, \ldots \qquad \text{(Paschen)}$$

$$\bar{v} = R_H \left(\frac{1}{4^2} - \frac{1}{n^2} \right), \; n = 5, 6, 7, \ldots \qquad \text{(Brackett)}$$

$$\bar{v} = R_H \left(\frac{1}{5^2} - \frac{1}{n^2} \right), \; n = 6, 7, 8 \ldots \qquad \text{(Pfund)}$$

We see that the wave numbers of hydrogen lines can be expressed as differences of two terms of the form $\frac{R_H}{n^2}$.

The next simplest spectra are of the 'monovalent' atoms of alkali metals Li, Na, K, etc. The lines in the spectrum of an alkali atom can be grouped unto four distinct series, a principal series' of intense lines a 'sharp series' of fine lines, a diffuse series' of comparatively broader lines and a 'fundamental series' which lies in the infra-red region.

The sharp and diffuse series line in the visible part and converge to a common limit. Rydberg represented the lines of a particular series by the formula.

$$\bar{v} = Z^2 R_A \left[\frac{1}{(m-\Delta_1)} - \frac{1}{(n-\Delta_2)^2}\right],$$

where R_A is the Rydberg constant for a particular element A, Z is the atomic number, m and n are positive integers, and Δ_1 and Δ_2 are constants for the particular series. Actually each line of an alkali spectrum is a close doublet. Again, we find that the wave numbers of the lines can be expressed as differences of two terms like $\frac{Z^2 R_A}{(n-\Delta)^2}$.

After alkali spectra, next in complexity are the spectra of 'divalent' atoms of alkaline-earths Be, Mg, Ca. In a typical alkaline-earth spectrum, we can distinguish two distinct systems of lines, a system of singlets and a system of distinguish two distinct systems of lines, a system of singlets and a system of triplets. Each system has a principal series, a sharp series, a diffuse series and a fundamental series.

As we proceed to atoms having several valence electrons, the spectra becomes more complex and the groupings of lines into series becomes less pronounced. Still regularities can be observed in complex spectra, and it is possible to express the wave number of any spectral line as the difference of two terms.

Rydberg-Ritz Combination Principle

The principle states *the wave numbers of spectral lines can be expressed by the difference of spectroscopic terms in such a way that other difference of those terms give also the wave numbers of lines in the same spectrum.* Suppose in a spectrum the wave numbers of two lines are given as

$$\bar{v}_a = T_2 - T_3 \text{ and } \bar{v}_d = T_1 - T_4,$$

Then lines of the following wave numbers are also expected in the same spectrum,

$$\bar{v}_b = T_2 - T4 \text{ and } \bar{v}_c = T_1 - T_3.$$

This means that constant differences exist between the wave numbers.

$$\bar{v}_b - \bar{v}_a = \bar{v}_d - \bar{v}_c$$

$$\bar{v}_c - \bar{v}_a = \bar{v}_d - \bar{v}_b.$$

Displacement Law

According to this law *the spectrum of any neutral atom of atomic number Z closely resembles the spectrum of the singly ionised atom of atomic number Z + 1.* For example, the spectrum of H^+ (Z = 2) closely resembles the spectrum of H (Z = 1) and can be represented by similar formula,

$$\bar{v} = 4\,R_{He}\left[\frac{1}{m^2} - \frac{1}{n^2}\right].$$

Other elements deprived of all but one electron, also produce hydrogen-like spectra which an be represented by the general formula

$$\bar{v} = Z^2 R_A\left[\frac{1}{m^2} - \frac{1}{n^2}\right].$$

In a similar way, the spectrum of a singly ionised alkaline-earth atom resembles the spectrum of an alkali atom. From this we conclude that it is the number of valence electrons in an atom which determines the qualitative character of the spectrum of that atom.

Hydrogen Spectrum

The spectrum of hydrogen atom consists of a number of lines. These lines have been grouped into a number of 'series'. The lines in each series are such that *their separation and intensity decrease regularly towards shorter wavelengths,* converging to a limit called the 'series'limit. The wavelengths in each series can be given by a simple empirical formula. The first such spectral series was observed by balmer in 1885 and is called the Balmer series of hydrogen. The first line with the longest wavelength (6563 Å) is named Ha. The next Hb, and so on. The series limit lies at 3646 Å, beyond which is a faint continuous spectrum. Balmer's formula for the wavelengths of the series is

$$\frac{1}{\lambda} = R\left(\frac{1}{2^2} - \frac{1}{n^2}\right), \; n = 3, 4, 5, \ldots \qquad \text{(Balmer)}$$

The quantity R is called the 'Rydberg constant' and has the value

$$R = 1.097 \times 10^7 \text{ meter}^{-1}.$$

The Ha line corresponds to n = 3, the H_β line to n = 4, and so on. The series limit experiment. The Balmer series contains only those spectral lines which fall in the visible part of the hydrogen spectrum. The lines falling in the ultraviolet and infrared parts form other series. The lines in the ultraviolet form the Lyman series whose wavelengths are give by

$$\frac{1}{\lambda} = R\left(\frac{1}{1^2} - \frac{1}{n^2}\right), \; n = 2, 3, 4, \ldots \qquad \text{(Lyman)}$$

In the infrared, three spectral series have been observed whose lines have the wavelengths given by the formulas

$$\frac{1}{\lambda} = R\left(\frac{1}{3^2} - \frac{1}{n^2}\right), \; n = 3, 5, 6, \ldots \qquad \text{(Paschen)}$$

$$\frac{1}{\lambda} = R\left(\frac{1}{4^2} - \frac{1}{n^2}\right), \; n = 5, 6, 7, \ldots \qquad \text{(Brackett)}$$

$$\frac{1}{\lambda} = R\left(\frac{1}{5^2} - \frac{1}{n^2}\right), \; n = 6, 7, 8, \ldots \qquad \text{(Pfund)}$$

DIFFERENT SPECTRAL LINES OF HYDROGEN

Every atom when excited emits radiations. The radiations form a line spectrum which is the characteristic of the emitter. Each atom has its own particular line spectrum which is regarded as the characteristic of that element to which the atom belongs. Like other elements, hydrogen possesses its own characteristic line spectrum. Hydrogen spectrum consists of a number of lines. These have been grouped into five series which are named after their discoverers, Fig 4. Many attempts were made by various workers to find a rule for underlying relationship which governed the wave lengths of these lines. We will discuss these by one.

Fig. 4

Calculation of Rydberg's Constant

The energy of the hydrogen atom when the electron is in the n_1th orbit.

$$En_1 = -\frac{me^4}{8\varepsilon_0{}^2h^2}\cdot\frac{1}{n_1{}^2}Z^2$$

and atom the energy of the atom when the electron is in the n_2th orbit

$$En_2 = -\frac{me^4}{8\varepsilon_0{}^2h^2}\cdot\frac{1}{n_2{}^2}Z^2$$

and the frequency of the photon emitted, when an electron jumps from n_2th to n_2th orbit is given by Bohr's third postulate, *i.e.*,

$$h\nu = En_1 - En_2 = -\frac{me^4}{8\varepsilon_0{}^2h^2}\left(\frac{1}{n_1{}^2}-\frac{1}{n_2{}^2}\right)Z^3$$

or

$$\nu = \frac{me^4}{8\varepsilon_0{}^2h^2}\left(\frac{1}{n_2{}^2}-\frac{1}{n_1{}^2}\right)Z^2.$$

Since the velocity of light $c = \nu\lambda$, we have

$$\frac{\nu}{c} = \frac{1}{\lambda} = \frac{me^4}{8\varepsilon_0{}^2ch^2}\left(\frac{1}{n_2{}^2}-\frac{1}{n_1{}^2}\right)Z^2 \qquad ...(1)$$

whcre

$$R = \frac{me^4}{8\varepsilon_0{}^2ch^2}.$$

On substituting the values, we get $R = 109737.302\ cm^{-1}$. Here ε_0 is a constant whose numerical value is equal to $8.854. \times 10^{-12}$. This value is found to be in agreement with the value obtained from the spectroscopic data of the Balmer's series which is $109677.67 cm^{-1}$.

Success of Bohr's Theory : This theory explains the following facts about atomic spectra.

FAILURES OF BOHR'S THEORY

(i) It failed to explain *Stark effect.* It is similar to Zeeman effect and is produced in the presence of external electrostatic field (Fig. 5).

(ii) It failed to explain Zeeman effect. When a substance emitting a line space trum is placed in a magnetic field, its lines would split up into a number of closely spaced lines. This is known as *Zeeman effect* (Fig. 5)

(iii) When spectral lines of hydrogen are observed very closely, each line is further made up of much closely spaced lines.

(iv) It failed to explain spectra of atoms other than hydrogen.

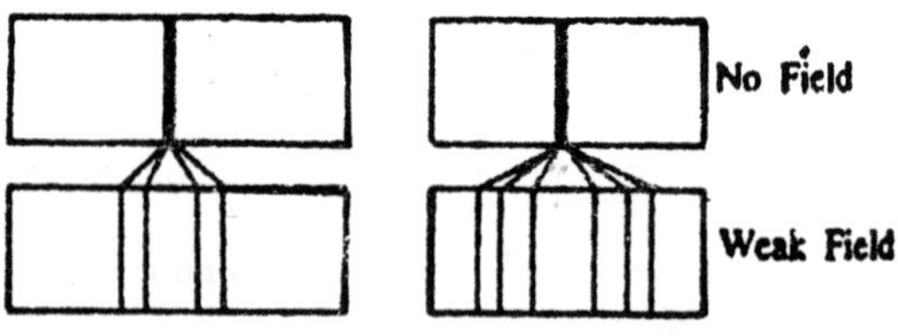

Fig. 5

(v) Another fundamental objection against Bohr's theory is that is used two theories which are opposed to each other, *i.e.*, *quantum theory* was used to account for the existence of stationary orbits and for frequencies of radiations emitted while motion of electron in its orbit obeyed *the law of classical mechanics.*

(vi) In the light of the Heisenberg's uncertainty principle, both the velocity and the position of the electron cannot be specified at a given time as Bohr did. Thus, it is another weakness of Bohr's theory.

(vii) Another weakness of Bohr's theory is that it did not throw light on the distribution and arrangement of electrons in atoms.

(viii) Bohr assumed that the nucleus is stationary and only electrons are revolving around it. Detailed facts have revealed that both the nucleus and the electrons move in closed orbits around their centre of mass. Therefore, it is major weakness of Bohr's theory.

LIGHT PHASE

Evidence in support of light phase. Hill and *Scrisbrick* found that isolated chloroplasts when illuminated produced oxygen from water. If certain hydrogen acceptors (oxidants) were present in the medium, they were reduced by the chloroplasts. The oxidants are called *Hill reagents* and the reaction as *Hill reaction.* Hill attributed the production of oxygen and hydrogen from water by light. *Arnonin* 1954, working on isolated chloroplasts reported that in addition to carrying out the Hill reaction, the chloroplasts could also synthesis ATP in the light from ADP and inorganic phosphate (Pt). This *photosynthetic phosphorylation* may be represented as.

$$ADP + P_t \xrightarrow{\text{Light energy}} ATP.$$

Reaction occurring in Light phase.

When chlorophyll molecule absorbs one quantum of light, it gets excited. The excited chlorophyll molecule brings about the following changes.

(i) Excited chlorophyll molecule interacts chemically with water to liberate oxygen.

$$4H_2O + \text{chlorophyll}^* \rightarrow 4[H] + 4[OH] + \text{chlorophyll}$$

$$4[OH] \rightarrow 2H_2O + O_2.$$

The nascent hydrogen reacts with NADP to form $NADPH_2$ which acts as a reducing agent.

$$NADP + 2H \rightarrow NADPH_2$$

(ii) Excited chlorophyll also brings about the reaction between ADP and inorganic phosphate Pt to form an energetic compound ATP.

$$ADP + P_i + \text{chlorophyll}^* \rightarrow ATP + \text{chlorophyll}.$$

The products obtained from light reaction enter the dark reaction of photosynthesis.

Mechanism of light phase reactions. When sun light falls on the leaves, the chlorophyll present in the leaves absorbs light photon and becomes excited. This photo excitation results in the displacement of an electron from the normal orbit (of chlorophyll) into a anew orbit. The hole left by the displaced electron is soon filled up by the return of the same electron or another one. If the same electron is returning the process is termed as *cyclic-transfer of electron.* However, if the different electron is returning, it is termed as non-cyclic transfer of electrons.

$$\text{Chlorophyll} + h\nu \rightarrow \text{Chlorophyll}^*$$

$$\text{Chlorophyll}^* \rightarrow (\text{chlorophyll})^* + e^-$$

Non-cyclic electron transfer. It involves the following steps.

(i) First of alkyl water dissociates into hydrogen and hydroxyl ions.

$$4H_2O \rightarrow 4H^+ + 4OH^-$$

(ii) When a quantum of short wavelength of light is received by a pigment II of chlorophyll molecule, it loses an electron. The loss of electrons is immediately compensated by the electrons of 4 OH^- ions, *i.e.,* converted into water and oxygen.

$$4\,OH^- \rightarrow 2H_3O + O_2 + 4e^-$$

(iii) The electron released from above process is raised from + 0.8 eV to zero potential at which it is captured by an electron carrier called *Plastoquinone.*

(iv) After this the electron travels downhill and falls back to + 0.4 eV in a dark reaction through a series of carriers (cytochrome b_6, cyt. f, and plastocyanine) to pigment system I. The energy released in the dowahill passage of electron from cytochrome b_6 to cytochrome f is utilised to convert ADP and inorganic phosphate into ATP.

$$ADP + P_i \rightarrow ATP$$

(v) The electron is pumped from + 0.04 to – 0.4 eV with energy of another quantum absorbed by the system. I. According to *Tagawa* and *Arnon* (1962) the electron from system I is captured by any iron-containing protein, called *ferredoxin* with a potential of – 4.32 eV (Fig. 2.31).

(vi) According to *Aron* (1967) the electron is finally passed on at – 0.32 e.v. to NAD which together with two H+ released from water, becomes reduced to NHDPH + H^+ (expressed as $NADPH_2$ for convenience).

$$NADP^+ + 2e^+ + 2H^+ \rightarrow NADPH + H^+$$

Energy from ATP helps to more the electron from $NADPP_2$ to phosphoglyceric acid into carbon cycle.

Cyclic electron transfer. The expelled electron from the chlorophyll molecule is first accepted by *ferredoxin* or by an other electron acceptor of the chloroplast.

The electron then traverses via cytochrome b_6 and cytochrome f, the two native cytochromes of chloroplasts, and ultimately reaches the original chlorophyll molecule from which it has been expelled. This scheme of electron transport is called *cyclic photophosphorylation.* In the cyclic photophosphorylation the electron, that is returned to the chlorophyll, is the same as the one that was expelled. In cyclic photophosphorylation energy is released during the transfer of electrons to cytochromes and the released energy is utilized to synthesize ATP.

Dark Phase

The present knowledge on the path of carbon in photosynthesis comes mainly from the work of *Melvin Calvin* and his associates,. They allowed the algae supplied with carbonlabelled CO_2 to photosynthesize for very brief periods of time and killed them instantaneously with boiling alcohol. Such studies using tracer technique revealed that products of CO_2 assimilation even for periods less than a minute were mainly sugars and amino acids shown in Fig. 6.

Calvin's group was interested to know the first stable product of photosynthesis. For this they allowed the algae to photosynthesis for 5 second and extracted the compounds after killing the organisms. The first stable product of very brief period of photosynthesis was found to be 3 carbon compound phosphoglyceric acid (PGA). Since PGA is a 3 carbon compound, it was thought that CO_2 is accepted by a 2-carbon molecule to form 3 carbon PGA. But no such compound was detectable in plants.

Later, *Benson* working in the same laboratory found that a 5-carbon sugar Ribulose diphosphate (RuDP) is the acceptor molecule of CO_2. On accepting CO_2 it was established that an unstable carbon intermediate is found and it cleaves into 2 molecules of PGA.

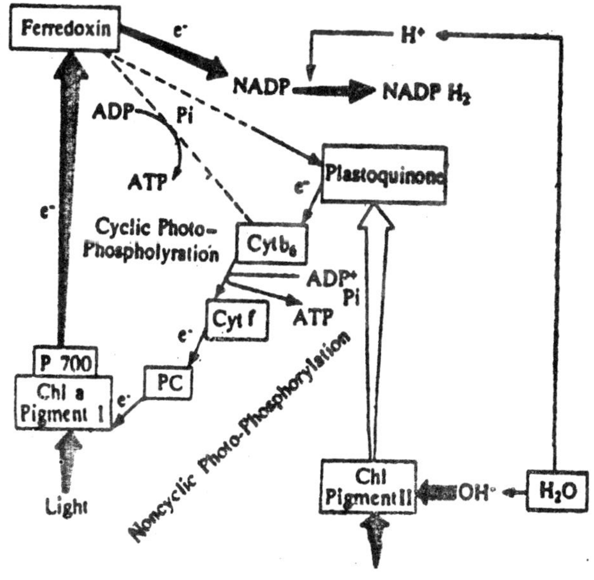

Fig. 6

The end products of photosynthesis, namely sugars or starches, are synthesized from PGA by reversed sequence of EMP pathways of glycolysis. It was also established that some of the PGA is used to regenerate RuDP by accepting some more CO_2 molecules. Thus, there is a regular cycle of reactions that ultimately produce the end product of photosynthesis.

The formation of hexose sugar in the dark reaction may be summarised as follows.

(a) 3 molecules of carbondioxide combine with 3 molecules of RuDP (Ribulose diphosphate) to form 6 molecules of phosphoglyceric acid (PGA).

$$3CO_2 + 3RuDP \rightarrow 6PGA$$

(b) 6 molecules of phosphoglyceric acid (PGA) combine with 6 molecules of ATP, which are formed in the light reaction, to form 6 molecules of diphosphoglyceric acid (DPGA).

$$6PGA + 6ATP \rightarrow DPGA + 5\ ADP$$

(c) 6 molecules of DPGA are reduced to 6 molecules of phosphoglyceraldehyde by 6 molecules of $NADPH_2$ which are produced in light reaction.

$$6\ DPGA + 6\ NADPH_2 \rightarrow 6\ \text{phosphoglyceraldehyde}\ H_2PO_4 + 6\ NADP$$

(d) Now one of the molecules of phosphoglyceraldehyde is utilised in the formation of a half molecule of hexose.

$$1\ \text{Phosphoglyceraldehyde} \rightarrow 1/2\ (\text{hexose}).$$

The five other molecules of phosphoglyceraldehyde are utilised for the continued regeneration of ribulose diphosphate.

$$5\ \text{Phosphoglyceraldehyde} \xrightarrow{\text{Changes}} 3\ \text{Ribulose Diphosphate.}$$

(i) The first molecule of phosphoglyceraldehyde is converted into its isomer, dihydroxy acetone phosphate. The latter molecule combines with second molecule of phosphoglyceraldehyde to form fructose 1, 6 diphosphate which in its turn gets dephosphorylated to form fructose monophosphate.

$$\text{Phosphoglyceraldehyde} \xrightarrow{\text{Isomerises}} \text{Dihydroxyacetone phosphate}$$

$$\text{Dihydroxy acetone phosphate} + \underset{\text{(II)}}{\text{phosphoglyceraldehyde}} \longrightarrow$$

$$\text{Fructose, 6 monophosphate} \xrightarrow{ADP \rightarrow ATP} \text{Fructose 1, 6 diphosphate.}$$

(ii) The fructose 6 monophosphate now combine with the third molecule of phosphoglyceraldehyde to give rise erthrose monophosphate (C_4) and xyluose monophosphate (C_5).

Fructose – 6 – monophosphate + Phosphoglyceraldehyde
(III)
⟶ Erythrose monophosphate (C_4)
+ Xylulose monophosphate (C_5)

(iii) The erythrose monophosphate now combines with the fourth molecule of phosphoglyceraldehyde to form sedoheptulose diphosphate which loses one phosphate to become sedoheptulose monophosphate. It then combines with the fifth molecule of phosphoglyceraldehyde to give rise to one molecule each of xylulose monophosphate and ribose monophosphate.

Erythrose monophosphate + Phosphoglyceraldehyde.
⟶ sedoheptulose diphosphate
↓ ADP – ATP

Xylulose monophosphate + Ribose monophosphate $\xleftarrow[\text{(V)}]{\text{Phosphoglyceraldehyde}}$ Sedoheptulose monophosphate

The ribose monophosphate gives rise to one molecule of ribulose monophosphate. The two molecule of xylulose monophosphate, one from this step and another form step (ii), are reversibly converted into its isomer ribulose monophosphate.

In this way five molecules of phosphoglyceraldehyde are converted into three molecules of ribulose monophosphate which in turn gets phosphorylated to form ribulose diphosphate. Now by doubling the number of CO_2 molecules of six, the path of carbon in dark phase of photosynthesis.

EXPLANATION OF THE DOUBLET STRUCTURE

When we study the spectra of elements other than hydrogen, we observe that many lines are actually multiples consisting of two, three or more lines close together.

In order to explain Uhlenbeck and Goldsmit (9135) suggested that this multiplicity of spectral lines is due to the spin of the electron. The contribution due to the spin is quantised and is expressed in terms of spin quantum numbers. The value of this number is + 1/2 and – 1/2. The resultant of the spin and azimuthal quantum number is given by j = l + s, where 'j' is called the *inner quantum number.* As 's' can have + 1/2 and + 1/2, it follows that for every value of 'l' there are two values of 'j' viz.,

$$j_1 \quad = l + 1/2 \text{ and } j2 = l - 1/2.$$

The above values hold good except when l is zero. In that case the two 'j' values are identical, *i.e.*, + 1/2, because it is not the + ve or – ve sign but is the numerical value only of 'j' which determines the momentum. It means that every 'l' level except l = 0 is consequently split into two levels with different energies. Let us now illustrate this discussion by applying to alkali metals.

(i) For an electron in an 'S' level, the value of l = 0. It means that the two values of j are numerically identical and hence it is a singlet level.

$$j_1 \quad = l + \frac{1}{2} \text{ and } j_2 = l - \frac{1}{2}$$

$$j_1 \quad = 0 + \frac{1}{2} \text{ and } j_2 = 0 - \frac{1}{2}$$

$$j_1 \quad = + \frac{1}{2} \text{ and } j_2 = - \frac{1}{2}$$

(ii) In the P level, l = 1 and so the corresponding values of 'j' are 3/2 and 1/2.

$$j_1 \quad = l + \frac{1}{2} \text{ and } j_2 = l - \frac{1}{2}$$

$$j_1 \quad = 1 + \frac{1}{2} \text{ and } j_2 = 1 - \frac{1}{2}$$

$$j_1 \quad = \frac{3}{2} \; j_2 = \frac{1}{2}.$$

Thus, each spectral line in 'P' level is doublet.

Similarly in the D and F levels, the values of 'j' are 5/2 and 3/2, 7/2 respectively. Again 'j' has two values and therefore each line will be a doublet. The doublet of the sharp series of spectra may be represented by the expressions

$$\bar{v} = aP_{1/2} - nS$$

and

$$\bar{v} = aP_{3/2} - nS$$

where S is a singlet and P is a doublet.

PHOTOCHEMICAL INHIBITION

There are certain substances which are able to retard the rate of a photochemical reaction when present in trace amounts. Such substances are known as *inhibitors* and the phenomenon is known as *photochemical inhibition.* Some examples are.

(i) Traces of nitric oxide and propylene lower the quantum yield of the photochemical combination of hydrogen and chlorine.

(ii) Traces of impurities like NH_3 which when present in the hydrogen and chlorine reaction lower its quantum yield from 10^6 to 10^4. In 1905, Chapman showed that the inhibitor action of NH_3 is due to the side reaction of NCl_3 which may result due to the reaction of NH_3 with Cl_2.

$$NCl_3 + 3HCl \rightleftharpoons NH_3 + 3Cl_3$$

$$NCl_3 + 3H_2O \rightleftharpoons NH_3 + 3HClO$$

At equilibrium the concentration of NH_3 is maintained and the actual formation of HCl is reduced,

(iii) SO_2 and O_2 were also found to be inhibitors in various photochemical reactions.

Explanation. It is generally accepted that photo-inhibitors interrupt the chain reactions by removing chain carrier atoms or radicals.

PERIOD OF INDUCTION

Many reactions are characterised by an initial period during which the process appears to be silent. This time is known as period of induction. Some examples are.

(i) An induction period is observed in hydrogen and chlorine reaction when certain impurities are present. With pure hydrogen and chlorine, no induction period is observed.

(ii) An induction period is also observed in the absorption of As_2O_2 by mercuric oxide in the presence of light.

Period of induction is due to presence of impurities like ammonia. These substances act as inhibitors by breaking these chains. When these impurities are completely converted into nitrogen and ammonium chloride during the reaction, then only normal photochemical reaction can take place.

PHOTOSTATIONARY STATE OR PHOTOCHEMICAL EQUILIBRIUM

A state of photochemical equilibrium is said to exist in a reaction when the rates of two opposing reactions of which at least one is light sensitive, become equal under the influence of light radiation.

Types : A number of cases of the photostationary state have been studied. These cases fall into two categories.

(a) In the First Category, only one reaction is light sensitive

$$A + B \underset{\text{dark}}{\overset{\text{light}}{\rightleftharpoons}} C + D$$

Some examples of such type are.

(i) *Dissociation of Nitrogen Dioxide*. When the vapour of nitrogen dioxide is exposed to light radiation of wavelength less than 3700Å, it dissociates to form nitric oxide and oxygen but the combination of these products is a dark reaction. Hence the equilibrium is

$$2NO_2 \underset{\text{dark}}{\overset{\text{light}}{\rightleftharpoons}} 2NO + O_2$$

When the reaction just starts, the pressure of the gas rises due to the decomposition of the nitrogen dioxide but becomes constant as soon as the stationary state it reached.

Dimerisation of anthracene. Another example of a photo-stationary state is the dimerisation of anthracene in solution in which the forward process is light sensitive but the back ward reaction is a thermal reaction, and so the equilibrium is

$$2_{14}H_{10} \underset{\text{dark}}{\overset{\text{light}}{\rightleftharpoons}} C_{28}H_{20}$$

Let us calculate the equilibrium constant of the reaction of category (a),

light $$A + B \underset{\text{dark}}{\overset{\text{light}}{\rightleftharpoons}} C + D$$

As the forward reaction is light sensitive, its rate will not depend upon the concentrations of A and B but upon the intensity of absorbed light. Hence.

Rate of the forward reaction = $k_1 I_{abs}$ where k_1 is constant for the forward reaction. As the backward reaction is a dark or thermal reaction, its rate will depend upon the concentrations of products C and D, *i.e.*,

Rate of backward reaction $= k_2$ [C] [D].

At photostationary state.

Rate of forward reaction = Rate of backward reaction

or $$\frac{k_1}{k_2} = \frac{[C][D]}{I_{abs}} \quad \text{or} \quad K = \frac{[C][D]}{I_{abs}}$$

where K is the equilibrium constant and is defined as the ratio of velocity constants of two opposing reactions.

The equilibrium constants for photochemical equilibria are constant only for a given light intensity, and vary as the latter is charged. Photochemical equilibrium constant is generally independent of temperature changes if the intensity of light is kept constant.

(b) In the Second Category, both the reactions are light-sensitive.

$$A + B \underset{\text{light}}{\overset{\text{light}}{\rightleftharpoons}} C + D$$

Few examples of this category are

(i) *Formation of sulphur trioxide.* It is an example of a photostationary state in which both the forward and back ward processes are light sensitive.

$$2SO_2 + O_2 \underset{\text{light}}{\overset{\text{light}}{\rightleftharpoons}} 2SO_3 .$$

The value of photochemical equilibrium constant of this reaction is almost independent of temperature from 500° to 800°C.

(ii) *Isomerisation of maleic acid into fumaric acid.* Another example is the isomerisation of maleic acid into fumaric acid in which both forward and backward reactions are light sensitive.

$$\begin{array}{c} H\text{—}C\text{—}COOH \\ \| \\ H\text{—}C\text{—}COOH \\ \text{Maleic acid} \end{array} \underset{\text{light}}{\overset{\text{light}}{\rightleftharpoons}} \begin{array}{c} H\text{—}C\text{—}COOH \\ \| \\ HOOC\text{—}C\text{—}H \\ \text{Fumaric acid} \end{array}$$

The value of equilibrium constant can be calculated in the similar manner as given in category (a).

Application. The concept of photostationary state has been used to explain the phenomenon of *vision.* When the light sensitive substance in eye called visual purple exposed to light, the latter gets bleached to form visual yellow.

Thus, an equilibrium state is established between visual purple and visual yellow. In dark the visual purple accumulates and eye becomes sensitive. When exposed to light further, it results in the phenomenon of *dazzling.*

BIOLUMINESCENCE

Certain living organisms emit light and show the phenomenon of chemiluminescence. It is known as bioluminescence. The phenomenon of bioluminescence was investigated by *E. N. Harvey* (1915). Some interesting examples are.

(i) The cold light produced by certain living organisms like fire-fly or glow worm is probably due to the oxidation of protein, luciferin, by the atmospheric oxygen in the presence of enzyme luciferase. The firefly emits light having a maximum intensity of wavelength of 5700Å. It can be easily judged by naked eye.

(ii) Certain marine animals also emit light. The protozoa Noctiluca of the sea produces phosphorescence. It has also been observed that many deep sea fishes have powerful organs which emit dazzling light.

PHOTOSYNTHESIS

The energy which supports the activities of most living organisms on the earth is derived directly or indirectly form the energy of sunlight through photosynthesis *Photo* = light, *synthesis* = to build up). *It is a process in which simple carbohydrates are synthesised from water and carbon dioxide in the chlorophyll containing tissues of plants in the presence of sunlight. Oxygen being a by-product is given out*

Kamen (1963) has defined photosynthesis as a *series of processes in which electromagnetic energy is converted into chemical free energy which can be used for biosynthesis.* By a single overall equation, it may be shown as below.

$$CO_2 + H_2O \xrightarrow[\text{chlorophyll}]{hv} 1/6\,(C_6H_{12}O_6) + O_2$$

In more general way, it is

$$CO_2 + H_2O \xrightarrow[\text{chlorophyll}]{hv} 1/n\,[CH_2O]_n + O_2$$

Scarce of Oxygen Liberated in Photosynthesis : The released oxygen (O_2) in photosynthesis comes from water [H_2O^{18}]. When green plants were supplied with water containing [H_2O^{18}]. The released oxygen was entirely of the O^{18} type, indicating that water is only source to release oxygen in photosynthesis. This can be represented by the following summary reaction.

$$6CO_2 + 12H_2O^{18} \rightarrow C_6H_{12}O_6 + 6H_2O + 6O_2^{\ 18}.$$

Photosynthesis and Pigments. The photosynthetic products are energy-rich organic compounds. The potential chemical energy of these compounds comes from the light energy. The light' energy to be effective in photosynthesis must be absorbed by a suitable pigment. The vital role is performed by the green pigment chlorophyll, in plants.

Chlorophyll pigment. There are at least seven types of chlorophylls known : chlorophylls *a, b, c, d* and *e.* bacteriochlorophyll and bacterioviridin. All these chlorophyll molecules contain a *tetrapyrrole* skelton formed into a ring with an atom of magnesium in the centre of the ring. A so-called pyrrole molecule contains a skeleton of five atoms, four carbon and one nitrogen and five are arranged in a ring.

Four such pyrroles arranged in a ring form the 'head' of a chlorophyll molecule. Attached to this *porhpyrin* ring at one point is an alcohol (phytol) "tail", a long chain of linked carbons. Relatively minor variations in the kinds and groupings of other atoms joined to this head and tail skeleton account for the differences among different kinds of chlorophylls.

Chlorophylls *a* and *b* are the two most abundant ones found in all the autotrophic plants except the pigment containing bacteria. Chlorophyll *b* is however, absent in blue-green, brown, and red algae. The other chlorophylls (c, d, e) are found only in algae in combination with chlorophyll a. Chlorophyll *a* possesses a $- CH_3$, a methyl group which is replaced by a – CHO, an aldehyde group, in chlorophyll *b.*

Path of Carbon Dioxide in Photosynthesis : The air contains 0.03% carbon dioxide. This gas diffuses into the air spaces between the cells of the mesophyll of the leaf through the open stomata. Here it combines with water to form carbonic acid.

Transport of carbon dioxide in the form of carbonic acid takes place from the spongy cells to the palisade cells until the main site of photosynthesis is reached. It is now proved that in photosynthesis carbon dioxide is not reduced as such but it first forms a complex with cellular compounds and is then reduced by some 'activated' compounds produced as a result of photochemical reaction.

Since isolated choloroplasts are unable to reduce carbon dioxide in the light, it is concluded that either carbon dioxide uptake must take place outside the chloroplast or some mechanism has been destroyed during extraction of the chloroplasts. The first alternative is more probable. *Frenkel* (1941) prepared CO_2 from radioactive (C^{14}) carbon.

He fed *Nitella* cells with this carbon dioxide ($C^{14}O_2$) in the dark and found that all the radioactive carbon was located in the cytoplasm. When exposed to light, 4/5 of radioactive carbon was found in the chloroplasts. The simplest interpretation of Frenkel's result is that carbon dioxide uptake occurs in the cytoplasm whereas reduction of carbon dioxide is associated with the chloroplasts.

Mechanism of Photosynthesis : Various theories have been propounded by various workers. But the earlier theories failed and still not theory is available which describes the mechanism satisfactorily. The uncertainty in the mechanism is die to.

(a) No products of the earlier stages of the process could be detected experimentally.

(b) When chlorophyll is extracted from plant, it fails to reproduce its photosynthetic property. So the photosynthesis cannot be studied under controlled experimental conditions.

(c) The complete structure of the chlorophyll was not known fully.

The mechanism of photosynthesis as currently understood is usually divided into three phases :

(1) The absorption of the light energy by the pigment system,

(2) Conversion of light energy into chemical energy by photo-phosphorylation, and

(3) Synthesis of organic compounds by the products of the photo-chemical reactions.

Enough evidence has accumulated in the recent past to show that photosynthesis consists of two stages, (a) light phase and a dark phase. The reaction of light phase is light sensitive and 'therefore' called a photochemical reaction. The reactions of the dark phase do not require light and are temperature sensitive. Such reactions are after called *Blackman reactions.*

$$\xrightarrow[\text{Photochemical reaction}]{\text{Step I}} \xrightarrow[\text{chemical reaction}]{\text{Step II}} \text{Photosynthetic Product}$$

(The component reactions of photosynthesis).

Evidences for the Existence of Light and Dark Phases : There are following evidences.

1. *Experiments with intermittent light.* When Chlorella cells were exposed to continuous illumination, *Warbarg* found that the rate of photosynthesis was less when compared to those exposed alternately to light and darkness. The increase in the rate of photosynthesis was enhanced greatly if the span of the light period was reduced. This is because light period followed by darkness allows the cells to utilize all the products of the light reaction for the reduction of CO_2 in the dark reaction. When the light was continuous or interrupted, the products of light phase accumulate because the pace of light reaction is far greater than that of dark.
2. *Temperature Experiments.* When the photosynthesis is carried out in the excess of carbon dioxide but in the presence of low light intensity, the rate of photosynthesis does not increase with the increase in temperature. This indicates that some reaction requires light and is not affected by rise in temperature. When the concentration of carbon dioxide is low and the intensity of light is high, the rate of photosynthesis becomes double for every 10°C rise in temperature. This shows that there must be a reaction which is not affected by light. This is the dark reaction of photosynthesis.

   Let us discuss these two stages separately.

LATENT IMAGE

The usefulness of present-day photographic materials is a result of the behaviour of the tiny individual particles of silver halide which are contained in photographic emulsion. In many cases these particles are extremely sensitive to light and store up the effect to an exceedingly small amount of light. The effect can, in turn, be multiplied many fold by the action of a developer. *This stored up effect of light is known as the latent image.*

It is so small that it cannot be developed itself. Consequently, the nature of the latent image and the nature of the development process are closely related.

Def. *When a photographic plate is exposed to light radiation for a certain time, an invisible image is formed on it. This invisible image is known as latent image.*

FACTS

(i) In the formation of latent image, no change in the emulsion has been detected.

(ii) The formation of latent image is very sensitive to traces of substances such as Ag, Ag_2 S etc.

(iii) When the photographic plate carrying latent image is exposed to mild reducing agent like pyrogallol, the latent image is developed to a negative plate.

Studies of the behaviour of the latent image under different conditions of development and exposure and its reactions with certain other chemicals have led to a fairly clear conception of its nature and the mechanism of its formation. An understanding of these processes requires knowledge of the structure of the silver halide grains themselves.

Theories of Latent Image Formation : When a film is exposed to light radiation, the absorbed energy ejects an electron from the bromide ion.

$$Br^- + h\nu \rightarrow Br + 1e^-.$$

The electron thus set free is captured by a silver ion to form a silver atom.

Thus, the overall reaction is the photochemical dissociation of silver bromide into silver and bromine by absorbing one quanta of light, *i.e.,*

$$AgBr + h\nu \rightarrow Ag + Br.$$

Thus during the formation of latent image, silver is formed and bromine atom being removed by gelatin.

Objection. The main problem is that light is absorbed all over the surface of the photographic film but silver atoms make their appearance only at a few isolated points within each grain to form a latent image. How does it happen ? How is the latent image formed ? Various theories regarding the formation of latent image are.

I. Paul's View : According to Paul there are several empty holes in the crystal. When a positive ion is missing at one point a negative ion will be missing at some other point. Thus, there exists some holes in the crystal.

When light is incident on a photographic plate, the electrons are given out and finally fall inside the holes. Thus,

$$X^- + h\nu \rightarrow \qquad X + e^-$$

the latent image was formed at these holes. When the exposed photographic plate is developed, negative image is formed at these holes. This theory does not explain the formation of latent image in complete manner.

+	–	+	–	+	–
–	+	–	+	–	+
+	–	hole	–	+	–
–	+	–	+	–	+
+	–	+	–	+	–
–	+	–	hole	–	+
+	–	+	–	+	–
–	+	–	+	–	+

Fig. 7 : Formation of latent image.

II. Webb's Theory : Recent advances in quantum mechanics and the applications of these principles to the structure and behaviour of crystals have contributed a great deal to knowledge of the problem of latent image formation. Webb (1936) first applied the principles of quantum mechanics to an explanation of the formation of the latent image.

Webb described the latent image as a concentration of electrons at a speck, or trap. The electrons are supposed to be transferred to a higher energy level by the absorption of light and in this energy level they could wander through the crystal at will as in the photo-conductance effect. The sensitivity specks were supposed to contain energy levels slightly lower than this conductances level, so that when an electron travels to such a specks, it gives up a bit of energy in going to this new level and is trapped. This gave an excellent explanation of the speck concentration of the effect of absorbed light but did not explain the formation of metallic silver as latent image substance.

III. Gurney and Mott's Theory : Gurney and Mott presented a theory for the formation of latent image.

1. The energy structure of a crystal of silver halide can be described as in Fig. 8. The Ag and Br represent the silver and bromide ions which are present alternately in a crystal. The crosshatches labelled as P represent the energy level of the electrons connected with

the bromide ions. This energy band is filled and electrons cannot move around in it. The clear energy band labelled S represents the energy level corresponding to the conductance level in metallic silver. In a crystal of silver halide all the valence electrons are associated with the bromide ions and none with silver so that this band is empty.

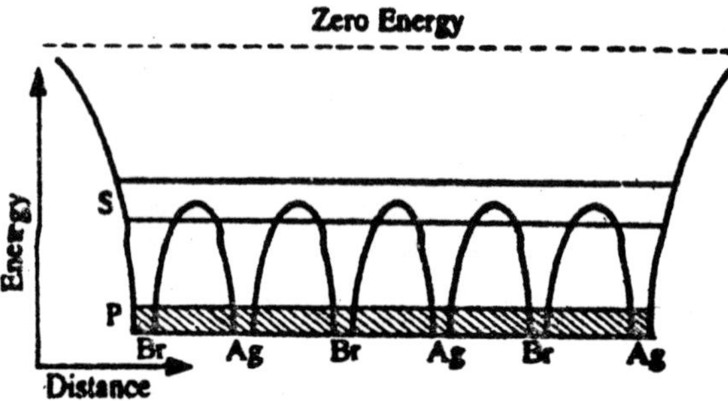

Fig. 8 : Quantum mechanical description of the energy level occurring in a crystal of silver bromide.

Electrons cannot have energy represented by the areas not included by these two bands. The photo-conductance phenomenon can be explained by this diagram–when light is absorbed, its energy is transferred to one of the electrons in the P and the electron is lifted to the S band where it is free of wander around just as the conductance electrons flow around in metallic silver.

2. Another concept which was applied to theory of photolysis by Gurney and Mott was the presence of interstitial ions in a crystal lattice. A perfect crystal would contain some ions which are not in their proper places but in the interstitial positions. Presumably these are caused by thermal motion of the ions in the crystals. This condition is shown in Fig. 9.

These ions and their movement through the crystal, either by going from the interstitial position to another or by moving to one of the holes which has been lifted by another interstitial ion, account for the every small electrical conductivity of crystals in the dark.

These ions might be considered as being in solution in a crystal just a salt is dissolved in water. The conductivity of such crystals follows the same laws as the conductivity of ionised salt solutions.

This conductivity decreases with decreasing temperature.

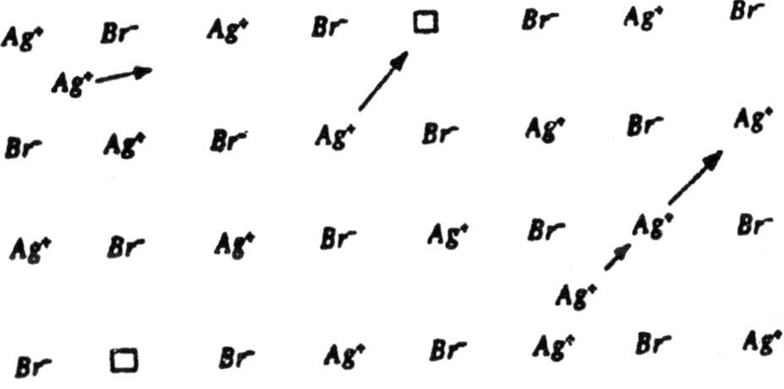

Fig. 9 : Interstitial silver ions in the silver bromide lattice.

3. Gurney and Mott also assumed that silver bromide contains particles of colloidal silver.

4. In order to apply his theory to the formation of latent image as well as to the photolytic formation of metallic silver he postulated that the sensitivity specks, silver sulphide or whatever their complete composition might be, can serve the same function as the colloidal silver specks, *i.e.,* they can act as traps for electrons in the conductance level of the crystal. The theory involves the following points.

(i) When a film is exposed to light radiation, the primary absorption of light occurs in Br– ions of silver bromide crystal.

$$Br^- + hv \rightarrow Br^- + le$$

producing an excited electron.

(ii) The bromine atom is termed as a positive hole and is transferred from its original position in the crystal lattice to the surface of the crystal by the following process.

$$Br + Br^- \rightarrow Br^- \ Br$$

On the surface, the positive hole is trapped by coming in contact with an impurity centre line Ag_2S, probably by absorbing an electron from the sensitizer.

(iii) The excited electrons [from (i)] are free to move on through the crystal lattice. The moving electrons come in contact with a specks of silver and the trapped. This charges the speck negatively and this negative charge acts at some of interstitial silver ions which migrate to speck.

The positive charges on the speck due to silver ions are neutralised by the excited electrons.

$$Ag^+ + 1e \rightarrow Ag.$$

Thus, the above process is repeated and a large number of silver atoms collect as an invisible speck at one point in a crystal. The same process is repeated in other silver bromide crystals on the film and ultimately leads to the formation of latent image.

Objections : These are.

(i) This theory had not taken sufficient account of the possibility of the recombination of photo released electrons with the positive holes which are formed at the same time.

$$\underset{\text{positive hole}}{Br} + 1e \rightarrow Br^-$$

(ii) *Gurney* assumed the existence of interstitial silver ions on a photographic film. *Mitchell* showed that a very small number of interstitial silver ions is found to exist on a photographic plate. This number is not sufficient to explain Gurney-Mott's theory.

IV. Mitchell's Theory (1957) : It involves the following steps.

(i) Upon the absorption of a quantum of light, a positive hole and photo-electron are formed.

$$Br^- \; h\nu \rightarrow \underset{\text{Positivee hole}}{Br} + \underset{\text{Photo electron}}{1e}$$

(ii) The positive hole is trapped at a sensitivity speck in the crystal and this liberates an interstitial silver ion.

(iii) The liberated interstitial silver ions absorbed at some distortion or kink in the crystal lattice.

Then, the absorbed positively charged silver ion absorbs the photo-electron.

$$Ag^+ + 1e \rightarrow Ag.$$

This single silver atom is called the latent pre-image.

(iv) Another interstitial silver ion released by the trapping of a positive hole is absorbed this latent pre-image.

$$\begin{array}{cccc} & & & Ag^+ \\ & & & | \\ Ag^+ & + Ag & \rightarrow & Ag \end{array}$$

Again, the silver ion traps a photo-electron to give a two-atom specks. This is called latent sub-image.

$$\begin{array}{cccc} Ag^+ & + 1e & \rightarrow & Ag \\ | & & & | \\ Ag & & & Ag \end{array}$$

(v) A third interstitial silver ion released by trapping of positive hole is absorbed on the latent sub-image and again, the absorbed silver ion traps a photo-electron, leading to the formation of latent image.

General Views : The series of experiments carried out in recent years have shown that the above theories are not satisfactory and they require further investigations.

THE SOMMERFELD MODEL

Inspite of many success, Bohr's theory was found to be inadequate to explain certain details in the spectrum of hydrogen. For instance, the Hx line of the Balmer series was found to contain several components. This fine structure of spectral lines could not be explained by Bohr's theory which assumed that there was only one orbit for each quantum number, whereas the observed fine structure suggested that for any given quantum number n there might be several orbits of slightly different energies.

Sommerfeld, in 1915, guided by the above suggestion modified Bohr's theory by introducing the following modifications.

(a) Concept of elliptical orbits, and

(b) Relativistic variation of the mass of electron.

We will discuss these modifications one by one.

(a) *Elliptical Orbits :* In order to explain the multiplicity of spectral lines. Sommerfeld introduced the concept of elliptical orbits. He postulated that :

(i) Since the electron is moving around and under the influence of a massive nucleus, like a planet around the central massive sun, it might describe elliptical orbits as well.

(ii) While retaining the first circular orbit suggested by Bohr, Sommerfeld assumed one additional elliptical orbit in the case of Bohr's second orbit and added two additional elliptical orbits to Bohr's third orbit and so on (Fig. 10).

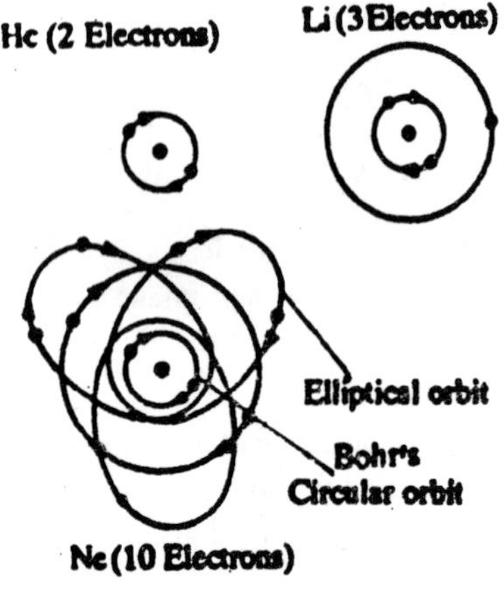

Fig. 10

(iii) The nucleus is one of the foci for all these orbits.

(iv) In an elliptical orbit, the major and minoraxes will differ in lengths, but as the orbit broadens, they will approach each other and becomes equal when the orbit becomes circular. Circular orbit is only a special case of the elliptical orbit. The orbit is defined by two quantum numbers, *i.e.*, n and k which correspond to the major and minor axes of the ellipse. They are related as follows.

$$\frac{\text{Principal quantum number}}{\text{Azimuthal quantum number}} = \frac{n}{k} = \frac{\text{length of major axis}}{\text{length of minor axis}}.$$

It is clear from the above that for any given value of n, k cannot be zero as in that case the ellipse would degenerate into a straight line passing through the nucleus. Further, k cannot be more than n since b is always less than a. When n = k the path becomes circular. This is a limit to the number of different orbits that an electron may have in any energy level.

Such possible paths are referred to as sub-levels. For a given value of n, k can have only n different values which mean that there may be only n elliptical orbits or sub levels with different eccentricity. For elliptical orbits the values of k would be (n – 1), (n – 2) etc., down to k = 1. When k = 0, the ellipse would be a straight line. When n = 3, k = 3, (circular), 2, 1, two of sommerfeld model lines in its subdivision of the original Bohr stationary levels into various sub-levels of slightly differing energies as given by difference in orbit shapes.

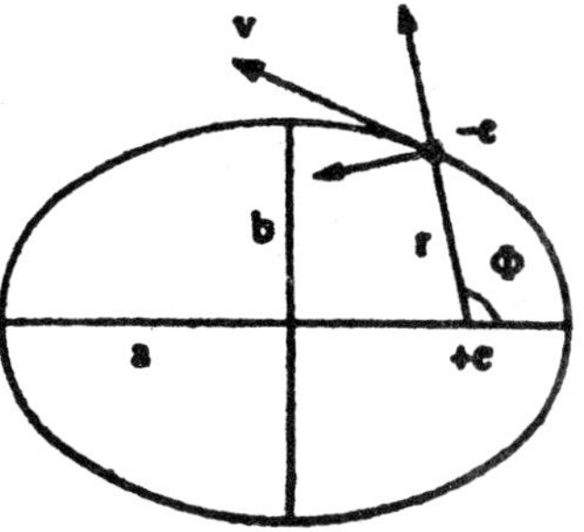

Fig. 11

Simple mathematical approach to Sommerfeld model : Let us consider the motion of an electron (– c) in an elliptical orbit as shown in (Fig. 11). Its position at any instant can be fixed in terms of polar coordinates, r and f where r is the distance of the electron from the nucleus (+ e) at one of the foci of the ellipse and f is the angle which the radius vector makes with the major axis of the ellipse. The tangential velocity v of the moving electron at any instant can be resolved into two components.

(i) One radial, *i.e., dr/dt* along the radius vector. Corresponding to this there will be a radial momentum Pr equal to m (dr/dt).

(ii) Other transverse, *i.e.*, m at right angles to the radius vector equal to r (df/dt). Corresponding to this there is angular or azimuthal momentum of equal to mr^2 (df/dt), where m is the mass of the electron.

As Sommerfeld considered the circular orbits to be special cases of elliptical orbits, he assumed that the elliptical orbits should satisfy the quantum condition of Bohr just as the circular orbits, *i.e.*,

$$\oint p_r \, dr = n_r h \quad ...(1)$$

$$\oint p_f \, d\phi = n_\phi l_t \quad ...(2)$$

Thus, the single n of Bohr's theory has been replaced by the two new quantum numbers n_r and n_ϕ, *i.e.*,

$$n = n_r + n_\phi \quad ...(3)$$

where n_r is the radial quantum number and nf is the angular or azimuthal quantum number. The total energy E is given by

$$E = \text{P.E.} + \text{K.E.}$$

$$= \text{P.E.} + \text{Radial K.E.} + \text{Angular K.E.}$$

$$= -\frac{e^2}{r} + \frac{1}{2}m\left(\frac{dr}{dt}\right)^2 + \frac{1}{2}mr^2\left(\frac{d\phi}{dt}\right)^2 \quad ...(4)$$

From equations (1), (2), (3) and (4), it can be shown that

$$1 - e^2 = \frac{b^2}{a^2} = \frac{{n_\phi}^2}{(n_\phi + n_r)^2}$$

where e is the eccentricity of the ellipse whose semi-major and semi-minor axes are a and b respectively.

and
$$E = -\frac{2\pi^2 mZ^2e^4}{h^2}\left(\frac{1}{n_\phi + n_r)}\right)^2 \quad ...(6)$$

From equation (5), we get

$$(1 - e^2)^{1/2n} = \frac{b}{a} = \frac{n_\phi}{n_\phi + n_r} = \frac{n_\phi}{n}.$$

Thus, equation (6) can be written as

$$E = \frac{2\pi^2 mZ^2e^4}{n^2h^2} \quad ...(7)$$

From equation (7), it follows that

(i) The elliptical orbits which have the same value of n, though of different eccentricities, have the same energy. It is not correct.

(ii) All orbits having the same values of the semi major axis possess the same energy, since the length for the semi-major axis is determined solely by the total quantum number n (= $n_f + n_r$). Hence the energy of any of these permitted elliptical orbits is identical with that of a circular Bohr orbit, whose radius is equal to the semi-major axis of the ellipse.

From the above it follows that the theory of elliptical orbits in spite of two new quantising conditions involved, introduces not new energy levels other than those given by Bohr's theory of circular orbits. No new spectral lines, which would explain the fine structure, are there fore predicted. Thus, Sommerfeld has to modify his own model by suggesting the variation of mass of electron with velocity called relativistic variation.

Relativistic Variation of the Mass of Electron : The velocity if an electron moving in an es path of the electron is found to be no longer a simple ellipse, it is indeed, no longer a closed figure, but is transformed into a complicated curve known as rosette –a processing ellipse (Fig. 12). The total energy E of the system corrected by the relativistic variation of the mass of electron can be shown to be

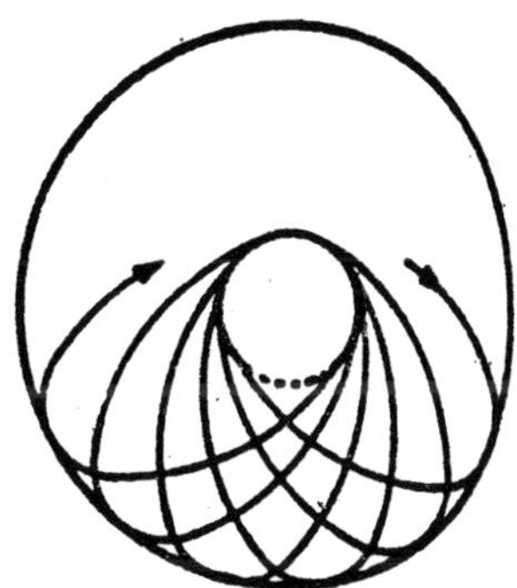

Fig. 12 : Sommerfeld elliptical orbit precessing about an axis through one of its focci.

$$E = -\frac{2\pi^2 m Z^2 e^4}{h^2}\left[\frac{1}{n^2} + \frac{4\pi^2 e^4 Z^2}{c^2 h^2}\left(\frac{n}{n\phi} - \frac{3}{4}\right)\frac{1}{n^4} + \ldots\right] \quad \ldots(8)$$

The relativistic correction, therefore, results in splitting up a given energy level E_α into n levels differing slightly from one another in energy. The splitting up of each energy levels gives rise to a fine structure of single spectral line, one application of the usual Bohr frequency conditions, When one explains the fine structure of lines, one should not take into account all theoretical possible transitions of the electron from one elliptical orbit to another that actually occur.

According to a principal known as the selection rule, transition can take place only between orbits for which the azimuthal quantum number changes by + 1 or – 1, *i.e.*, nf = ± 1.

Limitations of Sommerfeld model : (i) Though Sommerfeld's theory is fairly well verified yet it leads only to three components for the structure of Hx line, while there should be really five. Thus the relativistic atom model has met with a partial success.

(i) The concept of elliptical orbits due to Sommerfeld gives the correct total (n) of possible azimuthal quantum number, but the actual values are not correct. The experimental studies as well as theoretical treatment based on wave mechanics show that azimuthal quantum number can be zero, so that the values can be 0, 1, 2,... –1, thus making a total of n possibilities. The correct and new azimuthal quantum number is denoted by 'l' to avoid the confusion. Thus, l is equal to K – 1.

(ii) It provides no idea about the number of electron which can be accommodated in a particular orbit.

(iii) It provides macerate values for angular momentum.

(iv) Sommerfeld model could not explain Zeeman and Stark effect.

Explanation of fine details in line spectrum by Sommerfeld model : It can be explained as follows.

When an electron moves in an elliptical path there will be a displacement each time in its elliptical path and thus resulting in a small difference in energy. Thus, the path of the moving electron is no longer a simple closed ellipse but a complicated curve, known as rosette, which is made up of different elliptical orbits of slightly different energies. This *difference in energies of elliptical orbits explains the existence of components of spectral lines in the spectrum of higher elements.*

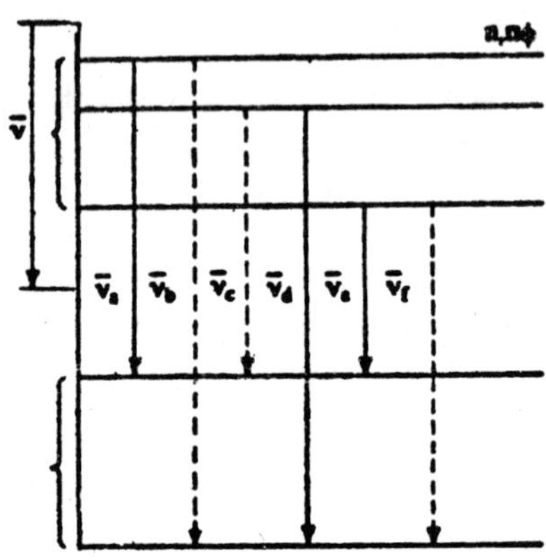

Fig. 13

Criticism of the theory : (1) Suppose there are two orbits with principal quantum numbers, n = 2 and n = 3, (Fig. 13) respectively.

Consider an electron which falls from third to the second orbit. There may be six possible transitions, each giving one fine line. But actual observations yield only five lines (Fig. 13).

Possibilities.

(i) $3K_3 \rightarrow 2K_2$ *i.e.*, $\Delta K = 3 - 2 = +1$

(ii) $3K_2 \rightarrow 2K_1$ *i.e.*, $\Delta K = 3 - 1 = +2$

(iii) $3K_2 \rightarrow 2K_3$ *i.e.*, $\Delta K = 2 - 2 = 0$

(iv) $3K_2 \rightarrow 2K_1$ *i.e.*, $\Delta K = 2 - 1 = +1$

(v) $3K_1 \rightarrow 3K_2$ *i.e.*, $\Delta K = 1 - 2 = -1$

(vi) $3K_1 \rightarrow 2K_1$ *i.e.*, $\Delta K = 1 - 1 = 0$

where ΔK indicates the change in azimuthal quantum number. Sommerfeld applied *selection rule* which limited the number of transitions. According to such rule only those transitions are possible for which the quantum number K changes by -1 or $+1$ *i.e.*, $\Delta K = \pm 1$. Hence transitions like ($3K_1 \rightarrow 2K_1$), ($3K_3 \rightarrow 2K_1$) and ($3K_2 \rightarrow 2K_2$) are forbidden. Therefore, according to this theory, the fine structure of H_α lines should made only of three lines. But actually it splits up into five lines. Sommerfeld theory fails to explain complicated systems.

(2) Sommerfeld's model could not explain Zeeman and Stark effect.

(3) The model gives no information regarding the relative intensities of lines, whose frequencies alone are predicted.

Wave mechanics and spectral lines : According to wave mechanics orbits do not exist in the atom. Thus, the interpretation of the emission of radiation due to a jump of the electron from outer to the inner orbits does not hold good.

In wave mechanics, place of orbits has been taken by stationary states of atoms with definite energies. When any state excited by the absorption of energy comes to the ground state, spectral lines are produced. Thus, *the frequency of each spectral line may be regarded as a 'beat' frequency between two states of the atom, which gives the same result as that of Bohr.*

As Ψ^2 represents the electrical charge density, then in the stationary state this remains constant and so also the charge density. But when any state becomes excited by the absorption of energy, Ψ is no longer constant and varies periodically and so also the charge density. *This periodic variation in charge density is accompanied by the emission of radiation.*

Excitation and Ionisation Potentials of an Atom : According to the Bohr's theory, in an atom there are certain discrete orbits only in which an electron can revolve without radiating energy. Each of these orbits of a given atom is characterised by a certain (quantised) energy. Hence we say that there are certain discrete energy levels in an atom.

When an electron in an atom absorbs sufficient energy from an outside source, it rises from its present energy level to an higher level. The atom is than said to be 'excited'.

This excited state lasts only for about 10^{-8} second after which the electron jumps back to the inner level, emitting the absorbed energy is emitted as electromagnetic radiations.

The energy required to excite or to ionise a given atom is perfectly definite.

For example, in case of hydrogen atom, the energy of the nth orbit is

$$E_n = -\frac{me^4}{8\varepsilon_0^2 h^2}\left(\frac{1}{n^2}\right).$$

Substituting the known values of

m (= 9.1×10^{-31} kg),

e (= 1.6×10^{-19}C),

h (= 6.62×10^{-34} J-s)

and ε_0 (= 8.85×10^{-12}C^2/N m^2), we get

$$E_n = -2.17 \times 10^{-18}\left(\frac{1}{n^2}\right) \text{ joule (J)}$$

$$= -\frac{2.17 \times 10^{-18}}{1.6 \times 10^{-19}}\left(\frac{1}{n^2}\right)$$

$$= -\frac{13.6}{n^2} \text{ eV.} \qquad [1 \text{ eV} = 1.6 \times 10^{-19} \text{ J}]$$

Putting n = 1, 2, 3, ... ∞, the energies of the 1st, 2nd, 3rd, ... ∞ energy levels come out to be – 13.6, – 3.4 – 1.51, ... 0 electron volts respectively. Hence the energy to be supplied to the atom to raise an electron from the first to the second orbit is (13.6 – 3.4) = 10.2 electron-volts, from first to the third orbit is (13.6 – 1.51) = 12.09 electron volts,* and to the ionized state is (13.6 – 0) = 13.6 electron volts.

Now, the atoms are usually excited or ionised by bombarding them with electrons accelerated through a potential. But the excitation or ionisation of an atom can take place only when the bombarding electron has the required amount of energy. *The minimum accelerating potential which imparts to the bombarding electron an energy sufficient to excite a given atom is called the 'excitation potential' of the atom. Similarly, the minimum accelerating potential which imparts to the bombarding electron an energy sufficient to ionise the atom is called the 'ionisation potential' of the atom.*

Franck-Hertz Experiment

Franck and Hertz, in 1914, performed a series of experiments to measure the excitation potentials of atoms of different elements. These experiments showed directly that in an atom discrete energy levels do exist.

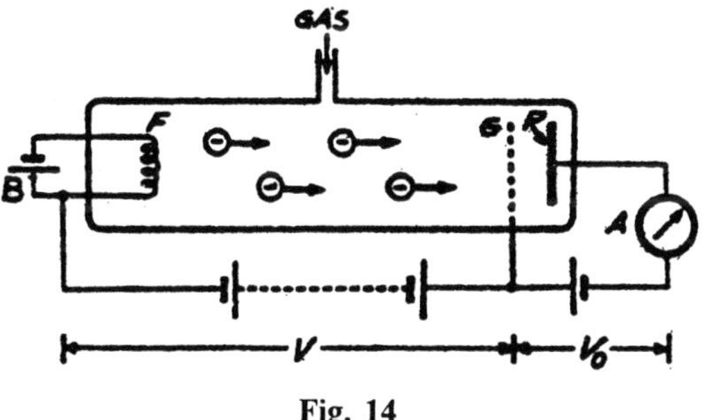

Fig. 14

Their apparatus is shown in Fig. 14. It consists of a glass tube in which are mounted a filament F, a grid G, and a plate P, as shown. The filament F is heated by a small battery B. An accelerating potential V is applied between F and G, and *a small fixed* retarding potential V_0 (about 0.5 volt) between G and P. The gas of the element whose atoms are to be studied is introduced in the tube at about 1 mm of mercury pressure.

The electrons emitted from the hot filament F are accelerated between F and G by the potential V, and retarded between G and P by the potential V_0. Thus, only electrons having energies greater than eV_0 at G are able to reach P. The current to P is recorded by an ammeter A.

The current is plotted against the *accelerating* potential V which is gradually increased from zero. The curve obtained shows a series of regularly spaced peaks (Fig. 15).

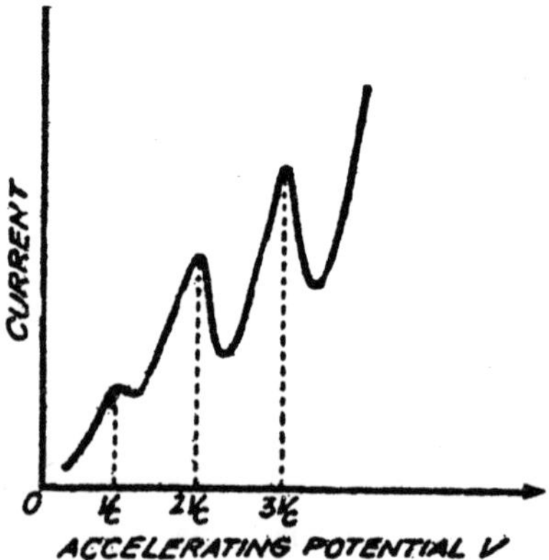

Fig. 15

Interpretation of the Curve

Electrons are emitted from F with a range of small energies. In the beginning, they acquire a small additional energy eV on reaching G. Those electrons whose energy is now greater than eV_0 reach P, and a current is obtained. As the accelerating potential V is increased, more and more of the electro.ıs arrive at the plate and the current rises. Between F and G the electrons collide with the gas atoms. But since the electrons energies are not enough to excite the atoms, the electrons do not lose energy in these (elastic) collisions.

When the accelerating potential V becomes equal to the first excitation potential V_c of the as atoms, the electron energy at G is eV_c. The electrons can now suffer *inelastic* collisions with gas atoms near G and excite them to an energy level above their ground state. The electrons which do so, lose their energy and are unable to reach P against the retarding potential. Hence the current drops sharply. Thus, the position of the first peak gives the excitation potential V_c of the gas atoms.

As the accelerating potential V is increased further, the electrons suffer inelastic collisions nearer and nearer to the filament, F, so that when they reach the grid G, once again they acquire enough energy to reach the plate P. The current thus begins to rise again. When V becomes equal to $2V_c$, a second inelastic collision occurs near the grid and the current drops again. Thus, the second peak gives $2V_c$. This process repeats as V is further increased.

The first peak occurs at a potential slightly less than V_c. This is because electrons are emitted from the filament with a finite velocity and hence with some energy. The true excitation potential V_c is obtained by measuring the difference between two successive peaks.

Limitations : Atoms have more than one excitation potentials and also an ionisation potential. Therefore, in an actual experiment the curve obtained is quite complicated and we cannot distinguish between which are excitation and which are ionisation potentials.

Demonstration of the Existence of Discrete Energy Levels : The experiment shows that electrons transfer energy to the atoms indiscrete amounts, and that they cannot excite atoms if their energy is less than eV_c. Franck and Hertz demonstrated it directly by observing the spectrum of the gas during electrons collisions. They showed that a particular spectral line does not appear until the electron energy reaches a *threshold* value. For example, in the case of mercury vapour, they found that minimum electron energy of 4.9 eV was required to obtain the 2536 Å line of mercury, and a photon of 2536 Å light has an energy of just 4.9 eV. This shows that discrete energy levels do exist in atom and the electrons of the atom can exist only in these levels.

Bohr's Correspondence Principle : The quantum theory gives results which are altogether different from those given by the classical theory in the microscopic world. Yet the two theories approach each other as the quantum number in question increases. This fact was pointed out by Bohr in 1932 who enunciated the following principle.

The predictions of the quantum theory for the behaviour of any physical system must coincide with the predictions of the classical theory in the limit in which the quantum numbers specifying the state of the system becomes very large. This is knows as 'Bohr's correspondence principle'. We establish it by means of an example.

According to classical theory, an electron moving in a circular orbit must emit radiation only of a particular frequency, namely, the frequency of revolution of the electron itself (or its integral multiples). According to quantum theory, on the other hand, radiation is emitted when the electron jumps from one orbit to the other and the frequency of radiation is determined by the difference in energy between the two orbits. Let us try to find a relation between the two frequencies.

The basic equations in the Bohr's theory of the hydrogen atom are the following.

$$\frac{mv^2}{r} = \frac{1}{4\pi\varepsilon_0}\frac{e^2}{r^2}$$

and $$mvr = n\frac{h}{2\pi}, \qquad n = 1, 2, 3, ...$$

where v is the velocity of the electron of mass m in the orbit of radius r. These equations give

$$v = \frac{nh}{2\pi m r}$$

and $$r = \frac{n^2h^2\varepsilon_0}{\pi m e^2}.$$

Hence the classical frequency of revolution of the electron is

$$f = \frac{\text{electron speed}}{\text{orbit circumference}} = \frac{v}{2\pi r}$$

$$= \frac{nh}{4\pi^2 mr^2} = \frac{nh}{4\pi^2 m\left(\frac{n^2h^2\varepsilon_0}{\pi m e^2}\right)^2}$$

$$= \frac{me^2}{4\varepsilon_0^2 n^3h^3} = \frac{me^4}{8\varepsilon_0^2ch^3}\frac{2c}{n^3}$$

$$= \frac{2Rc}{n^3} \qquad ...(i)$$

where $R\left(=\frac{me^4}{8\varepsilon_0^2ch^3}\right)$ is the Rydberg constant. This is the frequency which must be radiated by the moving electron classically.

Now, the frequency radiated on the basis of quantum theory is given by

$v = \frac{E_i - E_f}{h} = \left(=\frac{me^4}{8\varepsilon_0^2h^3}\right)\left(\frac{1}{n_f^2} - \frac{1}{n_i^2}\right)$, when the electron drops from the ni th orbit the nf the orbit. Again, introducing R, we have

$$v = Rc\left(\frac{1}{n_f^2} - \frac{1}{n_i^2}\right).$$

This may be written in the form

$$v = Rc \frac{(n_i - n_f)(n_i + n_f)}{n_f^2 n_i^2} \quad ...(ii)$$

Let us write $n_f = n$ and $n_i = n + 1$. Then the frequency of the emitted radiation for the transition $n + 1 \rightarrow$ (so that $\Delta n = 1$) is given by

$$v = Rc \frac{2n+1}{n^2 (n+1)^2}.$$

When n is very large, we can write $\frac{2n+1}{n^2 (n+1)^2} \simeq \frac{2}{n^3}$.

Under this condition, the emitted frequency is

$$v = \frac{2Rc}{n^3}. \quad ...(iii)$$

Equations (i) and (iii) yield

$$v = f.$$

If we consider transitions Dn = 2, 3, 4, ... we shall have

$$v = 2f, 3f, 4f,...$$

Thus, for very large quantum numbers, the quantum theory frequency v of the radiation is identical with the classical frequency of the revolution (or its harmonics) of the electron in the orbits. This is in accordance with Bohr's correspondence principle.

SPECTRA OF ALKALI METALS

Like the hydrogen atom, the spectra of the alkali metals consist of a series of lines with regularly decreasing separation and decreasing intensity. One important feature of the alkali-spectra is that the lines are mostly doublets though each and every line has not been resolved into two in the spectroscope

Quantum defect : The spectral lines obtained from alkali metals cannot be represented by a formula exactly analogous to Balmer's formula. As the lines converge to a limit, one is able to represent them by the difference of two spectral terms. Thus the values of the terms are not of Balmer's form *i.e.*, R/n^2 but of the form R/n^{*2}, where n is the principal quantum number and n* is the effective quantum number and its value is given by $(n + \alpha)$, where a is a small quantity and is known as the quantum defect.

The *Magnitude* of the *quantum defect* depends upon the penetration of the core by the orbit of the emitting electron. Increased penetration of the core depends upon the *shape of the orbit*. It is, therefore, expected that a series of different values of a will occur for different values of azimuthal quantum number.

There are *four* types of quantum defect (α) designated by S, P, D, and F. These words stand for the first letter of sharp, principal, diffuse and fundamental. The names of the series are attributed according to the order, as to which of them appears in the quantum defect of the running term. The principal series is most easily excited and is hence so calculate. The lines in the sharp series are well defined and those of the diffuse series are blurred. In emission spectra, other series in addition to the principal series may be observed for the alkali metals. These series partly overlap each other. The last named series viz., the fundamental of Bergmann series, occur in the infra-red.

However, the frequencies of the lines of the four series may consequently be represented by the following general expression.

Sharp series $\bar{v} = aP - nS$, Principal series $\bar{v} = aS - nP$,

Diffuse series $\bar{v} = aP - nD$, Fundamental series $\bar{v} = aD - nF$,

where a is a constant, and n is the principal quantum number.

***Explanation*:**

As earlier stated, the volume of n* is given by (n + α). Ωηεν we calculate the values of n and n* for the S, P, D and F terms of the different alkali metals, we get the following values.

	Terms Series	*n*(= n + α)*	*n*	*l*
Lithium	S	1.59	2	0
	P	1.96	2	0
	D	3.00	3	2
	F	4.00	4	3
	Terms Series	n* (= n + α)	*n*	*l*
Sodium	S	1.63	3	0
	P	2.12	3	1
	D	2.99	3	2
	F	4.00	4	3.

From the above table that the discrepancies between n* and n decrease in the order of S, P, D, and F of terms series. The discrepancies [n* – n (n + α) – n = α], also called quantum defect have been explained as follows.

If we compare the structure of an alkali metal with hydrogen, it may be regarded that its nucleus is a single unit and around which the additional electron rotates. The additional electron is known as *optical electron.* In hydrogen, the electron remains in its fixed orbits and therefore, definite spectral lines will be observed.

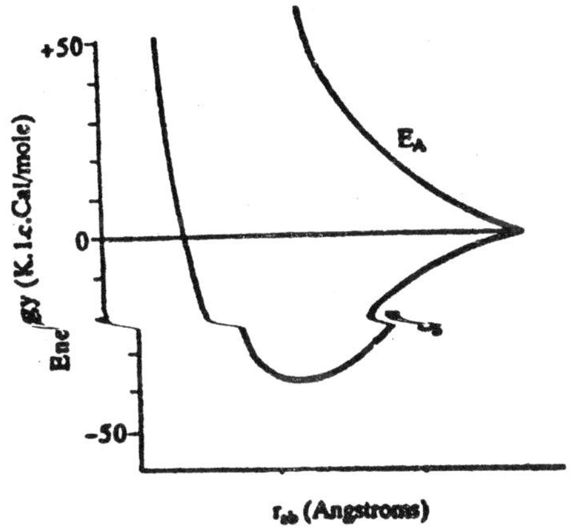

Fig. 16

In the case of an alkali metal, the optical electron does not always rotate in its own orbits. It has a tendency to penetrate into the other inner orbits and this penetration depends upon the eccentricity of the orbit. The greater the eccentricity of the orbit, the more closely the electron may approach the nucleus in the course of its rotation.

It is, therefore, more likely that greater the penetration of the inner levels, the greater will be divergence from hydrogen like behaviour. A simplified energy level diagram for sodium is shown in (Fig. 16), those for the other alkali metals are similar, except for the difference in the principal quantum number of the optical electrons.

The integers indicate the various values of the principal quantum number n in each spectral state, represented by S, P,D and ZF. It may be noted that the S, P, D and F terms are actually doublets.

SOLVED EXAMPLES

Example 1:

The average life-time of an electron in an excited state of hydrogen atom is about 10^{-8} s. How many revolutions does an electron in the n = 2 state make before dropping to the n=1 state? (R = 1.097 × 107 m^{-1}).

Solution:

Let v be the velocity of electron (mass m, charge e) in an orbit of radius r. This basic equations are.

$$\frac{mv^2}{r} = \frac{1}{4\pi\varepsilon_0}\frac{e^2}{r^2}$$

and $$mvr = n\frac{h}{2\pi}.$$

These equations give

$$v = \frac{nh}{2\pi mr}, \; r = \frac{n^2h^2\varepsilon_0}{\pi me^2}.$$

The number of revolutions of the electron in the orbit per second is

$$f = \frac{v}{2\pi r} = \frac{2Re}{n^3}$$

For n = 2 state, we have

$$f = \frac{2\times(1.097\times10^7\,m^{-1})\times(3\times10^8\,ms^{-1})}{8}$$

$$= 8.2 \times 10^{14}\ s^{-1}.$$

Hence the number of revolutions of the electron in its life-time of 0^{-8} second is $(8.2 \times 10^{14}) \times 10^{-8} = 8.2 \times 10^6$.

xample 2:

An orange photon of wavelength 600 nm is emitted from an atom. ind the difference in energy in the two atomic states involved. Find the ame result for the red 6563 Å line of hydrogen. (h = 6.63 × 10^{-34} j s, = 3 × 10^8 m s^{-1}, 1 eV = 1.6 × 10^{-19} J).

Solution:

By Bohr's postulate, the emitted frequency is given by

$$v = \frac{\Delta E}{h},$$

where ΔE is the difference in energy. But $v = c/\lambda$. Therefore

$$\frac{c}{\lambda} = \frac{\Delta E}{h} \text{ or } \Delta E = \frac{hc}{\lambda}.$$

Here $\lambda = 600 \text{ nm} = 600 \times 10^{-9}$ m.

$$\therefore \qquad \Delta E = \frac{(6.63 \times 10^{-34} \text{ Js}) \times (3.0 \times 10^{8} \text{ ms}^{-1})}{600 \times 10^{-9} \text{ m}}$$

$$= 3.31 \times 10^{-19} \text{ J}$$

$$= \frac{3.31 \times 10^{-19}}{1.6 \times 10^{-19}} = 2.07 \text{ eV}.$$

For $\quad \lambda = 6563$ Å $\times 10^{-10}$ m, we can show that

$$\Delta E = 3.03 \times 10^{-19} \text{ J} = 1.9 \text{ eV}.$$

Example 3:

What is the smallest wavelength in the spectral lines of Paschen series in the hydrogen atom ?

Solution:

The formula for Paschan series is

$$\frac{1}{\lambda} = R\left(\frac{1}{3^2} - \frac{1}{n^2}\right), \; n = 4, 5, 6, \ldots$$

For the smallest wave length, $n = \infty$.

$$\therefore \qquad \lambda = \frac{9}{R} = \frac{9}{1.097 \times 10^{7} \text{ m}^{-1}}$$

$$= 8.204 \times 10^{-7} \text{ m} = 8204 \text{ Å}.$$

Example 4:

The series limit of Balmer series is at 3646 Å. Calculate the wavelength of the first member Ha of this series.

Solution:

The formula for Balmer series is

$$\frac{1}{\lambda} = R\left(\frac{1}{2^2} - \frac{1}{n^2}\right), \; n = 3, 4, 5,$$

For series limit n = ∞ ad $\lambda = \lambda_\infty = 3646$ Å.

$$\therefore \qquad \frac{1}{3646\ \text{Å}} = \frac{R}{4} \text{ or } R = \frac{4}{3646\ \text{Å}}.$$

Again, for the first member n = 3.

$$\therefore \qquad \frac{1}{\lambda_1} = R\left(\frac{1}{2^2} - \frac{1}{3^2}\right) = \frac{5R}{36}$$

$$\text{or} \qquad \lambda_1 = \frac{36}{5R} = \frac{36}{5 \times \frac{4}{3646\ \text{Å}}} = 6563\ \text{Å}.$$

Example 5:

The wavelength of the first line of Balmer series is 6563 Å. Calculate Rydberg constant.

Solution:

The wavelength of the spectral lines of Balmer series are given by

$$\frac{1}{\lambda} = R_H\left(\frac{1}{2^2} - \frac{1}{n^2}\right),\ n = 3, 4, 5, \ldots$$

For the first ($\lambda = 6563 \times 10^{-10}$ m), n = 3.

$$\therefore \qquad \frac{1}{6563 \times 10^{-10}\ \text{m}} = R_H\left(\frac{1}{2^2} - \frac{1}{3^2}\right) = \frac{5}{36} = R_H$$

$$\text{or} \qquad R_H = \frac{36}{5 \times (6563 \times 10^{-10}\ \text{m})} = 1.097 \times 10^7\ \text{m}^{-1}.$$

Example 6:

Find the wavelength of the photon emitted when the hydrogen atom goes from n = 10 state to the ground state. ($R_H = 1.097 \times 10^{-3}$ Å^{-1}).

Solution:

Since the atom drops to the ground state (n = 1), the photon emitted belongs to the Lyman series. The wavelengths of the spectral lines in this series are given by

$$\frac{1}{\lambda} = R_H\left(\frac{1}{1^2} - \frac{1}{n^2}\right),\ n = 2, 3, 4, \ldots$$

For n = 10, we have

$$\frac{1}{\lambda} = R_H\left(1-\frac{1}{100}\right) = \frac{99}{100}\ R_H$$

or
$$\lambda = \frac{100}{99\,R_H}$$

$$= \frac{100}{99\times(1.097\times10^{-3}\ \text{Å}^{-1})} = 921\ \text{Å}.$$

Example 7(a):

The first line of the Balmer series in the spectrum of hydrogen has a wavelength of 6563 Å. Calculate the wavelength of the first line of Lyman series in the same spectrum.

Solution:

The wavelength of the spectral lines of hydrogen spectrum are given by

$$\frac{1}{\lambda} = R\left(\frac{1}{n_f^2}-\frac{1}{n_i^2}\right),$$

where R is Rydberg constant. For the first member of Balmer series $n_f = 2$ and $n_i = 3$.

$$\therefore \qquad \frac{1}{\lambda_1} = R\left(\frac{1}{2^2}-\frac{1}{3^2}\right) = \frac{5R}{36}.$$

For the first member of the Lyman series, $n_f = 1$ and $n_i = 2$.

$$\therefore \qquad \frac{1}{\lambda_1'} = R\left(\frac{1}{1^2}-\frac{1}{2^2}\right) = \frac{3R}{4}.$$

From (i) and (ii), we have

$$\frac{\lambda_1}{\lambda_1'} = R\ \frac{5R}{36}\times\frac{4}{3R} = \frac{5}{27}.$$

$$\therefore \qquad \lambda_1' = \frac{5}{27}\ \lambda_1 = \frac{5}{27}\times 6563$$

$$= 1215\ \text{Å}.$$

Example 7(b):

The first member of Balmer series of hydrogen has a wavelength of 6563 Å. Calculate the wavelengths of the second and fourth members.

Solution:

The wavelengths of the spectral lines of Balmer series are given by

$$\frac{1}{\lambda} = R\left(\frac{1}{2^2} - \frac{1}{n^2}\right), \ n = 3, 4, 5, ...$$

For the first member, n = 3.

$$\therefore \quad \frac{1}{\lambda_1} = R\left(\frac{1}{2^2} - \frac{1}{3^2}\right) = \frac{5R}{36} \quad ....(1)$$

For the second member, n = 4.

$$\therefore \quad \frac{1}{\lambda_2} = R\left(\frac{1}{2^2} - \frac{1}{4^2}\right) = \frac{3R}{16} \quad ...(2)$$

Dividing eq. (2) by (1), we get

$$\frac{\lambda_1}{\lambda_2} = \frac{3R}{16} \times \frac{36}{5R} = \frac{27}{20}.$$

$$\lambda_2 = \lambda_1 \times \frac{20}{27}.$$

But $\lambda_1 = 6563$ Å (given).

$$\therefore \quad \lambda_2 = (6563 \text{ Å}) \times \frac{20}{27} = 4861 \text{ Å}.$$

Similarly $\lambda_4 = 4102$ Å.

Example 7(c):

Calculate the (i) energy required to ionise hydrogen atom in the ground state (ii) limit of Balmer series. (Rydberg constant is $1.097 \times 10^7 \ m^{-1}$, $h = 6.63 \times 10^{-34}$ J s, c $3.0 \times 10^8 \ m \ s^{-1}$, 1 eV = 1.6×10^{-10} J).

Solution:

(i) The energy required to ionise, that is, to remove an electron from the hydrogen atom in the ground state (n = 1) to infinity (where the energy is zero) is numerically equal to the energy of the electron in the n = 1 orbit. Now, the energy of the electron in n th orbit of hydrogen is given by

$$E_n = -\frac{me^4}{8\varepsilon_0^2 h^2}\left(\frac{1}{n^2}\right)$$

$$= -\frac{Rhc}{n^2}. \qquad \left[\because R = \frac{me^4}{8\varepsilon_0^2 ch^3}\right]$$

For the ground (first) orbit, n = 1.

$$\therefore \quad E_1 = -Rhc$$

$$= -(1.097 \times 10^7 \text{ m}^{-1})(6.63 \times 10^{-34}\text{J})(3 \times 10^8 \text{m s}^{-1})$$

$$= -21.8 \times 10^{-19} \text{ J}$$

$$= -\frac{21.8 \times 10^{-19}}{1.6 \times 10^{-19}} = -13.6 \text{ eV}.$$

Hence the energy required to remove the electron from n = 1 orbit to infinity is 13.6 eV.

(ii) The Balmer series is represented by

$$\frac{1}{\lambda} = R\left(\frac{1}{2^2} - \frac{1}{n^2}\right), \; n = 3, 4, 5, \ldots$$

For the series limit n = ∞ so that

$$\frac{1}{\lambda} = \frac{R}{4}.$$

$$\therefore \quad \lambda = \frac{4}{R} = \frac{4}{1.097 \times 10^7 \text{ m}^{-1}}$$

$$= 3.646 \times 10^{-7} \text{ m} = 3646 \text{ Å}.$$

Example 8:

With Franck-Hertz type of experiment on sodium, the first spectral line to appear is the D-line, $\lambda = 5.89 \times 10^{-7}$ m. What is the first excitation potential of sodium ? Given : h = 6.63 × 10^{-34} J, s, c 3 × 108 m/s and 1 eV = 1.6 × 10^{-19} J.

Solution:

Let V volt be the excitation potential. Then the (excitation) energy imparted to the electron will be eV joule, where e coulomb is the charge on the electron. This energy is re-emitted as photon (radiation) when the electron returns to the normal state. If v be frequency of the emitted radiation, the photon energy will be hv. Thus

$$eV = hv$$

or $$V = \frac{h\nu}{e}.$$

If l be the wavelength of the emitted radiation, then $\nu = c/\lambda$.

$$\therefore \quad V = \frac{h\nu}{e\lambda}$$

$$= \frac{(6.63\times10^{-34}\,\text{J s})(3\times10^{8}\,\text{m s}^{-1})}{(1.6\times10^{-19}\,\text{C})(5.89\times10^{-7}\,\text{m})}$$

$$= 2.1\ (\text{J/C}) = 2.1\ \text{V}.$$

Example 9:

The first two excitation potentials of atomic hydrogen in Franck-Hertz experiment are 10.2 and 12.09 volts. Draw an energy level diagram and show all possible transitions for emission and absorption along their wavelengths.

($h = 6.63 \times 10^{-34}$ J s, c = 3.0 × 108 m/s, 1 eV = 1.6×10^{-19} J.)

Solution:

The energy level diagram, and the *three* possible transitions (a), (b) and (c) for emission are shown in the figure.

The frequency of radiation resulting from the transitions (a) is given by

$$\nu_a = \frac{E_2 - E_1}{h},$$

and the corresponding wavelength is

$$\lambda_a = \frac{hc}{E_2 - E_1}. \qquad [\because c = \nu\lambda]$$

Now, from the Fig., $E_2 - E_1 = 10.2$ eV $= 10.2 \times (1.6 \times 10^{-19})$ J. Therefore

$$\lambda_a = \frac{(6.63\times10^{-34}\,\text{J s})\times(3.0\times10^{8}\,\text{m s}^{-1})}{(10.2\times1.6\times10^{-19}\,\text{J})}$$

$$= 1.216 \times 10^{-7}\ \text{m} = 1216 \times 10^{-10}\ \text{m} = 1216\ \text{Å}.$$

Similarly, $$\lambda_b = \frac{hc}{(E_3 - E_1)}$$

$$= \frac{(6.63\times10^{-34})\times(3.0\times10^{8})}{(12.09\times1.6\times10^{-19})}$$

$$= 1.026 \times 10^{-7} \text{ m} = 1026 \times 10^{-10} \text{ m} = 1026 \text{ Å}.$$

Also, $$\lambda_c = \frac{hc}{(E_3 - E_2)}$$

$$= \frac{hc}{(E_3 - E_1) - (E_2 - E_1)}$$

$$= \frac{hc}{(12.09 - 10.2)\,\text{eV}} = \frac{hc}{1.89\,\text{eV}}$$

$$= \frac{(6.63 \times 10^{-34}) \times (3.0 \times 10^{8})}{(1.89 \times 1.6 \times 10^{-19})}$$

$$= 6.567 \times 10^{-7} \text{ m} = 6567 \times 10^{-10} \text{ m} = 6567 \text{ Å}.$$

In absorption, only the transitions starting from n = 1 shall be observed which correspond to 1216 Å and 1026 Å.

Example 10:

A positronium atom is a system consisting of a position and an electron. Calculate the reduced mass, the Rydberg constant and the wavelength of the first Balmer line for positronium. (Give $m = 9.1 \times 10^{-31}$ kg, $R_H = 1.09737 \times 10^{-3}$ Å^{-1} and $H_\alpha = 6563$ Å).

Solution:

The positron has the same mass m as the electron and has equal but positive charge. The reduced mass of the electron-positron atom is therefore

$$m = \frac{(m)(m)}{m + m} = \frac{1}{2}\, m = 4.55 \times 10^{-31} \text{ kg},$$

while the reduced mass of electron in hydrogen is very nearly m.

The Rydberg constant $\left(\frac{\mu e^4}{8\varepsilon_0^{\,2} ch^3}\right)$ for positronium is therefore half that for hydrogen (with infinitely heavy nucleus). Thus

$$R_P = \frac{1}{2}\, R_H = 0.54868 \times 10^{-3} \text{ Å}^{-1}.$$

The wavelength of first Balmer line (H_α) for hydrogen atom is given

$$\frac{1}{\lambda_H} = R_H \left(\frac{1}{2^2} - \frac{1}{3^2}\right),$$

while that for positronium atom is $\frac{1}{\lambda_p} = R_P \left(\frac{1}{2^2} - \frac{1}{3^2}\right)$.

Thus $$\frac{\lambda_P}{\lambda_H} = \frac{R_H}{R_P} = 2.$$

$$\therefore \quad \lambda_P = 2\lambda_H = 2 \times 6563 = 13126 \text{ Å}.$$

Example 11:

Find the recoil speed of hydrogen atom after it emits a photon in going from $n = 3$ to $n = 1$ state. Electron mass is 9.11×10^{-31} kg and $h = 6.626 \times 10^{-34}$ J s.

Solution:

The energy of the hydrogen atom in the n th state is given by

$$E_n = -\frac{Rhc}{n^2},$$

where R is Rydberg constant. From this, we get

$$E_1 - E_3 = -\frac{Rhc}{1^2} + \frac{Rhc}{3^2} = -\frac{8}{9} R h c.$$

The energy of the emitted photon is

$$\Delta E = E_1 \sim E_3 = \frac{8}{9} R h c.$$

The momentum of the photon is

$$p = \frac{\Delta E}{c} = \frac{8}{9} R h.$$

By conservation of momentum, the recoil momentum of the hydrogen atom will be equal (and opposite) to the momentum of the emitted photon. The recoil speed of the atom is

$$v = \frac{\text{momentum}}{\text{mass}} = \frac{8}{9}\frac{Rh}{m_H}.$$

But $m_H = 1836$ m, where m is electron mass.

$$\therefore \quad v = \frac{8}{9}\frac{Rh}{(1836\,m)}.$$

Putting the given values of h, m and using $R = 1.097 \times 10^7 \text{ m}^{-1}$, we get

$$v = \frac{8}{9}\frac{(1.097 \times 10^7 \text{ m}^{-1})(6.626 \times 10^{-34} \text{ Js})}{1836\,(9.11 \times 10^{-31} \text{ kg})}$$

$$= 3.86 \text{ m/s}.$$

Example 12:

The ionisation potential of hydrogen tom is 13.6 volt. Find the wavelength of the Lyman series limit.

Solution:

When an atom absorbs energy so that its two stationary states becomes existed simultaneously, two superimposed sets of radiations are produced to give a 'beat' variation. If v¢ and v¢¢ represent the frequencies of two superimposed vibrations, the frequency v of the emitted beat is

$$v = v' - v''.$$

The wave mechanical vibrations are related to the energies of the corresponding state by the usual quantum expressions $E' = hv'$ and $E' - E'' = hv''$ so that

$$E' - E'' = h\,(v' - v'')$$

$$\Delta E = 13.61\left(\frac{1}{(1)^2} - \frac{1}{\infty}\right) = 13.61 \text{ eV}.$$

Therefore, ionisations potential = 13.61 eV.

Example 13:

A beam of monochromatic photons of energy 9 eV is incident on hydrogen gas all of whose atoms are in the ground state. It is found that the beam is fully transmitted without absorption. Why ? The ground state energy of an electron in the hydrogen atom is $E_1 = -13.6$ eV.

Solution:

The minimum energy that can be absorbed by ground-state hydrogen atom is $E_1 \sim E_2$, which would excite it to the next state (n = 2). Now,

$$E_1 \sim E_2 = E_1 \sim \frac{E_1}{4} \qquad \left[\because E_n = \frac{E_1}{n^2}\right]$$

$$= \frac{3}{4} E_1$$

$$= \frac{3}{4} \times 13.6 = 10.2 \text{ eV}.$$

Hence photons of energy 9 eV cannot be absorbed by hydrogen atoms.

Example 14(a):

Calculate the ionisation potential of hydrogen from the following date.

Solution:

According to Bohr's theory, the energy of an electron in the nth orbit of hydrogen atom is

$$E = -\frac{2\pi^2 e^4 m}{n^2 h^2}$$

When the atom is in the normal state, the only electron in the hydrogen atom in the first orbit, *i.e.*, n = 1.

$$\therefore \quad E = -\frac{2\pi^2 m e^4}{h^2}$$

or $$E = \frac{2 \times (3.14)^2 \times 9 \times 10^{-28} \times (4.8 \times 10)^4}{(6.6 \times 20^{27})^2} = 2.165 \times 10^{-11} \text{ eg.}$$

To remove the electron from first orbit to infinity 2.165×10^{-11} erg of energy must be supplied. The amount of energy is called the ionisation potential of hydrogen atom.

Ionisation potential of hydrogen atom

$$= 2.165 \times 10^{-11} \text{ erg}$$

$$= \frac{2.165 \times 10^{-11}}{1.6 \times 10^{-12}} \text{ electron volts} = 13.53 \text{ eV.}$$

Example 15:

Energy in a Bohr's orbit is given to be equal to – B/n^2 *where* B = 2.179×10^{-11} *erg. Calculate the frequency of radiation and also the wave number when the electron jumps from the third orbit to the second* (h = 6.62×10^{-27} *erg sec) orbit.*

Solution:

Energy of a Bohr's orbit is given by

$$E = \frac{B}{n^2} = \frac{2.179 \times 10^{-11}}{n^2} \text{ erg.}$$

For the third orbit, n = 3

$$E_2 = \frac{2.179 \times 10^{-11}}{9} = -\ 0.2421 \times 10^{-11} \text{ erg.}$$

For the second orbit, n = 2

$$E_2 = \frac{2.179 \times 10^{-11}}{4} = -\ 0.54475 \times 10^{-11} \text{ erg.}$$

$$E_3 - E_2 = (0.54475 \times 10^{-11}) - (0.2421 \times 10^{-11})$$
$$= 0.30265 \times 10^{-11} \text{ erg.}$$

When electron jumps from the third to the second orbit, a photon is emitted, the energy of which is given by

$h\nu = E_3 - E_2$ where ν is the frequency

$$\nu = \frac{E_3 - E_2}{h} = \frac{0.30265 \times 10^{-11}}{6.62 \times 10^{-27}} = 4.572 \times 10^{14} \text{ sec}^{-1}.$$

The corresponding wave number is given by

$$\bar{\nu} = \frac{1}{\lambda} = \frac{\nu}{c} = \frac{4.572 \times 10^{14}}{3 \times 10^{-10}} = 15240 \text{ cm}^{-1}$$

Example 16:

Give using spectral notation for the following states of the atom.

(i) n = 4, L = 2, S = 0

(ii) n = 4, L = 1, S = 1, J = 0 and

(iii) n = 3, L = 2, multiplicity 2.

Solution:

(i) Multiplicity (2S + 1) = 1, J = L + S = 2

∴ State will be 4 $^1D_2$.

(ii) Multiplicity (2S + 1) = 3 ∴ State will be 4 $^3P_0$.

(iii) Multiplicity will be (2S + 1) = 2 or S = 1/2

As L = 2, J = 5/2 or 3/2.

The two states which are positive are 3 $^2D_{5/2}$ or $^2D_{3/2}$. *Give using spectral notation for the following states of the atom.*

(i) n = 4, L = 2, S = 0

(ii) n = 4, L = 1, S = 1, J = 0 and

(iii) n = 3, L = 2, multiplicity 2.

Solution:

(i) Multiplicity (2S + 1) = 1, J = L + S = 2

$\therefore$ State will be 4 $^1D_2$.

(ii) Multiplicity (2S + 1) = 3 $\therefore$ State will be 4 $^3P_0$.

(iii) Multiplicity will be (2S + 1) = 2 or S = 1/2

As L = 2, J = 5/2 or 3/2.

The two states which are positive are 3 $^2D_{5/2}$ or $^2D_{3/2}$.

Example 17(a):

Calculate the energy in calories per mole or per Einstein for radiations of wavelength 100 Å..

Solution:

Energy per quantum = hv

But h = 6.62 × 10^{-27} erg-sec.

$$\therefore \quad V = \frac{c}{\lambda} = \frac{3\times10^{10}}{1000\,\text{Å}} = \frac{3\times10^{10}\,\text{cm.sec}^{-1}}{1000\times10^{-8}\,\text{cm}} = 3\times10^{15}\ \text{sec}^{-1}$$

Energy per quantum = hv

$= (6.62 \times 10^{-27}$ erg. sec) $(3 \times 10^{15}$ sec$^{-1})$

$= 19.86 \times 10^{-12}$ ergs.

But N = 6.02 × 10^{28} molecules/mole

and hv = 19.86 × 10^{-12} erg.

Therefore, the energy per mole or per Einstein

= N hv

= (6.02 × 10^{23} molecules mole^{-1}) 19.86 × 10^{-12} ergs/mole

= 11.94 × 10^{12} ergs/mole

$= \frac{11.94\times10^{12}}{10^{7}}$ Jule/mole [$\because$ 1 Joule = 10^7 ergs]

$$= \frac{11.94 \times 10^{12}}{4.184 \times 10^{7}} \text{ cal/mole} \qquad [\because 1 \text{ cal} = 4.184 \text{ Joules}]$$

$$= \frac{2.8590 \times 10^{5}}{10^{3}} \text{ K cal/mole } [\because 1 \text{ K cal} = 10^3 \text{ cal}]$$

$$= 285.90 \text{ K cal/mole.}$$

$$= \frac{285.90}{23.06} \text{ electron volt } [\because 1 \text{ eV} = 23.06 \text{ K cal/mole}]$$

$$= 12.390 \text{ electron-volts.}$$

Example 17(b):

$$B \rightarrow C$$

1.0×10^{-5} mole of B was formed on absorption of 6.62×10^7 ergs at 3600 Å. Calculate the quantum yield or efficiency.

Solution:

No. of moles reacting

$$= 1.0 \times 10^{-5} \times 6.02 \times 10^{23} \text{ molecules}$$

$$= 6.02 \times 10^{18} \text{ molecules}$$

No. of quanta absorbed

$$= \frac{\text{Total energy absorbed}}{\text{Energy of one quantum}} = \frac{6.62 \times 10^{7} \text{ ergs}}{hv}$$

$$= \frac{6.62 \times 10^{7}}{hc/\lambda}$$

$$= \frac{6.62 \times 10^{7} \times \lambda}{hc} \qquad \left[\because v = \frac{c}{\lambda}\right]$$

But $\quad l = 3600 \text{ Å} = 3600 \times 10^{-8}$ cm.

$c = 3 \times 10^{10}$ cm/sec, $h = 6.62 \times 10^{-27}$ erg/sec.

$$\text{No. of quanta absorbed} = \frac{6.62 \times 10^{7} \times 3600 \times 10^{-8}}{6.62 \times 10^{-27} \times 3 \times 10^{10}} = 1.2 \times 10^{19}$$

$$\text{Therefore, equation yield} = \frac{\text{No. of molecules reacting}}{\text{NO. of quanta absorbed}}$$

$$= \frac{6.02 \times 10^{18}}{1.2 \times 10^{19}} = 0.506.$$

Example 17(c):

The bond energy in a molecule is 142.95 cal/mole. What is the longest wave-length of light capable of dissociating this molecule ?

Solution:

Energy per mole = N hv

where N= 6.02×10^{23} molecules, h = 6.62×10^{-27} erg-sec

$$v = \text{frequency} = \frac{c}{\lambda} = \frac{3.0 \times 10^{10}}{\lambda}$$

$$\therefore \quad \text{Energy per mole} = \frac{6.02 \times 10^{23} \times 6.62 \times 10^{-27} \times 3 \times 10^{10}}{\lambda} \quad \text{...(A)}$$

The energy per mole = 142.95 cal/mole

$= 142.95 \times 10^3$ cal/mole

$= 142.95 \times 10^3 \times 4.184 \times 10^7$ ergs/mole ...(B)

Equating eqs. (A) and (B), we get

$$\frac{6.02 \times 10^{23} \times 6.62 \times 10^{-27} \times 3 \times 10^{10}}{142.95 \times 10^3 \times 4.184 \times 10^7}$$

$$= 142.95 \times 103 \times 4.184 \times 107$$

or

$$\lambda = \frac{6.02 \times 10^{23} \times 6.62 \times 10^{-27} \times 3 \times 10^{10}}{142.95 \times 10^3 \times 4.184 \times 10^7} \text{ cm}$$

$$= 2000 \times 10^{-8} \text{ cm} = 2000 \text{ Å}.$$

Example 18:

The dissociation energy of hydrogen is 102900 cal/mole. If H_2 is dissociated by illumination with radiation of wave-length 2537 Å. What fraction of the radiant energy will be converted in to kinetic energy ?

Solution:

Dissociation energy= 102900 cal/mole

$$= \frac{102900}{6.023 \times 10^{23}} = 1.708 \times 10^{-19} \text{ cal/molecule.}$$

Also, one quantum of light is able to dissociate one molecule of hydrogen. Therefore, Energy of one quantum = hv = $\frac{hc}{\lambda}$.

But $h = 6.625 \times 10^{-27}$ erg sec,

$c = 3 \times 10^{10}$ cm/sec

$\lambda = 2537$ Å $= 2537 \times 10^{-8}$ cm

$\therefore$ Energy of one quantum $= \dfrac{6.625 \times 10^{-27} \times 3 \times 10^{10}}{2537 \times 10^{-8}}$ ergs

$$= \frac{3 \times 6.625 \times 10^{-9}}{2537} \text{ ergs}$$

$$= \frac{3 \times 6.625 \times 10^{-9}}{2537 \times 4.18 \times 10^{7}} \text{ cal}$$

[$\because$ 1 cal 4.18 $\times 10^7$ ergs]

But the dissociation energy per mole $= 1.708 \times 10^{-19}$ cal

$\therefore$ Energy converted into kinetic energy

$= (1.874 \times 10^{-19} - 1.708 \times 10^{-19})$ cal

$= 0.166 \times 10^{-19}$ cal

Hence, fraction converted into K.E. $= \dfrac{0.166 \times 10^{-19}}{1.874 \times 10^{-19}}$.

Example 19:

Calculate the time taken by the electron to traverse the first orbit in the hydrogen atom. Electron mass and charge are 9.1 $\times$ 10^{-31} kg and 1.6 $\times$ 10^{-19} C and h = 6.63 $\times$ 10^{-34} J-s.

Solution:

The radius of the n the Bohr orbit is

$r_n = 4\,\pi\,\varepsilon_0 \dfrac{n^2 h^2}{4\pi^2 m e^2}$ and the velocity of electron in the n th orbit is

$$v_n = \frac{1}{4\pi\varepsilon_0} \frac{2\pi e^2}{n h}.$$

The time taken by the electron to traverse the n the orbit is therefore

$$T_n = \frac{2\pi r_n}{v_n}$$

$$= 2\pi\,(4\,\pi\,\varepsilon_0)^2 \frac{n^2 h^2}{4\pi^2 m e^2} \frac{n h}{2\pi e^2}$$

$$= \frac{4\varepsilon_0^2 h^3 n^3}{me^4}.$$

For the first orbit, n = substituting the given values of h, m, e and $\varepsilon_0 = 8.85 \times 10^{-12}$ $C^2/N\text{-}m^2$, we get

$$T_1 = \frac{4.(8.85\times10^{-12})^2\,(6.63\times10^{-34})^3}{(9.1\times10^{-31})\,(1.6\times10^{-19})^4} = 1.5 \times 10^{-16}\text{ s.}$$

Example 20:

Find an expression for the radius of the electron orbit in the hydrogen atom in its n th state. What will be the approximate quantum number n for an electron in an orbit of radius 1 mm ? (Take required values from problem 1).

Solution:

The basic equation in the Bohr's theory of hydrogen atom are.

$$\frac{mv^2}{r} = \frac{1}{4\pi\varepsilon_0}\frac{e^2}{r^2} \qquad ...(1)$$

and

$$mvr = \frac{nh}{2\pi}. \qquad ...(2)$$

Squaring (2) and dividing by (1), we get

$$r = 4\,\pi\,\varepsilon_0\,\frac{n^2h^2}{4\pi^2me^2}.$$

Substituting the given values, we get

$$r = \frac{1}{9\times10^9\text{ Nm}^2/\text{C}^2}\times$$

$$\frac{n^2\,(6.63\times10^{-34}\text{ Js})^2}{4\times(3.14)^2\times(9.1\times10^{-31}\text{ kg})\times(1.6\times10^{-19}\text{ C})^2}$$

$= 0.53 \times 10^{-10}$ n^2 meter

$= 0.53$ n^2 Å.

For r = 1 mm = 10^7 Å, we have

$$10^7 = 0.53\ n^2.$$

$$\therefore \qquad n^2 = \frac{10^7}{0.53} = 18.87 \times 10^6 \text{ or n; } 4350$$

Example 21:

The wavelength of the first line of Balmer series of hydrogen is 6562.8 Å. Calculate (i) the ionisation potential and the first excitation potential of the hydrogen atom. ($h = 6.63 \times 10^{-34}$ J-s,

$c = 3\times 10^{8} m\ s^{-1}$).

Solution:

(i) The wavelengths of the Balmer lines are given by

$$\frac{1}{\lambda} = R\left(\frac{1}{2^2} - \frac{1}{n^2}\right), \qquad n = 3, 4, 5, ...$$

For the first line n = 3 and l = 6562.8 Å = 6562.8×10^{-10} m.

$$\therefore \quad \frac{1}{6562.8\times 10^{-10}} = R\left(\frac{1}{2^2} - \frac{1}{3^2}\right) = \frac{5R}{36}$$

or
$$R = \frac{36}{5\times 6562.8\times 10^{-10}\,\text{m}} = 1.097 \times 10^{7}\ \text{m}^{-1}.$$

This ionisation potential of an atom is numerically equal to the (ionisation) energy required to remove an electron completely from the atom in the normal state. In hydrogen atom, the electron stays in the first orbit (n = 1). Hence the energy required to remove this electron to infinity (where the energy is considered to be zero) is numerically equal to the energy of the electron in the first orbit. We know that electron energy in hydrogen atom is given by

$$E_n = -\frac{Rhc}{n^2}.$$

In the ground state, n = 1.

$$\therefore \quad E_1 = -Rhc$$

$$= -13.6 \text{ eV}. \qquad \text{(as obtained in the leas problem).}$$

Here the ionisation potential of the hydrogen atom is 13.6 volts.

(ii) The first excitation potential of the atom is the energy required to shift the electron from n = 1 to n = 2 orbit, that is, $E_1 \sim E_2$. Now,

$$E_1 = -13.6 \text{ eV}.$$

and
$$E_2 = \frac{E_1}{4} = -3.4 \text{ eV}. \qquad \left[\because E_n \propto \frac{1}{n^2}\right]$$

$\therefore \quad E_1 \sim E_2 = 13.6 - 3.4 = 10.2$ eV.

Hence the first excitation potential is 10.2 volts.

Example 22:

The wavelength of a yellow line of sodium is 5896 Å. Calculate its wave number and frequency.

Solution:

The wave number $\bar{v}$ is the reciprocal of wave-length λ, that is,

$$\bar{v} = \frac{1}{\lambda}.$$

Here λ = 5896 Å = 5896 × 10^{-8} cm.

$$\therefore \quad \bar{v} = \frac{1}{5896 \times 10^{-8}\,\text{cm}} = 16960 \text{ per cm.}$$

The frequency v is related to the wavelength l by

$$c = v\lambda,$$

where c is the speed of light. Thus

$$v = \frac{c}{\lambda} = \frac{3 \times 10^{8}\ \text{cm/s}}{5896 \times 10^{-8}\ \text{cm}} = 5.1 \times 10^{12}\ \text{s}^{-1}$$

2

Neutron Proton Scattering at Low Energies

INTRODUCTION

Two kinds of the reactions can be involved in neutron proton interaction: One scattering and other radiative capture. The latter has low probability and cross section for high energy neutrons, as the cross section for the competing radiative capture reaction decreases with 1/v, where v is the neutron velocity. In practice protons are bound in molecules. The chemical binding energy of the proton in a molecule is about 0.1 eV. Thus for neutron energies = > 1eV the proton can be assumed as free. *This sets a lower limit to the neutron energy*. If the neutron energy is less than 10 MeV, only the S-wave overlaps with the nuclear potential and is scattered.

In the centre of mass system, the Schrodinger equation for the two body (n-p system) problem is

$$\Delta^2\phi + \frac{M}{\hbar^2}[E - V(r)]\,\psi = 0 \qquad ...(1)$$

where,

M = Proton or neutron mass = 2 × Reduced mass of the system.

E = Incident kinetic energy in C-M system = (incident K. E. in L-co-ordinates)

and V(r) = Inter-nucleon potential energy.

At large distances from the centre of scattering the solution of this equation is expected to be of the form

$$\psi = e^{ikz} + \frac{e^{ikr}}{r} f(\theta) \qquad ...(2)$$

The term e^{ikz} represents a plane wave describing a beam of particles moving in the z-direction towards the origin (scattering centre). The second term represents the scattered wave. The complex quantity $f(\theta)$ is the scattering amplitude in the direction θ and is to be evaluated in terms of k. In the case of a spherically symmetric potential the entire arrangement is axially symmetric about the incident direction and hence does not depend on the azimuthal angle ϕ. The 1/r dependence is necessary for the conservation of particles in the outgoing wave. The volume of a spherical shell, between r and r + dr is $4\pi r^2$ dr and hence the density of particles in it or the probability of finding one particle in the spherical shell must vary with $1/r^2$ which is proportional to the square of the amplitude of the scattered wave in the shell. Hence the amplitude of the scattered wave must vary with 1/r.

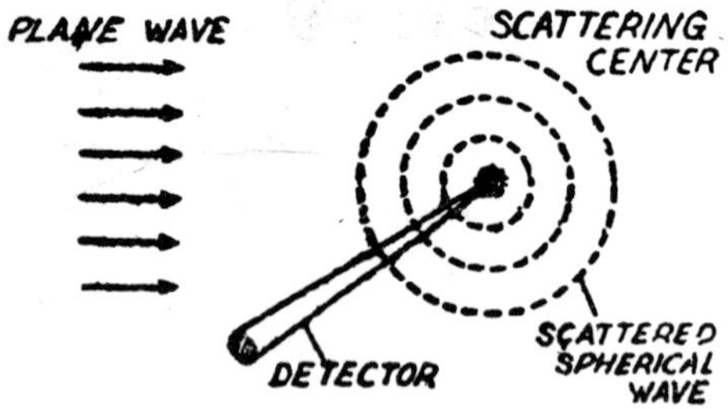

Fig. 1 : Scattering process.

To compute the differential scattering cross section, we must find the number of particles dN scattered in unit time by one target nucleus into a solid angle $d\Omega$ and the incident flux F. If v is the speed of an incoming particle with respect to the scatterer, then

Incoming flux of particles $F = \psi_{in}^* \psi_{in} v = v$

Similarly, dN is equal to the flux of scattered particles $\psi_{sc}^*\psi_{sc}v$ multiplied by the area $r^2 d\Omega$ cut out by $d\Omega$ on a spherical surface of radius r and is given by

$$dN = \psi_{sc}^*\psi_{sc}\, vr^2 d\Omega = |\, f(\theta)\,|^2\, v d\Omega$$

$\therefore$ The differential cross section $d\sigma = |\, f(\theta)\,|^2 v d\Omega = |\, f(\theta)\,|^2 \Omega$

or $$\sigma = \int |f(\theta)|^2 d\Omega = 2\pi \int |f(\theta)|^2 \sin\theta d\theta \qquad ...(3)$$

First of all let us consider the wave equation (1) in the absence of a scattering centre [V(r) = 0 for all values of r].

$$\Delta^2\phi + \left(\frac{ME}{\hbar^2}\right)\psi = 0 \qquad ...(4)$$

This has the solution $\psi = e^{ikz}$ ...(5)

where $k = \frac{1}{\lambda\!\!\!^{-}} = \frac{\sqrt{(ME)}}{\hbar}$

Rayleigh proposed that this type of wavefunction can be expanded into a series in terms of spherical harmonic functions. Thus, eqn. (5) can be written as an infinite series

$$\psi = e^{ikz} = e^{ikr\cos\theta} = \sum_{i=0}^{\infty} R_l(r)\, Y_l, (\theta), \qquad ...(6)$$

where l is the integer representing the number of the partial waves. It, as usual, signifies the orbital angular momentum of the system. The radial functions $R_l(r)$ are solutions of the radial part of equation (4).

$$\frac{1}{r^2}\frac{d}{dr}\left(r^2\frac{dR}{dr}\right) + \left(k^2 - \frac{l(l+1)}{r^2}\right)R = 0 \qquad ...(7)$$

This equation has two solutions, one is not finite at the origin and cannot represent the plane wave. The other is finite at origin and can be represented in terms of spherical Bessel functions as

$$R_l(r) - i^l \sqrt{[4\pi(2l+1)]}\, j_l(kr) \qquad ...(8)$$

The square of this gives the r-dependence of the probability density for each partial wave in expression (6). The values of first few spherical Bessel functions are given as

$$j_0(kr) = \frac{\sin kr}{kr},\ j_1(kr) = \frac{\sin kr}{(kr)^2} - \frac{\cos kr}{kr},$$

$$j_2(kr) = \left(\frac{3}{(kr)^3} - \frac{1}{kr}\right)\sin kr - \frac{3\cos kr}{(kr)^2},$$

The spherical harmonic function is given by

$$Y_{l,\,0}(\theta) = \frac{(2l+1)^{1/2}}{(4\pi)^{1/2}}\, Pt\,(\cos\theta), \qquad ...(9)$$

where P_l (cos θ) is the Legendre polynominal of order *l*. The square of the spherical harmonic function gives the angular dependence of the probability density. The values of the first few Legendre polynomials are

$$P_0 (\cos\theta) = 1,\ P_1 (\cos\theta) = \cos\theta,\ P_2 (\cos\theta) = \frac{1}{2}(3\cos^2\theta - 1).$$

For incident neutrons kinetic energy less than 10 MeV (in the lab-system), the only partial wave involved in scattering is the $l = 0$ or S-wave. The scattering is then spherically symmetric in the centre of mass system. The higher the *l*-value, the larger the impact parameter has to be for a particle with given linear momentum. In the absence of a scattering potential equation (39) can be written as

$$\psi = R_0(r) Y_{0,0}(\theta) + \sum_{l=1}^{\infty} R_l(r) = \frac{\sin kr}{kr} + \left(e^{ikz} - \frac{\sin kr}{kr}\right) \quad ...(10)$$

The averaged value of the quantity within the brackets over all directions in space is zero. The first term corresponds to the spherically symmetric partial wave (S-wave). For S-wave scattering, only the first term is affected and can therefore be written as ψ_s in the presence of the scattering potential V(r), We can write it as $\psi_s = u(r)/r$.

Thus, as $r \to \infty$, the solution u(r) assumes the form c sin $(kr + \delta_0)$, where c is an arbitrary constant and δ_0 is some phase angle. Thus the complete wave function outside the scattering potential is

$$\psi = \frac{c\sin(kr + \delta_0)}{r} + \left(e^{ikz} - \frac{\sin kr}{kr}\right)$$

$$= e^{ikz} + \frac{c}{r}\frac{e^{ikz}e^{i\delta_0} - e^{0kr}e^{i\delta_0}}{2i} - \frac{1}{kr}\frac{e^{ikr} - e^{-ikr}}{2i}$$

$$= e^{ikr} + \frac{e^{ikr}}{r}\left(\frac{ce^{i\delta_0} - 1/k}{2i}\right) - \frac{e^{-ikr}}{r}\left(\frac{ce^{-\delta_0} - 1/k}{2i}\right)$$

This scattered wave must contain no incoming wave. Therefore, we can write the coefficient of e^{-ikr} as zero. Thus, we have

$$ce^{-i\delta_0} - \frac{1}{k} = 0 \quad \text{or} \quad c = \left(\frac{1}{k}\right)e^{i\delta_0},$$

hence $$\psi = e^{ikz} + \frac{e^{2i\delta_0}}{r}\frac{e^{2i\delta_0} - 1}{2ik} \quad ...(11)$$

Comparing this with the standard solution (2), we get

$$f(\theta) = \frac{e^{2i\delta_0} - 1}{2ik} = \frac{e^{i\delta_0}}{k} \cdot \frac{e^{i\delta_0} - e^{-i\delta_0}}{2i} = \frac{e^{i\delta_0}}{k} \sin \delta_0, \quad ...(12)$$

The total elastic scattering cross section is given as

$$\sigma_0 = 2\pi \int_0^{\pi} \frac{\sin^2 \delta_0}{k^2} \sin\theta \, d\theta = \frac{4\pi}{k^2} \sin^2 \delta_0$$

$$= 4\pi \lambda\!\!\!^{-2} \sin^2 \delta_0 \quad ...(13)$$

The analysis here is carried through only for $l = 0$ scattering. Higher orbital angular momentum waves also have to be considered at higher energies. The total cross-section can be written as a sum of partial cross-sections, one for each l-wave. The partial cross-sections are

$$\sigma_l = 4\pi \lambda\!\!\!^{-2} (2l + 1) \sin^2 \delta_l \quad ...(14)$$

Scattering Length

For neutrons of very low energy scattered free protons, $\lambda\!\!\!^{-}$ is very large and hence k is very small. It can seen from eqn. (12) that as $k \to 0$, δ_0 must also approach zero, otherwise $f(\theta)$ would become infinite. Thus for low energy neutrons $f(\theta)$ can be written as

$$f_0 = \lim_{\delta_0 \to 0} \frac{e^{i\delta_0} \sin \delta_0}{k} = \frac{\delta_0}{k} = -a \quad ...(15)$$

where the quantity +a is called the *scattering length* in the convention of Fermi and Marshall. Hence, for low energy neutrons

$$u(r) = c\,(kr + \delta_0) = ck\,(r - a) \quad ...(16)$$

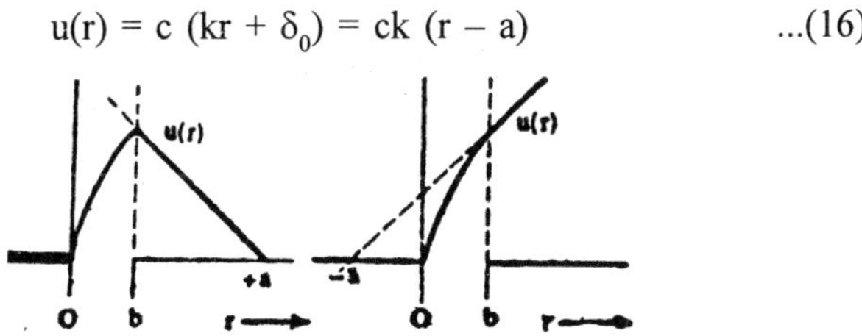

Fig. 1 : (left) Positive scattering length (bound state); (right) Negative scattering length (unbound state).

This is the equation of a straight line intersecting the r-axis at $r = a$, and is obtained by extrapolating the radial wave function u(r) from the point just beyond the range of the nuclear force. Scattering from a potential giving a bound state produces a positive a. If the potential gives only a virtual state, the slope of the inner wave function at r = b is positive and a is negative.

From eqns. (13) and (15) the zero energy scattering cross-section becomes

$$\sigma_0 = 4\pi a^2 \qquad ...(17)$$

This is identical with the scattering cross-section of an impenetrable sphere of radius a, in the limit of zero energy. The measurement of σ_0 *determines the magnitude of the scattering length a but not its sign.*

Determination of the Phase Shift δ_0

We shall now attempt to determine the phase shift δ_0 for low energy neutron-proton scattering by solving also the Schrodinger's equation in the region where the interaction between the two particles takes place. For this we make the simple assumption of a square well for the nuclear potential. Inside the well of depth V_0 and radius b the radial wave equation for particles whose total energy has the positive value E is

$$\frac{d^2u}{dr^2} + \frac{M}{\hbar^2}[E + V_0]\,u(r) = 0 \qquad ...(18)$$

Inside the square well this equation has the simple solution

$$u(r) = A \sin k_1 r, \text{ where } k_1 = \sqrt{[M(E + V_0)]}/\hbar \qquad ...(19)$$

Outside the square well, the solution can be written as

$$u(r) = B \sin (kr + \delta_0) \qquad ...(20)$$

At the edge of the rectangular well (r = b), the two solutions and their derivatives with respect to r must be continuous.

$$\therefore \qquad A \sin k_1 b = B \sin (kb + \delta_0)$$

and $k_1 A \cos k_1 b = Bk \cos (kb + \delta_0)$

Hence $\qquad k_1 \cot k_1 b = k \cot (kb + \delta_0) \qquad ...(21)$

This relationship is analogous in form to equation

$$K \cot Kb = \alpha$$

which describes the binding energy B of the deuteron in terms of the same rectangular well (V_0, A).

Here we have $K = \sqrt{[M (V_0 - B)]}/\hbar$ and $\alpha = \sqrt{[MB)}/\hbar$

For low energy neutrons ($E < V_0$), we may assume $K = k_1$ (as $V_0 > B$), hence *the wavefunction u(r) inside the well is nearly the same for the deuteron and the n-p-scattering system.* Thus, for approximation we have

$$\frac{\sqrt{(ME)}}{\hbar} \cot\left(\frac{\sqrt{(ME)}}{\hbar} b + \delta_0\right) = \frac{\sqrt{(ME)}}{\hbar}$$

As the scattering length a is much larger than the range b of the potential, thus for very low energy neutrons kb can be neglected in comparison to δ_0.

$$\therefore \qquad \cot \delta_0 = -\sqrt{\left(\frac{B}{E}\right)} \text{ or } \sin \delta_0 = \frac{E}{E + |B|} \qquad ...(22)$$

Substituting this value of $\sin^2 \delta_0$, we obtain the approximate value of total scattering cross-section as

$$\sigma = 4\pi\hbar^2 \frac{E}{E + |B|} = \frac{4\pi\hbar^2}{M} \cdot \frac{1}{E + |B|} \qquad ...(23)$$

The Spin Dependence of Nuclear Forces

At very low energies the situation is very different Numerical substitution of B = 2.22 MeV In equation (23) gives a predicted value for zero energy neutrons ($E = 0$), $\sigma_0 \simeq 2.3$ barns. Which is in violent disagreement with the measured value $\sigma_0 = 20.36 \pm 0.10$ barns. This disagreement is a sign of some fundamental error in our assumptions. This point was cleared up by E. P. Wigner in 1935. He suggested that the scattering occurs not only in the triplet state (^3S), but in the singlet state (^1S) as well.

Experimentally the total angular momentum of the deuteron nucleus is unity. The spins are, therefore, correlated in the ground state of the deuteron. The situation is different in n-p scattering experiment. When unpolarized neutrons strike randomly oriented protons, their uncorrelated spins add up to unity in three fourths of the collisions and to zero in one fourth of the collisions. In other words we can say that the triplet state (S = 1) has three times the statistical weight (2S + 1) of a singlet (S = 0) state. The total n-p scattering cross-section will then be

$$\sigma = \frac{3}{4}\sigma_t + \frac{1}{4}\sigma_s,$$

where σ_t and as indicate the cross-sections in triplet and singlet states respectively.

One test of Wigner's hypothesis is by measurement of the cross-section over the range 0 to 5 MeV, where the theoretical expression for a, should hold. We can use the calculated value σ_t = 2.3 barns and the experimental information on σ_0 (zero energy cross-section) to calculate σ_s. To fit the experimental data, we must have (in barns)

$$20.36 = \frac{3}{4}(2.3) + \frac{1}{4}\sigma_s$$

$$\therefore \qquad \sigma_s = 74 \text{ barns.} \qquad ...(24)$$

Introducing the singlet and triplet scattering lengths as and a_t, where $\sigma_s = 4 + 4\pi a_s^2$ and $\sigma_t = 4\pi a_t^2$, we obtain the rough estimates $a_t \sim 4.3$ fermi and as $a_s \sim 24.3$ fermi.

COHERENT SCATTERING OF SLOW NEUTRONS

We can find the magnitude, but not the sign, of the scattering length a The most important evidence is the scattering of neutrons by *ortho and para hydrogen*. An experimental comparison of the coherent scattering from ortho and para hydrogen was first suggested by Teller in 1936 to test the spin dependence of the neutron proton interaction. Not only we are able to verify that the scattering is spin dependent but also will be able to show that singlet state scattering length is negative. In para-hydrogen molecules the proton spins are anti-parallel. The molecules can rotate with energies given by

$$E_J = J(J+1)\frac{\hbar}{2g}, \qquad (J = 0, 2, 4,...),$$

where g is the moment of inertia. In ortho hydrogen proton spins are parallel. The rotational states are given by J = 1, 3, 5..... If temperature is low enough practically all the para hydrogen will be in the state J = 0 and all the ortho hydrogen will be in state J = 1.

We shall now derive an expression for the scattered intensity from a molecule of ortho or para hydrogen when the incident neutron energy is so small that λ_n is much greater than 0.78×10^{-10}m the distance between the atoms in hydrogen molecule. The theoretical expressions for total cross-section for the special case of neutron energy of 0.001463 eV and a gas temperature of 19.5° K, as derived by Schwinger and Teller in 1937, are

$$\sigma_{para} = 6.69\ (3a_t + a_s)^2 \quad ...(1)$$

$$\sigma_{ortho} = 6.69\ [(3a_t + a_s)^2 \text{ or } (a_t - a_s)^2] + 1.74\ (a_t - a_s)^2 \quad ...(2)$$

where a_t and a_s are triplet and singlet scattering lengths. The last term in σ_{ortho} was added to take into account inelastic scattering by conversion of ortho to para. This process is energetically possible but its cross-section is small. It is clear from above relations that σ_{para} and σ_{ortho} become identical if $a_t = a_s$. In this case the total nuclear forces are said to be spin independent.

According to the experimental results of Sutton (1947), $\sigma_{para} = 4.0$ barns and $\sigma_{ortha} = 12.5$ barns. This great difference between these cross-sections indicates that the n-p force is strongly spin dependent.

$| 3a_t + a_s |$ is very much smaller than $| a_t \times a_s |$ when $a_s < 0$ but very much larger when $a_s > 0$. Since $\sigma_{para} < \sigma_{ortho}$, hence $| 3a_t + a_s |$ is smaller than $| a_t - a_s |$ or $a_s < 0$. This –ve value, of a_s indicates that the *singlet state of the deuteron* is virtual or unbound. We can now solve equations (1) and (2) for a_t and a_s. There are four possible pairs of a' and a values. Two can be discarded because they have –Ve a_t values. A third pair can be shown to be inconsistent with n-p scattering data. The remaining solution is $a_t = 0.52 \times 10^{-14}$m and $a_s = 2.3 \times 10^{-11}$m. Therefore, the total cross-section will be

$$\sigma_0 = \frac{3}{4}\sigma_t + \frac{1}{4}\sigma_s = \frac{3}{4} \times 4\pi a_t^2 + \frac{1}{4} \times 4\pi a_s^2 = 19.8 \text{ barns,}$$

which is in fair agreement with experimental value 20.36 barns.

Spin of Neutron

The large observed value of $\sigma_{ortho}/\sigma_{para}$ relates to the spin of the neutron. The ground state of the deuteron ha I = 0 and I = 1, if

$$S_r + S_p = \frac{1}{4} + \frac{1}{2} = 1 \qquad \text{(spin parallel)}$$

$$S_n + S_p = \frac{3}{2} - \frac{1}{2} = 1 \qquad \text{(spin anti-parallel).}$$

In the second case the relative statistical weights for n-p collisions with free protons would change from their values of 3/4 and 1/4 to values of 5/8 and 3/8. This will change the scattering cross-sections.

$$\sigma_{ortho}/\sigma_{para} \simeq 2.$$

This is against experimental results, hence the neutron has spin 1/2 not 3/2.

SHAPE INDEPENDENT EFFECTIVE RANGE THEORY

While defining the scattering length $a = -\delta_0/k$ we make the use of scattering of S-wave by using a square well potential. We shall discuss here a more general method, which gives a very good approximation for phase shift δ_0 and at the same time does not require a specified form of potential V(r). It is possible to expand the quantity cot δ_0 in a simple power series of k.

This will simplify the procedure of computing phase shifts. In the energy range 0 to 15 MeV, two terms in the power series suffice. The first term contains the scattering length and the second term contains a quantity called the effective range, to be defined below.

In developing the effective range theory we begin by writing Schrodinger's equation for an S-wave and positive energy for the relative motion. Let u and u_0 be the radial wavefunctions corresponding to two energies E and E_0. Then we have

$$\frac{d^2u}{dr^2} + \frac{M}{\hbar^2}[E - V]\,u = 0 \qquad ...(1)$$

$$\frac{d^2u_0}{dr^2} + \frac{M}{\hbar^2}[E_0 - V]\,u_0 = 0 \qquad ...(2)$$

where V is the nuclear potential of any shape. Multiply eqn. (1) by u_0 and eqn. (2) by u, subtract and integrate from 0 to ∞. Then

$$\left[uu'_0 - u_0u'\right]_0^\infty = \left(k^2 - k_0^2\right)\int_0^\infty uu_0\, dr \qquad ...(3)$$

Consider a new function v which is identical with u outside the potential well and extrapolates the behaviour of u to the origin. The functions are given as $v = c \sin(kr + \delta)$. These are normalized in such a way that $v \to 1$ at the origin. Thus

$$v = \sin(kr + \delta)/\sin\delta \qquad ...(4)$$

We get a relation like eqn. (3) for the wave functions v and v_0. Thus

$$\left[vv'_0 - v_0v'\right]_0^\infty = \left(k^2 - k_0^2\right)\int_0^\infty vv_0\, dr \qquad ...(5)$$

We may now subtract eqn. (3) from eqn. (5). Since v = u and v_0 = u_0 outside the potential well and $u = u_0 = 0$ for r = 0, hence

$$-\left[vv'_0 - v_0v'\right]_{r=r_0} = \left(k^2 - k_0^2\right)\int_0^\infty (vv_0 - uu_0)\, dr \qquad ...(6)$$

As relation (4) gives

$$v' = k \cos (kr + \delta)/\sin \delta$$

$$\therefore \qquad (v')_{r=0}\ k \cot \delta \text{ and } (v)_{r=0} = 1$$

Similarly $(v'_0)_{r=0} = k_0 \cot \delta_0$ and $(v_0)_{r=0} = 1$

Substituting these values in equation (6), we have

$$-k_0 \cot \delta_0 + k \cot \delta = (k^2 - k_0^2)\int_0^\infty (vv_0 - uu_0)\, dr \qquad ...(7)$$

Consider now the case when $k_0 = 0$ (the case of zero energy). By definition of scattering length $k_0 \cot \delta_0 = -1/a$ when $k_0 \rightarrow 0$. Thus eqn. (7) becomes

$$k \cot \delta = -1/a + \frac{1}{2} k^2 \rho(0, E) \qquad ...(8)$$

where $\frac{1}{2}\rho\,(0, E) = \int_0^\infty (vv_0 - uu_0)\, dr$

Since u and v differ only inside the range of nuclear forces, therefore, the integrand vanishes outside the nuclear force range. Inside the range, the wavefunctions are almost independent of the energy, because the potential energy is much larger than the energy of the system E. We can thus write

$$v \simeq v_0 \text{ and } u \simeq u_0$$

and $\frac{1}{2}\rho\,(0, 0) = \int_0^\infty \left(v_0^2 - u_0^2\right) dr = \frac{1}{2} r_0$ ...(9)

The quantity r_0 is constant and is known as *effective range* of nuclear potential. It is one of the average properties of the potential to be determined from the scattering. The factor 1/2 is included in the definition to make r_0 approximately equal to the radius. This justifies the name "effective range" for r_0. Thus eqn (8) becomes

$$k \cot \delta = -1/a + \frac{1}{2} r_0\, k^2. \qquad ...(10)$$

$$\therefore \text{ Cross-section } \sigma = \frac{4\pi a^2}{\left\{\left[1-\frac{1}{2}ar_0k^2\right]^2 + a^2k^2\right\}} \quad ...(11)$$

Whatever the shape of the potential, the scattering can be described by the two parameters a and r_0. The experimental values of δ and k serve to determine a and r_0. Any reasonable potential shape can be made to give matching values of a and r_0 by suitable choice of its depth V_0 and range b. Hence the two experimentally determined lengths a and r_0 do not determine the shape of the nuclear potential, but if the shape is chosen then a and r_0 fix the depth and range. For this reason above relation is known as the *shape independent approximation.* Triplet effective range r_0 is obtained from a and has value $1.70 \pm 0.03 \times 10^{-15}$m. Singlet effective range r_0 is obtained from a_s and has the value $2.5 \pm 0.2 \times 10^{-15}$m.

PROTON-PROTON SCATTERING AT LOW ENERGIES

The p-p scattering is the only way to get direct evidence on proton-proton force. There are two essential differences between p-p and n-p scattering. First, the p-p scattering is caused not only by the nuclear forces but also by the Coulomb force. Second, the scattering and scattered particles are identical and obey the Pauli exclusion principle and, therefore, the wavefunction describing the two protons must change sign on the interchange of the two particles.

Hence a symmetric space wavefunction (S, D states, etc.) can only be associated with an anti-symmetric (singlet) spin wavefunction, whereas a symmetric (triplet) spin wavefunction is required for an anti-symmetric space wavefunction (P, F states, etc.). For incident protons of energy below 10 MeV, only the S state interaction is of any importance in the scattering, since protons in higher orbital angular momentum states stay apart from each other beyond the range of the nuclear force.

Experimentally the study of p-p scattering, is capable of much higher accuracy than n-p scattering, for the following reasons:

1. Protons are easily available over a wide range of energies.
2. Protons can be made monoenergetic.
3. Protons can be produced in well collimated beam.
4. Protons can be easily detected by their ionizing properties.

5. Protons undergo Coulomb scattering simultaneously with nuclear scattering. This increases the sensitivity in case one of the scattering probabilities is small and also gives sign of the phase shifts resulting from nuclear scattering.
6. The proton combination obeys Fermi Statistics. This simplifies the analysis of proton-proton scattering.

We are now interested in obtaining a theoretical expression for the differential elastic scattering cross-section of protons by protons. The theory is more complicated than that of n-p scattering because the Coulomb potential appreciably distorts the incident wave eve at finite distances. The radial wave equation for the proton-proton system is

$$\frac{d^2u}{dr^2}+\frac{2m}{\hbar^2}\left[E-V(r)-\frac{e^2}{4\pi\varepsilon_0 r}-\frac{l(l+1)\hbar^2}{2mr^2}\right]u=0$$

where V(r) is the square well nuclear potential. For low energies, the nuclear potential will affect only the $l = 0$ partial wave, but the Coulomb potential produces higher l-value scattering. Rutherford first investigated the scattering by the Coulomb field from the classical stand point. The Rutherford cross-section for Coulomb scattering of particle of charge Z_1e by a particle of charge Z_2e in the centre of mass system is

$$d\sigma=\left[\frac{(Z_1e)^2\ (Z_2e)^2}{4(4\pi\epsilon_0)^2\, m^2v^4}\operatorname{cosec}^4\frac{\theta}{2}\right]2\pi\sin\theta\, d\theta, \qquad \text{...(1)}$$

where v is the velocity of the incident particle, m is the reduced mass of the system and θ is the angle of scattering in the centre of mass system. For p-p scattering $Z_1 = Z_2 = 1$, $m = M/2$, $\theta = 2\theta_1$ and $\sin\theta\, d\theta = 4\sin\theta_1\cos\theta_1\, d\theta_1$ (where θ_1 is in laboratory system). Thus, in laboratory system, the cross-section becomes

$$d\sigma=\frac{e^4}{(4\pi\epsilon_0)^2\, E_0^2}\left[\frac{1}{\sin^4\theta_1}+\frac{1}{\cos^4\theta_1}\right]\cos\theta_1\ 2\pi\sin\theta_1\, d\theta_1, \qquad \text{...(2)}$$

where $E_0=\frac{1}{2}Mv^2$ (kinetic energy of incident proton in the laboratory system). The term containing $\cos^4\theta_1$ is added because each proton at angle θ_1 in the L-system is accompanied by a recoil proton potential is repulsive or attractive, as δ_0 is positive corresponding to attractive potentials and negative to repulsive potentials For large energies the last term becomes the most important because of the v_2 coefficient.

The differential cross section is a complicated function of the energy E_0, the angle of scattering θ_1 and the phase shift δ_0. Differential cross-section for p-p elastic scattering for 2.4 MeV protons is shown in Fig. 2. At small scattering angles the scattering is essentially pure Coulomb scattering. The nuclear scattering predominates in the central region of angles, near a C.M. angle of 90° (lab angle of 45°). At forward and backward angles, the dips are caused by interference between nuclear and Coulomb scattering.

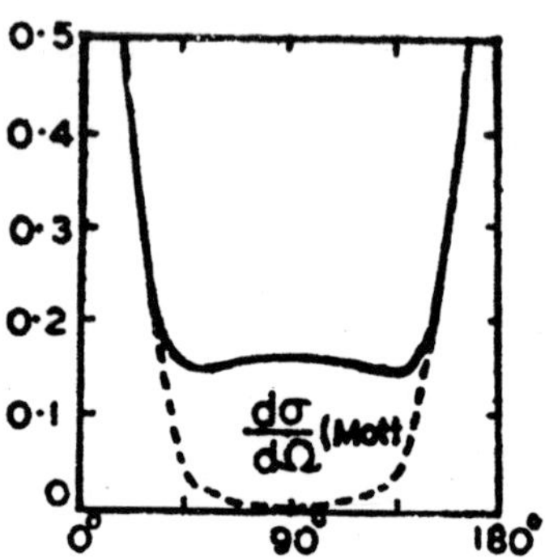

Fig. 2 : Differential cross-section for p-p scattering.

Experiments on p-p Scattering

The experimental data are analyzed further along lines similar to those used in n-p scattering but are complicated considerably by the presence of the Coulomb force. In the interpretation of the p-p scattering results, the phase shifts δ_0 are the common meeting ground of experiment and theory. Jackson and Blatt showed that the observed variation of δ_0 with incident proton energy can be matched by any of the conventional shapes of potential well, with suitable choices of depth and range. A shape independent theory, therefore, his a natural attractiveness. Such a theory, for proton interaction, contains as parameters, the effective range r_0 and scattering length a_p. The best fit to the data compiled by Jackson and Blatt is obtained with

$$r_0 = 2.65\ (1 \pm 0.03) \times 10^{-15}\ \text{m and } a_p = 7.7\ (1 \pm 0.07) \times 10^{-15}\ \text{m}.$$

The fact that a_p is negative indicates that the S^1 state for the p-p system cannot be bound. Therefore the di-proton or He^2 can not exist, The exclusion principle prevents any ^3S state, which would be the analog of the stable deuteron. The absolute value of a_p requires further interpretation because it includes Coulomb as well as specifically nuclear effects. The latter can be isolated by treating the Coulomb force, with in the range of the nuclear force, as a small perturbation. Thus a/i is increased to the value equal to

$$a_{eq} = -17 \times 10^{-15} \text{ m.}$$

We know that for neutron-proton estate $a_s = -24 \times 10^{-15}$m and $r_{0s} = 2.5 \times 10^{-15}$m. The effective ranges do not differ significantly in view of their large experimental uncertainties. The scattering lengths are different significantly. Such a difference might arise from the magnetic moment interaction, as the interinsic magnetic dipole moments of the neutron and proton are of opposite sign. Schwinger has evaluated a net attractive magnetic interaction in the ^{3}[np) force and a net attractive magnetic interaction in the ^{1}(pp) force, at an angle $(90 - \theta_1)$ and it is not possible to distinguish between the two protons.

Equation (2) does not agree with experiment even at low energies where nuclear scattering is negligible. Mott introduced a third term from wave mechanical consideration. This places symmetry requirements on the wave-function and represents the interference between the wave functions of protons. Thus, eqn. (2) becomes

$$d\sigma = \frac{e^4}{(4\pi \epsilon_0)^2 E_0^2} \left[\frac{1}{\sin^4 \theta_1} + \frac{1}{\cos^4 \theta_1} - \frac{\cos\{e^2/4\pi \epsilon_0 \hbar v) \log \tan^2 \theta_1}{\sin^2 \theta_1 \cos^2 \theta_1} \right]$$

$$\cos \theta_1 \; 2\pi \sin \theta_1 d\theta_1$$

The negative sign is because the protons obey Fermi Statistics. Second and third terms drop out for unlike charged particles. For proton, energies above 1 MeV, $\frac{e^2}{4\pi \epsilon_0 \hbar v} > \frac{1}{7}$ and the cosine term is nearly unity, unless θ_1 is close to zero or $\pi/2$. Thus away from $\theta_1 = 0$ or $\pi/2$, for $E_0 > 1$ MeV, the Mott term is simplified and above equation reduces to

$$d\sigma \; \frac{e^4}{(4\pi \epsilon_0)^2 E_0^2} \left[\frac{1}{\sin^4 \theta_1} + \frac{1}{\cos^4 \theta_1} \frac{1}{\sin^2 \theta_1 \cos^2 \theta_1} \right]$$

$$\cos \theta_1 \; 2\pi \sin \theta_1 d\theta_1 \qquad \text{...(3)}$$

The main difference between proton and neutron seems to be of electric charge and the nuclear force apparently does not arise from charge. It is reasonable to assume that the nuclear potential between two protons has the some characteristics as that between proton and neutron. Therefore, in p-p scattering at low energies it is expected that only the $l = 0$ scattering processes will be affected by the nuclear potential, just as in n-p scattering. When the effects of S wave-nuclear scattering are included we obtain scattering term. For S wave-nuclear scattering the complete differential scattering cross-section in laboratory system is given as

$$\frac{d\sigma}{d\Omega}=\frac{e^4}{(4\pi \in_0)^2 E_0^2}\left[\frac{1}{\sin^4\theta_1}+\frac{1}{\cos^4\theta_1}-\frac{1}{\sin^2\theta_1\cos^2\theta_1}-\frac{8\pi \in_0 \hbar v \sin\delta}{e^2}\right.$$

$$\left\{\frac{\cos\left[\delta_0+\dfrac{e^2}{4\pi \in_0 \hbar v}\log\ \sin^2 1/2\theta\right.}{\sin^2(\theta/2)}+\frac{\cos\left[\delta_0+\dfrac{e^2}{4\pi \in_0 \hbar v}\log\cos^2 1/2\theta^2\right.}{\cos^2(\theta/2)}\right\}$$

$$\left.+\left(\frac{8\pi\varepsilon_0 \hbar v}{e^2}\right)^2 \sin^2\delta_0\right]\cos\theta_1 \qquad ...(4)$$

In this equation 1st, 2nd and 3rd terms are due to Coulomb repulsion, the 5th term due to nuclear scattering and the fourth term due to interference between the Coulomb and nuclear scattering effects. This relation reduces to equation (74) for a pure Coulomb field when nuclear phase shifts δ_0 is zero *i.e.*, when there is n nuclear scattering.

The fourth term is useful in determining δ_0 when δ_0 is small, because it is linear in $\sin^2 \delta_0$ and it is large compared to the last term which involves $\sin^2 \delta_0$. The linearity in δ_0 of the interference term also permits determination of whether the nuclear. *This equality of the parameters of the (pp) and (np) singlet interactions gave rise to the hypothesis of the charge independence of the nuclear forces.* This hypothesis states that the forces between two nucleons in the same spin and orbital angular momentum state are independent of the charge of the two nucleons.

SIMILARITY BETWEEN (NN) AND (PP) FORCES

(i) Similar to diprotons, di-neutrons are not found in a stable stationary state. The scattering length for the (nn) interaction appears to have the same (negative) sign as that for the (pp) interaction.

(ii) The study of lightest pair of mirror nuclei, H^3 and He^3 provides us the direct evidence on the equivalence of ^{1}(nn) and ^{1}(pp) forces. The possible interactions between two particles in these nuclei can be written as

$$H^3 \rightarrow {}^3(np) + {}^1(np) + {}^1(nn),\ He3 \rightarrow {}^3(np) + {}^1(np) + {}^1(pp)$$

These are all S-state interactions because orbital angular momentum is zero for each nucleus. The observed binding energy difference is 0.76 MeV. This corresponds to the Coulomb energy of two discreate protons

$$\left[\frac{3}{5}\left(\frac{e^2}{4\pi \in_0 R}\right)z(z-1)=0.75 \text{ MeV}\right].$$

This proves that (nn) = (pp) force.

(iii) Inelastic scattering of deuteron by neutrons has given quantitative support to the similarity of (nn) and (pp) force.

(iv) The differential cross-sections at ~ 10 MeV for the H^3 (d, n) He^4 and He^3 (d, p) He^4 reactions show very similar asymmetries. These observations support the equality of (nn) and (pp) force.

(v) Strong confirmation for equality of n-n and p-p forces comes also from the steady of heavier mirror nuclei, such as ${}_{16}S^{31}$ and ${}_{51}P^{31}$.

NON-CENTRAL FORCES

(a) Experimental Evidence for the Existence of Non-Central Forces

A serious discrepancy was discovered between theory and experiment, performed by Kellogg in 1939. The experimental results showed a fine structure in the radio frequency magnetic resonance spectrum of deuterium, but it was different from the theoretical prediction. The quadrupole moment established by this method was 2.82×10^{-31} m^2. It was assumed that n-p force was a central force. The ground state of deuteron must be a S-state whose quadrupole moment should be zero and the magnetic moment should he the algebraic sum of the magnetic moments μ_p, and μ_n. The existence of quadrupole moments in heavier nuclei can be explained by assuming the force between neutron and proton as partly a central force and partly noncentral (*or tensor*) force.

From the small deviation of the deuteron wavefunction from spherical symmetry, we can calculate that the non-central part of the force is also very small compared to fie central force part. The tensor force resembles a magnetic interaction between two dipoles. It results in a lower potential energy for n and p when the line joining them is | | to the spin direction than when it is ⊥.

(b) General Form of the Non-Central Force

Let us now see whether we can construct a potential which depends not only on the relative position vector of the particles r, but also on their spin orientations $\vec{\sigma}_1$ and $\vec{\sigma}_2$ and which can account for the quadrupole moment. This potential must be invariant under rotations and reflections of the co-ordinate system which we use to describe the relative motion of the particles and hence must be a scalar.

Wigner in 1941 has showed that, if the interactions are assumed to be invariant with respect to displacement, rotation and inversion of the observer's co-ordinate system and independent of the particle velocities, the general potential can be written as

$$V(r) = V_1(r) + V_2(r)\,\vec{\sigma}_1 . \vec{\sigma}_2 + V_3(r) S_{12} \qquad ...(1)$$

where the potentials V may depend on the orbital momentum of the two particle system as well as on the charge of the particles. The reasons of this limited choice are:

1. The vector r changes to –r under reflection and thus mutt occur in even powers only.
2. Derivatives of r cannot occur since forces are velocity independent.
3. Since $\vec{\sigma}_1$ and $\vec{\sigma}_2$ are not invariant against rotation and $\vec{\sigma}_1 \cdot \vec{\sigma}_2$ is invariant.
4. $\vec{\sigma} \cdot \vec{r}$ is invariant against rotation, but not against inversion, since $\vec{r} \rightarrow -\vec{r}$ and $\vec{\sigma} \rightarrow \vec{\sigma}$ on inversion. Because of this, only even power of $(\vec{\sigma} \cdot \vec{r})$ may occur such as $(\vec{\sigma}_1 \cdot \vec{r})(\vec{\sigma}_2 \cdot \vec{r})$.

The first term is corresponding to central force. The second term corresponding to a spin dependent but central force. The value of $\sigma_1 . \sigma_2$ for a triplet state and –3 for a single state. The last term is corresponding to the interaction potential which depends not only on the separation between neutron and proton but also on the angle which their spins make with the line joining the two particles. The non-central character of the interaction is contained in the tensor operator.

$$S_{12} = 3(\vec{\sigma}_1 \cdot r)(\vec{\sigma}_2 \cdot r)/r^2 - \vec{\sigma}_1 . \vec{\sigma}_2 \qquad ...(2)$$

The first term gives the depnndence of the interaction on spin angles. The second term has been subtracted so that the average of S_{12} over all directions r is zero.

(c) Properties of the Non-Central Force

Since S_{12} is a scalar, the tenser force potential is a scalar. It can be shown that whin $V_3(r)$ is finite, the total angular momentum **I** and the parity remain good quantum numbers but that the orbital angular momentum **L** and the spin angular momentum S are not constants of motion. The magnitude of the spin angular momentum S^2 is still a constant of motion for the two body systems.

The tensor operator S_{12} is defined in such a way that its average over all directions is zero. In the singlet spin state there is no preferred direction for the spin orientation hence total spin angular momentum

$$S = \frac{1}{2}(\vec{\sigma_1} + \vec{\sigma_2}) = 0. \text{ Thus, we get}$$

$$S_{12} = -3(\vec{\sigma}_1 \cdot \vec{r})(\vec{\sigma}_2 \cdot r)/r^2 - \vec{\sigma}_1 . \vec{\sigma}_2$$

This results that the tensor force is zero in the singlet state and all previous results for singlet states are unaffected entirely.

We also know that a triplet states of a given total angular momentum **I** can be written as a linear superposition of triplet states with orbital angular momenta L = I – 1, I + 1. Since parity of a wave function is $(-1)^L$, hence the state with L = I has opposite parity to the two states with L = I – 1 and L = I + 1. Hence the I = 1 states of even parity are mixtures of $^3S_1$ and $^3D_1$ and the I = 1 states of odd parity are pure $^3P_1$.

(d) The Ground State of the Deuteron

Magnetic Moment

We know that the small discrepancy between the sum of the magnetic moment of proton and neutron and the measured value for the deuteron can be interpreted as a contribution of the orbital motion of the proton in the D-state in the deuteron ground state. If nuclear forces are supposed to be central forces, the difference will be zero. This contribution can appear only with the non-central forces. The operator describing the magnetic moment of deuteron is

$$\vec{\mu} = \vec{\mu}_p \times \vec{\sigma}_p + \vec{\mu}_n \times \vec{\sigma}_n + \vec{L}r \qquad ...(3)$$

where $\vec{\mu}_n$ and $\vec{\mu}_p$ are the magnetic moments of neutron and proton measured in nuclear magnetons, $\vec{\sigma}_n$ and $\vec{\sigma}_p$ their unitary spin operators and $\mathbf{L}_p$ is the orbital angular momentum of the proton. The uncharged neutron cannot contribute any magnetic moment by orbital motion alone. In the centre of mass system the orbital angular momentum of the proton is half of the combined orbital angular momentum **L**. Thus the above equation can be written as:

$$\mu = (\mu_n + \mu_p).\frac{1}{2}(\sigma_n + \sigma_p) + \frac{1}{2}(\mu_n - \mu_p)(\sigma_n - \sigma_p) + \frac{1}{2}L.$$

Since the operator $(\sigma_n - \sigma_p)$ in the second term vanishes for a triplet state, and I = L + S, hence tne magnetic moment operator becomes

$$\vec{\mu} = (\mu_n + \mu_p)\,\mathbf{I} - (\mu_n + \mu_p - \frac{1}{2})\mathbf{L} \qquad ....(4)$$

The observed value of μ is the expectation value of this expression in the state with $\mathbf{I}_z$ = **L**. We, therefore, can replace L by L_z.

$$\therefore \quad \mathbf{L} \to \mathbf{L}_z = \frac{\mathbf{LI}}{\mathbf{I}^2}I_z \; \frac{I(I+1)+l(L+1)-S(S+1)}{2I(I+1)}I_z \qquad ...(5)$$

For, the deuteron, I(I + 1) = S(S + 1) = 2 and in a mixture of S and D states, $[L(L + 1)]_{av} = 0 \times p_S + 6 \times p_D = 6p_D$. Here p_D and p_S represent the D-and S-state probabilities respectively. For the deuteron I = 1, the value of I_z in the state with I_z = I is unity.

$$\therefore \quad \mu \text{ of deuteron} = \mu_n + \mu_p - \frac{3}{2}(\mu_n + \mu_p - \frac{1}{2})p_D \qquad ...(6)$$

The last term gives a direct measure of the D-state probability. The experimental results imply a D-state probability of about 4%. The above relation does not give accurate results. There are various other causes which can give corrections, especially of relativistic effects. Hence the measured magnetic moment gives only a rough estimate of the D-state probability. One may expect that p_D is really between 2 and 8%.

Quadrupole Moment

The quadrupole moment Q was defined as the average value of $(3z^2 - r^2)$ in the state with m = I. When evaluating the quadrupole moment of the deuteron we must remember that only the proton contributes to the quadrupole moment and its distance from the centre of gravity is half of the proton neutron separation r.

$$\therefore \qquad Q = \frac{1}{4}(3z^2 - r^2) = \frac{1}{4}r^2\,(3\cos^2\theta - 1) \qquad ...(7)$$

Since the quadrupole moment is estimated from the S and D waves beyond the potential well, hence the expectation value of this operator Q is given by

$$\therefore \qquad (\psi, Q\psi) = (\psi s, Q\psi s) = (\psi_D, Q\psi_D) + 2\,(\psi s, Q\psi_D) \qquad ...(8)$$

The first term is zero because the S state is spherically symmetrical and cannot have a quadrupole moment. The second term is a pure D-state term and is smaller than the cross term (as $p_S >> p_D$), hence only the cross term contributes. The measured quadrupole moment Q is, therefore, be written as

$$Q = \frac{1}{\sqrt{(50)}} \int_0^\infty r^2 u(r)\,\omega(r)\,dr \qquad ...(9)$$

where the constant comes from the spin sum and angular integration u(r) is the deuteron ground state S-wave function outside the range of nuclear forces and can be written as $u(r) = Ns\, e^{-kr}$.ω(r) is the deuteron D-wave function at distances beyond the range of the specific potentials and can be written as

$$\omega(r) = N_D e^{-kr}\,(1 + 3/kr + 3/k^2r^2).$$

Here N_S and N_D are normalization constants and $k = \sqrt{(MB)}/\hbar^2$.

The rough estimate of Ns can be obtained by neglecting the small D-state probability compared to unity by substituting

$$p_s = \int_0^\infty u^2 dr = 1 \qquad N_s = \sqrt{(2k)} \qquad ...(10)$$

Since the weighing factor r^2 flavours the contribution of the outside wave function, hence we can estimate the quadrupole moment by substituting the values of u(r) and ω(r) into the equation (9). The result will be

$$Q = \frac{N_S N_D}{\sqrt{8k^2}} \qquad ...(11)$$

This relation can be used for a good estimate of N_D. If value of N_S is taken from eqn. (10), we have

$$Q = \frac{N_D}{2k^{5/2}} \qquad \text{or} \quad N_D = 2Qk^{5/2} \qquad \text{...(12)}$$

Thus, we see that the function ω(r) outside the range of the forces is determined completely by the quadrupole moment. This result implies that the D-state probability p_D depends strongly on the tensor force range RT and increases rapidly as R_T is made shorter. The integral of ω^2 from the radius R_T on out is given by

$$\int_{R_T}^{\infty} \omega^2 \, dr = \frac{N_{D^2}}{R_{T^3} k^4} \qquad \text{...(12)}$$

To take into account the contribution from $r < R_T$ we can roughly double this and thus get the physically important quantity

$$P_D = \int_0^{\infty} \omega^2 \, dr = 2\int_{R_T}^{\infty} \omega^2 \, dr = \frac{6N_{D^2}}{R_{T^3} k^4} = \frac{24Q^2 k}{R_{T^3}} \qquad \text{...(13)}$$

This equation implies that the tensor force cannot have an arbitrarily small range otherwise the ground state would become a predominantly D state rather than predominantly S state. The experimental value of Q is + 2.73(e × 10^{-31}m^2).

A rough measurement of p_D is obtained from the deuteron magnetic moment (eqn. 6). This value of p_D leads to a fair estimate of R_T, since it is in the cube of R_T which occurs in eqn. (13). This gives that the tenser force range R_T falls nest 3 × 10^{-15} m, which is almost independent of the range of the central force and is slightly larger than the range of central forces.

(e) Neutron-Proton Scattering Below 10 MeV

In principle the analysis of the scattering is very much altered by non-central forces. The separation of the various partial waves is no longer complete, since the orbital angular momentum *l* is not a constant of the motion. At energies below 10 MeV, it is advantageous to decompose the wave function into spherical harmonics by the use of the spin angular function. The spin S can have values 0 and 1 corresponding to the singlet and triplet states.

The S-scattering in the singlet state (S = 0) is unchanged by the presence of tensor forces since they do not act in that state. Therefore, we have to discuss only the triplet scattering. We know that the tensor force couples the $^3D_1$ state with the $^3S_1$ state. Thus both states contribute to the scattering even at low energies.

If the ingoing wave is a pure $^3S_1$ wave, the outgoing wave is a mixture of S and D waves owing to the action of the tensor force during the impact. Similarly, an ingoing pure $^3D_1$ wave emerges as a mixture outgoing wave. The waves, which have the same mixture ratio of S to D states before and after the scattering event, are called *eigenstates for the scattering*. There are two such eigenstates: One is called the α-wave in which S wave is predominant. Other is called the β-wave in which D wave is predominant. Since we are dealing with the same angular momentum state $^3S_1$ and $^3D_1$ as the ground state of the deuteron, the asymptotic form of perturbed u(r) and ω(r) can be written in the form

$$u(r) = a \sin (kr + \delta)$$

$$\omega(r) = b \sin (kr + \delta - \pi),$$

where the angle δ represents the phase shift compared to the asymptotic form of the unperturbed plane wave. The solution u, ω of the wave equation is an eigenstate of the scattering provided that the phase shift δ is the same for both u(r) and ω(r). There are two (α and β) solutions satisfying this condition, with two different phase shifts (δ_α and δ_β) and two different values of the ratio b/a which determine ω(r)/u(r).

If we neglect the phase shifts of all the waves except the α-wave, the differential scattering cross-section in C-M system becomes

$$\frac{d\sigma}{\delta\Omega} = \frac{\sin^2\delta_\alpha}{k^2}\left[\left(1+\frac{\tan\in}{\sqrt{8}}\right)\sin^2 (2\in)P_2(\cos\theta)\right] \quad ...(14)$$

where P_2 (cos θ) is the second Legendre polynomial and $\varepsilon = \tan^{-1} b_\alpha/a_\alpha$, *i.e.*, the asymptotic ratio of a to u, can be determined by the quadrupole moment as

$$\in \simeq \tan\in \simeq b_\alpha/a_\alpha \simeq \sqrt{(2)}Qk^2. \quad ...(15)$$

If we substitute the numerical value of Q in this eqn. and evaluate the angle dependent term in eqn. (14), it gives that the coefficient of $P_2(\cos\theta)$ in the angular distribution is less than 0.01 even at 10 MeV. Hence, this tensor effect in the scattering is much too small because in this energy region the experimental errors for angular distribution measurements are at present of the order of 10%. P-wave scattering can be expected to contribute terms. The triplet parameters a_t and r_{0i} are not much affected by the tenser forces, if the potentials have similar shape.

SATURATION OF NUCLEAR FORCES

Thus, far we have discussed mainly the character of the forces acting between pairs of nucleons. When we come to consider systems comprising more than two particles, we must expect some complications to enter. We shall deal with certain quantitative aspects of nuclear forces which become important when not only the two, body problem but also the properties of other nuclei are studied. There are two important facts: the density of nucleons is roughly equal for all nuclei *(saturation of density)*, and the binding energy per nucleon is roughly equal for all nuclei (*saturation of binding energy*).

In the case of two body problem, following assumptions have been made:

(a) The nuclear force acts between the pair of nucleons and does not influenced by the presence of neighbouring nucleons.

(b) The nuclear force between the nucleons is velocity independent, attractive, independent of the type of the nucleons and central.

These assumptions do not correspond to reality. The basic force law would have to be modified, if we add nucleon to a nucleus. The force between two nucleons is attractive for distances $r > d$, and is repulsive otherwise. Hence d is of the order of the distance between nearest neighbours in the nuclei.

(A) Exchange Forces

In 1932, Heisenberg proposed, in order to explain the saturation of nuclear forces, that nuclear forces were *Exchange Forces* which would depend explicitly on the symmetry of the wave function. At the time of Heisenberg's idea of exchange forces, mesons were known and it was known that π-meson was being exchanged between nucleons, by any of the following processes:

$$\begin{array}{rclclclc} n, p & ; & p, n & ; & p, p & ; & n, n \\ p + \pi^-, p & ; & n + \pi^+, n & ; & p + \pi^o, p & ; & n + \pi^o, n \\ p, \pi^-, p & ; & n, \pi^+ + n & ; & p, p + \pi^o & ; & n, n + \pi^o \\ p, n & ; & n, p & ; & p, p & ; & n, n, \end{array}$$

The exchange of a pion is thus equivaler t to charge exchange. We can think of the nucleons as exchanging their space and spin co-ordinates. The wave equation of the two body system for an ordinary central force is

$$[(\hbar^2 M)\nabla^2 + E]\Psi(r_1, r_2, \sigma_1, \sigma_2) = V(r)\Psi(r_2, r_1, \sigma_1, \sigma_2). \quad ...(1)$$

The force is known as no *exchange, ordinary* or *Wigner force*. The interaction does not cause any exchange. Exchange forces are classified as:

1. *Majorana Forces*

The Majorana interaction is that in which it is assumed that two particles attract one another if the wave function describing the entire system does not change sign when the special co-ordinates of the two particles are interchanged and that they repel if the wave function changes sign, *i.e.*,

$$[(\hbar^2 M)\nabla^2 + E]\psi(r_1, r_2, \sigma_1, \sigma_2) = V(r)\psi(r_2, r_1, \sigma_1, \sigma_2). \quad ...(2)$$

This interaction reflects the coordinates and replaces r by –r in the wave-function. Hence eqn. (2) may be written as

$$[(\hbar^2/M)\nabla^2 + E]\psi(r) = (-1)^l\, V(r)\psi(r). \quad ...(3)$$

This eqn. indicates that the force is always attractive for states of even l(S, D, G,...) and always repulsive for states of odd l.

2. *Burtlett Forces*

This involves the exchange of the spin hut not the position coordinates of the two interacting nucleons. For such an interaction, the Schrodinger eqn is

$$[(\hbar^2 M)\nabla^2 + E]\psi(r_1, r_2, \sigma_1, \sigma_2) = V(r)\psi(r_1, r_2, \sigma_2, \sigma_1). \quad ...(4)$$

The wave function of two particles is symmetric if the total spin S = 1 and anti-symmetric if S = 0. Thus eqn. (4) gives

$$[(\hbar^2/M)\nabla^2 + E]\psi(r) = (-1)^{s+1}\, V(r)\psi(r). \quad ...(5)$$

This relation is equivalent to an ordinary potential which changes sign between S = 0 and S = 1. The nuclear force can not be totally of the Bartlett type because it is clear from neutron proton scattering data that both the ^3S and ^1S potentials are attractive.

3. *Heisenberg Forces*

In the type of interaction there is an exchange of both the position and the spin co-ordinates of the two nucleons. For such an inter Action, the Schrodinger eqn. is

$$[(\hbar^2 M)\nabla^2 + E]\psi(r_1, r_2, \sigma_1, \sigma_2) = V(r)\psi(r_2, r_1, \sigma_2, \sigma_1). \quad ...(6)$$

Since the wave function of two particles is symmetric for $(l + S)$ even and anti-symmetric for $(l + S)$ odd, hence eqn (6) may be written as

$$[(\hbar^2 M)\nabla^2 + E]\ \psi\ (r) = (-1)^{l+s+1}\ V(r)\psi(r) \quad ...(7)$$

This relation indicates that sign of ordinary potential is positive if $(l + S)$ is odd and is negative if $(l + S)$ is even. This gives that the force is attractive for even l triplet states and odd l singlet states, but is repulsive in odd l triplet and even l singlet states.

The three types of exchange operators P^M, P^B and P^H are used to construct these three types of exchanged forces. The reversal of sign between ^3S and ^1S states indicates that the nuclear force cannot be wholly of the Heisenberg type. The difference between the n – p interactions in these states can be explained by assuming that the interaction is roughly 25 per cent Heisenberg or Bartlett and 75 per cent Wigner or Majorana.

The three types of exchange operators can be related as

$$P^H = P^M P^B \text{ and } (P^M)^2 = (P^B)^2 = (P^H)^2 = 1. \quad ...(8)$$

This shows that each operator has only two eigenstates +1 and –1.

The Majorana exchange operator is +1 for the states of even l and –1 for the states in odd l. The Bartlett exchange operator gives +1 in triplet states and –1 in singlet states, independent of l. In the various states of the two particle systems, the exchange operators have the values given below:

Operator	**Even parity states**		**Odd parity states**	
	Triplet	**Singlet**	**Triplet**	**Singlet**
P^H	1	–1	–1	1
P^M	1	1	–1	–1
P^B	1	–1	1	–1

The most general potential of the exchange type is written in the form

$$V_W(r) + V_M(r)P_M + V_B(r)P^B + V_H(r)P^H \qquad ...(9)$$

This potential also has tensor operator S_{12} term for mixtures of Wigner and Majorana forces.

(B) Isotopic Spin Formalism

In this formalism we are considering the proton and the neutron as different quantum states of the same particle, the nucleon. The total wave-function is written as a product of the space part, a spin part and an isospin part. The nucleus must obey Fermi-statistics in order to the consistent with the ordinary theory. Thus, the total wave-function for the two or more particles

$$\psi = \psi(\text{space})\ \psi\ (\text{spin})\ \psi\ (\text{isotopic spin}) \qquad ...(10)$$

must be anti-symmetric with respect to interchange of all co-ordinates of two nucleons. In the ground state of the deuteron, for example, ψ (space) is symmetric, as it a mixture of an S-state and a D-state, ψ (spin) is symmetric (the two spins are parallel), so that the ψ (isotopic spin) must be anti-symmetric and thus $T = 0$ (the two isotopic spins are oppositely oriented); The lowest state of deuteron in which the two nucleon spins are opposed, ψ (spin) is then antisymmetric, is the lowest one in which $T = 1$.

The concept of the T multiplet has been applied to β-decay, γ-decay and to nuclear reactions. *The success of these applications supplies additional support for the hypothesis of the charge independence of nuclear forces.*

In order to confirm with isospin conservation, the Hamiltonian describing the interaction between two nucleons must be rotationally invariant in isospace. Thus it must contain scalar quantities formed with the isospins $\vec{\tau}_1$ and $\vec{\tau}_2$. The product of operators $\vec{\tau}_1 \cdot \vec{\tau}_2$ gives + 1 when applied to isotopic spin triplet states and –3 when applied to isotopic spin singlet states. We can use $\vec{\tau}_1 \cdot \vec{\tau}_2$ to define an isotopic spin operator P^r analogous to spin operator σ.

$$P^r = \frac{1}{2}(1 + \vec{\tau}_1 \cdot \vec{\tau}_2) \qquad ...(11)$$

The operator gives + 1 when applied to the (symmetric) isotopic spin triplet states, –1 when applied to the (anti-symmetric) isotopic spin singlet states. Hence, it is equivalent to simple exchange of the isotopic spin co-ordinates η_1 and η_2 of the two particles. Thus, we can write

$$P^r\psi(r_1, \zeta_1, \eta_1; r_2, \zeta_2, \eta_2) = \psi(r_1, \zeta_1, \eta_2; r_2, \zeta_2, \eta_1) \qquad ...(12)$$

where r_1 denotes the position of the one particle and ζ_1 its spin direction.

We have required complete anti-symmetry under the full exchange of all the co-ordinates of the two particles. Since Heisenberg exchange operator P^H exchanges position mechanical spin and the isotopic spin operator P^r exchanges the isotopic spin co-ordinates, hence we can write.

$$P^H P^r \psi = -\psi \qquad ...(13)$$

This is a condition on ψ not an operator entity. We multiply equation (13) by P' on both sides and have

$$P^H\psi = -P^r\psi \qquad ...(14)$$

because equation (12) shows that $(P^r)^2 = 1$.

It is clear from equation (14) that we can replace the Heisenberg exchange operator by the isotopic spin operator $-P^r$. As the Bartlett exchange operator ^{PB} can be written as

$$P^B = \frac{1}{2}(1 + \vec{\sigma}_1 \cdot \vec{\sigma}_2) \qquad ...(15)$$

Thus, we can replace the Majorana exchange operator by the combination

$$P^M = \frac{1}{4}(1 + \vec{\sigma}_1 \cdot \vec{\sigma}_2)(1 + \vec{\tau}_1 \cdot \vec{\tau}_2) \qquad ...(16)$$

Equations (14), (15) and (16) show the expressions for the three linearly independent exchange operators in terms of the mechanical and isotopic spin operators. From the point of view of the isotopic spin formalism it is more convenient to use three other linearly independent operators $(\vec{\sigma}_1 \cdot \vec{\sigma}_2)$ $(\vec{\tau}_1 \cdot \vec{\tau}_2)$ and $(\vec{\sigma}_1 \cdot \vec{\sigma}_2)(\vec{\tau}_1 \cdot \vec{\tau}_2)$.The values of these products in states of the two particle system are listed below:

Spin Product	**Even Parity States**		**Odd Parity States**	
	Triplet	**Singlet**	**Triplet**	**Singlet**
$\vec{\sigma}_1 \cdot \vec{\sigma}_2$	1	–3	1	–3
$\vec{\tau}_1 \cdot \vec{\tau}_2$	–3	1	1	–3
$(\vec{\sigma}_1 \cdot \vec{\sigma}_2)(\vec{\tau}_1 \cdot \vec{\tau}_2)$	–3	–3	1	9

INSULATING CORE TRANSFORMERS

Transformers are very well known to us. They consist of two coils, called primary and secondary. But increasing the number of turns in secondary, output ac potential can be increased easily. The transformers consisting iron core cannot deliver voltages in the million volt range. The limitation is due to insulation required between the secondary coil (high potential) and the core, which is not far from ground potential.

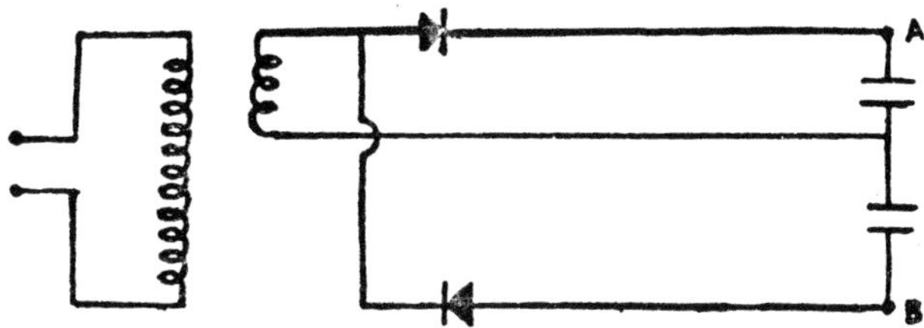

Fig. 3 : Insulating core transformer (Equivalent circuit).

In the insulating core transformer the iron core has not been dispensed with, but is made out of sections which are electrically insulated from one another. The segments form a central column surrounded by an outer shell of magnetic material, where the primary coil is wound. Each section in the central core is surrounded by a secondary winding, where an ac voltage of about 30 kV is induced. The output from each secondary is rectified by means of the voltage doubling circuit. By connecting in series all the dc power supplies thus formed, we can obtain high voltage. The first insulating core transformer, designed in 1960, was successfully operated at one million volts and 25 milliamperes with a power conversion efficiency of about 90%.

LINEAR ACCELERATORS

In a linear accelerator (often abbreviated as linac) the energy of a charged particle is increased steadily or in stops as it travels in a straight line. There are two principal devices which make use of direct acceleration in a straight tube.

1. Drift Tube Accelerators

The idea of accelerating positive ions by the aid of an alternating radio-frequency field was first suggested by Ising in 1925. By such a method D.H. Sloan and Lawrence in 1931 produced a beam of mercury ions emerging with energy 1.25 MeV, using initially a 30 mega cycles per second radio frequency supply of only 40 kilo volts. The principle is shown in Fig. 4. The linear accelerator consists of a number of cylindrical electrodes of increasing length arranged in a straight line.

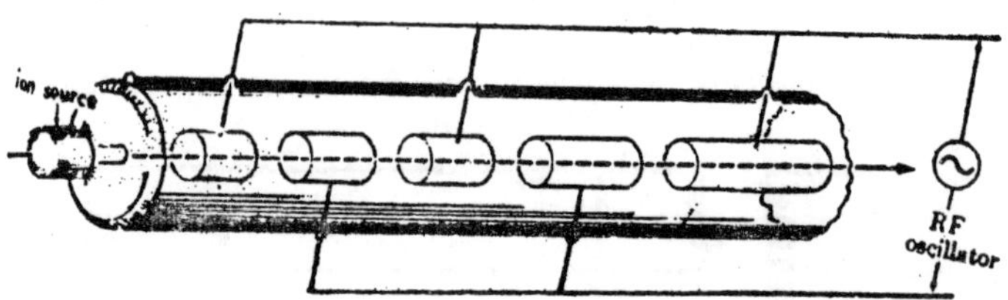

Fig. 4 : Schematic diagram of a linear accelerator.

Alternate cylinders are connected together, the cylinders numbered 1, 3, 5, etc., being joined to one terminal and the remaining cylinders to the second terminal of the high frequency power supply. At any instant, therefore, alternate electrode carried opposite electrical potentials. All the electrodes which are at positive potential in a particular half cycle becomes negative in the next half cycle.

Suppose positive ions from the ion source move from left to right along the common axis of the cylindrical electrodes. While passing through first electrode, the ions receive no acceleration, since they are within a uniform potential. If the first is positive and the second is negative, the positively charged particles will be accelerated in the gap between these electrodes.

The positive ions then enter the second electrode and travel through it at a constant but higher speed. The length of the second electrode is such that just as the ions reach the gap between it and the third electrode the potential of these electrodes are reversed. Now, the second becomes positive and third negative, thus the positive ions are accelerated in this gap.

The length of a cylindrical electrode plus a gap or the separation between two consecutive gaps is the distance traversed by the particles during one half cycle of the applied field of frequency f and is given by

$$L_n = v_n T_2/2 = v_n/2f \qquad ...(1)$$

where v_n is the velocity of the particle in this n^{th} electrode. If V is the peak value of the potential difference between two terminals, then the energy acquired by the ions in passing one gap will be qV. If there are n such gaps, the total kinetic energy acquired by the ions will be qV_n. The velocity of the ions in the nth electrode is, thus, given by the relation

$$\frac{1}{2}Mv_n^2 = nqV + K, \qquad ...(2)$$

where M is the mass of the particle and q its charge. The constant K allows for the fact that the injection energy is greater than zero and for the fact that the first gap has only half the accelerating voltage V if the oscillator is balanced with respect to ground. By combining eqns. (1) and (7) we obtain the length in the nth electrode

$$L_n = \frac{1}{2f}\left[\frac{2(nqV+K)}{M}\right]^{1/2} \qquad ...(3)$$

This relation indicates that in order to get a large energy the peak voltage of the oscillator should be large and number of cylinders should be as large as possible. The length of the cylinders can be reduced by increasing the frequency of the oscillator.

Relativistically, for particles of high energy ($v \rightarrow c$) the distance between gaps is constant, *i.e.*,

$$L = c/2f = \lambda/2. \qquad ...(4)$$

Hence, each energy increment is mainly represented by an increase in mass rather than velocity. The cylinders then approach uniformly in size and spacing and the distance between two adjacent gaps is a half wave length of the wave emitted by the electric oscillator.

Radial Focusing

The ion beam during its passage through the gaps is not only accelerated but also focussed radially due to the curved nature of the electric lines of force in the gap. The Fig. 5 shows the gap between two coaxial cylinders when a difference in potentials exists between them. In the first half of the gap, any oncoming ion with a velocity Vi will receive an increment of velocity ΔV forward and ΔV_c toward the axis of the cylinder.

During the latter half of the gap it will receive an equal increment in its forward motion but the radial component of force is directed outwards. This latter velocity increment V'_c is not as great

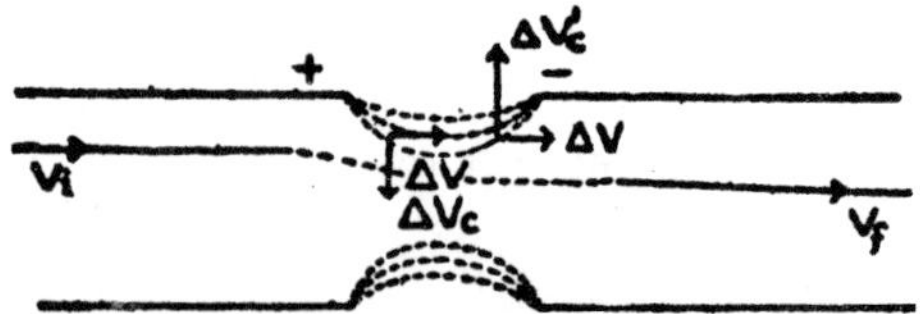

Fig. 5 : Focusing due to electric field.

as that toward the axis, because ion is travelling faster and hence is acted upon by a force for a shorter time. Thus, there is a net displacement of ions towards the axis of the cylinders. The defect of this instrument is that of its larger size and it is very difficult to maintain vacuum in large chamber.

Phase Focusing

In the derivation of relation (3), we have assumed a constant potential difference V applied between consecutive cylinders. As alternating potential Vo sin wt is applied by the radio-frequency source, hence the ion should cross the gaps at times O, T, 2T,...., T = 2π/ω or the ions should approach in same phase at successive gaps. Particles which arrive at each gap at the instant corresponding to the point indicated by P in the figure will always receive the appropriate impulse. A particle which

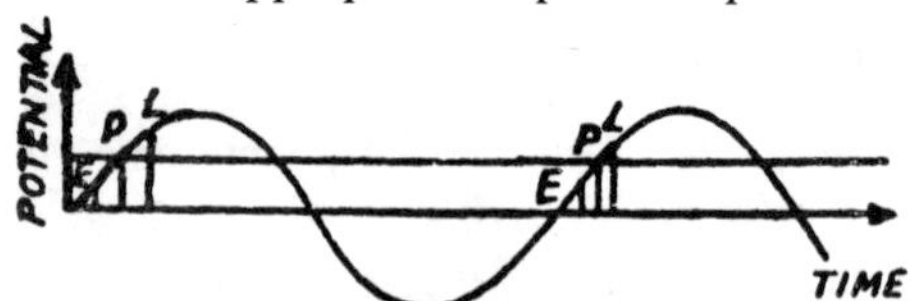

Fig. 6 : Principle of phase stability.

has more energy than that called for at a particular gap will necessarily have greater velocity and will have phase such as E. It will thus be accelerated less than normal and will arrive at the next gap with a phase closer to P. An ion that arrives at a gap with less than the characteristic energy will arrive late, phase L, It will be accelerated more than the normal and will gradually shift its phase toward. Thus all ions within a certain phase range will be trapped and will oscillate about the point of *stable phase* (point P).

2. Wave Guide Accelerators

Modern accelerators make use of waveguides to establish the electric field. A waveguide is a pipe of conducting material in which an oscillating electromagnetic field is established. It electromagnetic waves with frequencies around 10^4 or 10^5 megacycles per sec are fed into one end of a waveguide, the waves and the associated oscillating electric and magnetic fields travel along the length of the guide. It may at first appear surprising that the rate of propagation of the wave-front in a waveguide, the *phase velocity*, in a simple waveguide, is usually greater than the velocity of light. The reason is that the electromagnetic waves approach the walls of the waveguide at an angle so that they are reflected back and forth and travel along the guide in a zig-zag path. The interference of these waves leads to a new wave pattern in which the wavelength is greater than in free space. The phase velocity v_p is equal to the product of this wavelength and the normal frequency which is constant and is given by the relation

$$v_p = c[1 - (\lambda/\lambda_c)^2]^{-1/2} \qquad ...(5)$$

where λ is the wavelength of the applied r-f in free space and λ_c is the cut off wavelength which depends on the type of wave and the dimensions of the guide. $\lambda_c = 2.615$ a for the circular guide of radius a and for TM_{01} (or E_{01}), *transverse magnetic field pattern* waves. If the electromagnetic wave is completely absorbed at the end of the waveguide, the result is a travelling wave.

But if there is an electrical conductor at the end, part of the wave will be reflected back along the guide, and the combination will be a standing wave. The radiofrequency signal is produced by master oscillator and amplified at each feeding station by klystrons. The speed of the electromagnetic wave is synchronised with the speed of the particle so that the particle gains energy from the electric component of the wave.

Linear accelerators have been developed for both electrons and protons. The design considerations of proton and electron linear accelerators are basically different for the fundamental reason that the less massive electron travels most of the time with a velocity very close to that of light. On the other hand, the protons gain speed along the total light path. The chief drawback of linear accelerator is its very high power consumption. A linear accelerator of length l excited by a field whose free space wavelength is λ and accelerating particles to a final energy of T eV, requires a total power

$$P = CT^2\lambda^{1/2}/l \text{ watts}, \qquad ...(6)$$

where $C \simeq 3 \times 10^{-8}$ for proton machines and $\simeq 10^{-7}$, for electron machines.

The ratio of the square of the maximum electric field on the axis to the power dissipated per meter of length is called the *shunt impedance*. It is given by

$$Z = 10^{-6}\, E_z^2/P \text{ megohms/m.} \qquad ...(7)$$

Electron Linear Accelerator

If a bunch of electrons injected along the axis of a guide carrying an E_{01} travelling wave it may be possible for the electrons to be accelerated by the axial electric field of the wave and thereby reach high energies. In order for an electron to be accelerated in a wave guide system, the phase velocity must be essentially equal to that of the accelerating electron. The phase velocity can be reduced by placing metal discs with holes in the centre, at intervals along the guide. The resulting system is known as an *iris loaded* or *disc loaded wave guide.* The beam of electrons to be accelerated passes through these holes. Near the injection end of the loaded guide the electron velocity may be of the order of 1/2c (for 79 keV) and the phase velocity of the wave must be adjusted to this value. After travelling a short distance the electron velocity is very close to c and the phase velocity is made equal to c and remains same over the remaining length of the accelerator.

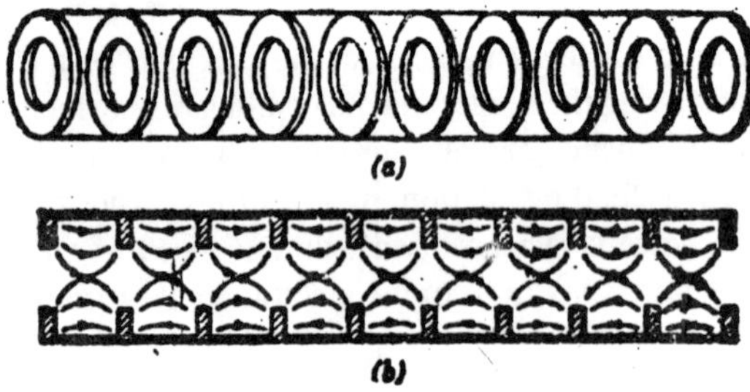

Fig. 7 : (a) A disc loaded wave guide. (b) The electric field configuration.

The tendency for the electron bunch to diverge under the influence of the transverse field components can be compensated in the initial stage of acceleration by an axial external dc magnetic field. As the electron reaches an energy of a few MeV, the defocusing effects become ineffective. The radii of the hole in the iris a and the guide radius b are not independent. For any choice of a there exists a value of b. As a is increased b also increases rather slowly. For relatively small values of a, b is represented by the relation

$$b/\lambda = 0.383\ [1 + 20\ (a/\lambda)^3]. \qquad ...(8)$$

The flow of energy is strongly influenced by the size of the apertures. The energy distribution along the wave guide follows from following relations:

Group velocity v_g = energy flow/energy density.

Energy dissipated = $\dfrac{d}{dz}$ (rate of energy flow)

Q-factor of the waveguide = $\dfrac{\omega\,(\text{energy stored})}{\text{rate or energy dissipation}}$

or Rate of energy dissipation $\propto e^{(\omega z/vgQ)}$.

Since the energy dissipation is proportional to the square of the electric field E_z, hence

$$E_z = E_{zo}\ e^{-(\omega z/2vgQ)}.$$

If the distance between two feedpoints is l the total energy gain of a particle reaching from one feed point to the next will be

$$e\int_0^l E_z dz = \frac{2ev_g QE_{z0}}{\omega}[1 - e^{(wl/2v_g Q)}] \qquad ...(9)$$

The first travelling wave accelerator was constructed by D.W. Fry and his associates in England in 1946 In the following year Hansen and his colleagues completed a similar device to produce 6MeV electrons. Electrons with energies up to 1BeV have been obtained with the 220ft Stanford linear accelerator constructed at Stanford University.

This accelerator is divided into 10 sections, each of which is fed with high frequency power from a separate klystron amplifier and producing a travelling wave in the tube having a frequency of 3×10^9 cycles/sec. The accelerator is operated at the rate of 60 pulses/sec, each pulse lasting one microsecond and produces about 5×10^{10} electrons per pulse. These electrons have a narrow energy spread of 0.5 per cent about the mean energy BeV.

A linear electron accelerator, 2 miles long, was constructed at Stanford Linear Accelerator Centre (SLAG) and was completed in 1966. It consists of 960 sections of copper tubing, each 10 feet long and 4 inches external diameter. With 240 klystrons in operation, the maximum electron energy is 20 BeV. It is planned eventually to increase the number of klystrons to 960, the electron energy should then be 40BeV.

Proton and Heavy ion Linear Accelerators

Above technique cannot be applied to proton acceleration. The main difference arises because of the much lower injection velocity of protons (4 MeV proton has a velocity of ~0.1c). Disc loaded-wave guide is impracticable for phase velocities below about 0.4 c.

To accelerate protons a long cylindrical resonant cavity is excited in a TM_{010} mode with an infinite phase velocity. The electric field lines are parallel to the axis of the cavity and terminate on the end walls of the cavity. Since an ion takes many periods of the rf-field to travel from one end of the cavity to the other, some device must be introduced to shield it from the field while the field direction is such as to decelerate rather than to accelerate.

It is done by introducing a series of tubes along the axis through which the beam passes. These tubes are known as drift tubes, as ions experience no forces while they travel through beam and cross the gap between two *drift tubes* when the electric field is increasing with time. Their length increases along the accelerator as the ion velocity increases. The tubes decrease in diameter to keep the resonance frequency uniform along the length of the accelerator.

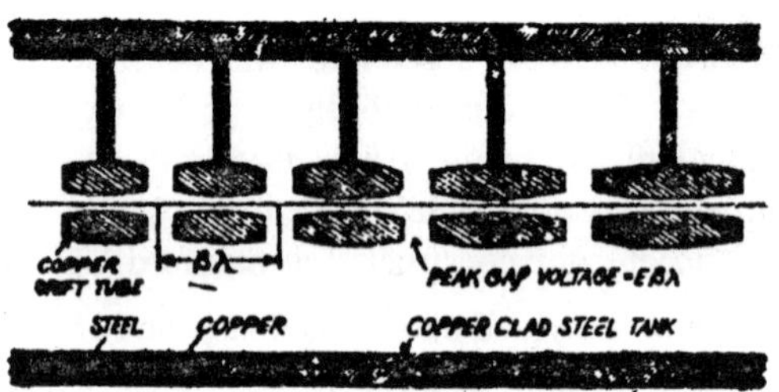

Fig. 8 : Drift tube linear accelerator.

One of the difficult problems was rf-defocusing of the beam. At velocities low compared with c a phase stable particle will be lost very rapidly by the radial components of rf-field. The radial field at the entrance to the drift tube is removed by placing a grid just infront of the drift tube. However, this focusing grid or foil in-front of each drift tube intercepts or scatters some of the protons and results in the considerable loss in beam intensity.

The use of electric quadrupole lenses along the axis of the accelerator is an alternative approach to the focusing problem.

In 1965, the highest energy attained with a proton linear accelerator was 70 MeV. Several linear accelerators have been built follower energies. At Brookbaven a design study has been made for a 500 MeV linac and at Los Alamos design study has been made for an 800 MeV linac.

LOW ENERGY CIRCULAR ACCELERATORS

Instead of building up a high potential, as is done in electrostatic accelerators, a generator of relatively low voltage is used repeatedly to add energy to the ions in cyclic accelerators. In these accelerators the length of path can be increased without increasing the length of the apparatus. Examples are the cyclotron and betatron.

1. Cyclotron (Fixed Frequency)

Lawrence realized that, instead of using cylinders of gradually increasing lengths, the particles could be accelerated by using one pair of electrodes only if the particles were compelled to cross the same accelerating gap repeatedly by means of a magnetic field. This instrument was named as *magnetic resonance accelerator*. By 1936, the name *cyclotron* has come into general use. In 1939, Professor Lawrence was awarded the Nobel Prize in Physics in recognition of this achievement in the conception and development of the cyclotron.

In its simplest form, the cyclotron consists of two flat, semicircular metal boxes, called dees because of their shape. These hollow chambers have their diametric edges parallel and slightly separated from each other. A radio-frequency alternating potential of the order of megacycles per second is applied between the dees, which act as electrodes.

These dees are surrounded by a closed vessel, containing gas like hydrogen, helium, deuterium at low pressure and the whole apparatus is placed between the poles of a strong electromagnet which provides a magnetic field perpendicular to the plan of the dees. The positively charged particles arc produced at the center between two dees, as shown by S in Fig. 9. Suppose that at any particular instant the alternating potential in the direction which makes D_1 positive and D_2, negative.

A positive ion starting from the source S will be attracted by the dee D_2. Since there is a uniform magnetic field B acting at right angles to the plane of the dees, the ion of charge q and mass M will move in a circular path of radius $r = M_v/Bq$. In the interior of the dee, the speed of the ion remains constant. After it has traversed half a cycle, the ion comes to the edge of D_2.

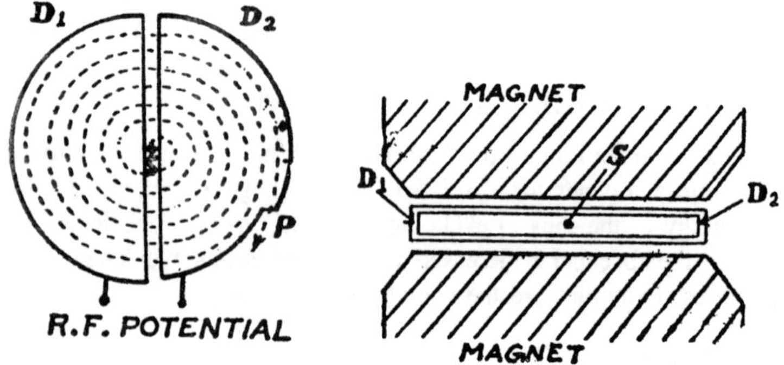

Fig. 9 : Cyclotron (Magnetic Resonance Accelerator).

If in the meantime, the potential difference between D_1 and D_2 has changed direction so that D_2 is now positive and D_1 negative, the positive ion will receive an additional acceleration while going across the gap between the dees, and then travel in a circular path of larger radius inside D_1 under the influence of magnetic field. After traversing a half cycle in D_1, it will reach the edge of D_1 and receive an additional acceleration between the gaps because in the meantime the potential difference between dees has changed.

The ion will continue travelling in a semi-circle of increasing radii, each time it goes from D_1 to D_2 and from D_2 to D_1. The time taken by the charged particle to traverse the semi- circular path in the dee is given by $t = \pi r/v = \pi M/Bq$ and hence the time for a complete circular path

$$\tau = 2t = 2\pi M/Bq \quad \text{...(1)}$$

Thus, for any given value of M/q, τ is determined by the magnetic field intensity B. By adjusting B the time can be made the same as that required to change the potentials. The frequency of the oscillations required to keep the ion in phase is given by the relation

$$f = 1/\tau = Bq/2\pi M. \quad \text{...(2)}$$

If f and B are adjusted to keep the charged ion always in phase, each time the ion crosses the gap it receives an additional energy and at the same time it describes a flat spiral of increasing radius. Eventually, the ion reaches the periphery of the dee, where it can be brought out of the chamber by means of a deflecting plate charged to a high negative potential. This attractive force draws the ion out of its spiral path and thus can be used easily. If R is the radius of the dee, the kinetic energy of the ion emerging from fie cyclotron is thus given by

$$T = \frac{1}{2}M(BqR/M)^2 = q^2B^2R^2/2M = 2\pi^2R^2f^2M. \qquad ...(3)$$

This relation indicates that the maximum energy attained by the ion is limited by the radius R of the dees, magnetic field B or the frequency of the alternating potential f. It is also clear that the maximum energy acquired by the charged particle in a particular cyclotron is independent of the alternating voltage. It can be explained by the fact that when the voltage is small the ion makes a large number of turns before reaching the periphery, but when the voltage is high the number of turns is small. The total energy remains same in both the cases, provided B and R are unchanged.

Above equations are also true for relativistic velocities, providing $M = M_0(1 - v^2/c^2)^{-1/2}$. The quantity Br, often called *rigidity of a beam* of ions is thus given as

$$Br = \frac{M_v}{q} = \frac{W_v}{qc^2} = \frac{1}{qc}(W^2 - W_0^2)^{1/2} \qquad ...(4)$$

If the ion has Ze charge, and T and W_0 are in MeV then above relation gives

$$Br = \frac{1}{300Z}(T^2 + 2TW_0)^{1/2} \text{ weber/meter} \qquad ...(5)$$

If there are accelerations and the average potential difference between D's is V, the final energy can also be written as

$$T = NqV. \qquad ...(6)$$

Eliminating T from eqns. (3) and (6), we get

$$R = \frac{1}{B}\left(\frac{2MV}{q}\right)^{1/2} N^{1/2} \qquad ...(7)$$

This shows that the radii of successive paths increase as $N^{1/2}$, getting closer together as N becomes large.

Electric and Magnetic Focusing

The electric vertical forces are analogus to those already discussed in the linear accelerator. If the particles cross the gap when the field is rising the focusing field they experience when they approach the gap is over compensated by the stronger defocusing field that exists after the gap is passed. If the field is decreasing as they cross the gap, a net focusing field experienced. If the ions are to be kept in the vicinity of the median plane there must be forces returning them to this plane, whenever they leave it. These forces may result by shaping the magnetic field so that the vertical component B_0 decreases with increasing distance from the centre of the dees and by adjusting electric field between the dees. Without such restoring forces, it would be likely to drift towards the inside of the dees and be lost. The magnetic field varies as

$$B = B_0 (r_0/r)^n, \qquad (8)$$

where B_0 is the field at some fixed radius r_0, the index n is obtained by differentiation- It is zero at the centre, rises almost linearly upto the point where fringing effects start, and increases rapidly to exit-slit location. The frequency of the axial oscillations is given by

$$f_z = f_0 n^{1/2} \qquad ...(9)$$

where f_n is the cyclotron resonance frequency. The radial oscillations also occur, with a frequency

$$f_r = f_0(1 - n)^{1/2} \qquad ...(10)$$

Let us discuss the main components of a cyclotron:

(a) *Vacuum Chambers:* The vacuum inside the machine is obtained with the help of diffusion pumps. Oil-diffusion pumps of large diameter which have two or three pumping stages and automatic fractionation of the oil are best suited to this purpose. The baffles are used to prevent back streaming of the oil vapour, so that the oil deposition within the vacuum chamber is very small.

(b) *Deflector:* The deflector is used to pull the ion beam out of its circular path and direct it in a linear path. The deflector system may be divided as dc, rf, and dc-plus rf-types. In the dc-type the dc-electric field is applied between two curved plates forming a cylindrical condenser. The inner one is called the *septum* and the outer the *deflector*. The blade of the deflector is made of Cu with oil cooling arrangement and the septum is made of very thin tungsten plate. As a chosen maximum radius R a thin septum is inserted having a larger radius of curvature R + ΔR, which splits the beam and allows a fraction of ions to pass into a channel behind the septum. If V_d is the deflector voltage and d is the spacing, then we have

$$\frac{mv^2}{R} = Bqv$$

and
$$\frac{mv^2}{R + \Delta R} = Bqv \frac{V_d q}{d}$$

$$\therefore \qquad \frac{V_d}{d} = \frac{2T}{qR} \frac{\Delta R}{R + \Delta R} \qquad ...(11)$$

This equation determines the maximum permissible value of radius.

(c) *Oscillator:* To produce a high frequency potential difference of 10 to 100 kv in the frequency range between 10 to 20 megacycles/ sec between the dees, resonant circuits are required. The resonant circuit is equivalent to a pair of quarter wave coaxial transmission lines with the dees supported on the ends of the inner conductors. The frequency is controlled by *Cu-shorting bars* or *spiders* reaching from stem to outer cylinder or in shielded pair structure from stem to stem.

(d) *Magnets:* The main magnet of a cyclotron has a basic structure of two cylindrical tapered poles, to rectangular yokes and two rectangular uprights. The purpose of the taper is to keep flux density approximately constant. The pole base has diameter greater than that of D's to keep the flux density in the base of the pole (about 52" for 42" cyclotron). A stabilized current is supplied to the magnet coils.

To tune up a cyclotron to preserve resonance and maintain beam intensity, shimming coils are used. These are flat-wound coils of many concentric turns with individual leads brought out radially from the inner turns.

(e) *Dees:* The dees are made of Cu, preferably the oxygen free high conductivity variety, and are supported in cantilever fashion by *dee stems* of 3" diameter. Few cyclotrons have only one dee, in which case the stem is located so as to be ⊥ to the open edge of the dee. This yield a symmetrical voltage pattern across the edge. To obtain a vertically symmetrical field pattern, a *dummy dee* (consists of a Cu frame) faces the active electrode.

(f) *Ion Source:* Recent sources are of the hooded arc variety. A hollow cylinder or chimney is closed at the top but with a hole in its side at mid-plane height, facing one dee. The cathode is formed of heavy tungsten or tantalum rod, formed as a short stemmed 'V' or 'U' clamped in terminating lugs and using large diameter water cooled leads. The chimney of the ion source may be moved in x-y-z direction to place the hole to any desired position so as to maximise the output current. The filament may be replaced without the loss of vacuum in the main chamber through air locks.

Cyclotrons are usually described in terms of the diameter of the pole faces of magnet. The first machine constructed by Lawrence and Livingston in 1932 used a magnet with pole faces 11 inches in diameter and produced 1.2 MeV protons. Consequently, cyclotrons with larger and larger magnets were built by Lawrence and his associates at Berkeley and later by others at several laboratories in the United States and aboard. Among them are several of about 60" pole-face diameter that accelerate deuterons to about 20 MeV and helium ions to about 40 MeV. The relativistic mass increase has limited the energies achieved with standard cyclotrons to about 25MeV for deuterons and 50 MeV for alpha particles. Although the cyclotron is capable of producing beams of energetic ions that are more intense than those from the electrostatic accelerators, the particle energies are much less constant. Hence, cyclotrons have found their widest use for studies requiring particles of high energy but where essential to know the exact value of energy. These high energy particles are used for the study of nuclear reactions, for the production of protons neutrons and radionuclides.

2. The Betatron

The cyclotron principle is inapplicable for the acceleration of electrons to high energies because of the large relativistic increase of mass at low energies. The voltage multiplier and the Van de Graaff electrostatic generator can both be used to accelerate electrons but the energies are limited to a few MeV. High energies X-rays, used in biological and medical research as well as in atomic research, can be secured by bombarding a target with high energy electrons. The first important machine for producing high energy electrons was the *betatron*. The first *magnetic induction accelerator* or *betatron* was constructed by D. W. Kerst at the University of Illinois in 1940, to accelerate electrons to an energy of 2.3 MeV. Several other betatrons have been constructed, the largest being the one at the University of Illinois, completed in 1950. It yields electrons of 300 MeV energy.

The action of the betatron depends on the same fundamental principle as that of the transformer in which an alternating current applied to a primary coil induces an alternating current usually with higher or low voltage in the secondary coil. In the betatron secondary coil is replaced by a doughnut shaped vacuum chamber. Electrons produced in the doughnut from a hot filament, are given a preliminary acceleration by the application of potential difference of 20 to 70 kV. When an alternating magnetic field is applied parallel to the axis of the tube, two effects are produced: *an electromotive force is produced in the electron orbit by the changing magnetic flux that gives an additional energy to the electrons; a radial force is produced by the action of magnetic field whose direction is perpendicular to the electron velocity which keeps the electron moving in a circular path.* Instead of spiralling, as in the cyclotron, conditions are arranged such that the increasing magnetic field keeps the electrons in a circular orbit of constant radius.

Using Faraday's Law of induction, the work done on an electron of charge e in one revolution $W = e d\Phi/dt$. If F is the tangential force acting on the electron, the work done W can be expressed as $W = 2\pi RF$.

$$\therefore \qquad F = \frac{e\, d\Phi}{2\pi R\, dt} \qquad ...(12)$$

For an electron moving in a circular orbit

$$Bev = Mv^2/R \quad \text{or} \quad Mv = BeR \qquad ...(13)$$

This relation indicates that the magnetic field B at the orbit must increase as the electron energy increases, otherwise radius R will not be constant. If R is kept constant, then

$$\frac{d}{dt}(Mv) = eR\frac{dB}{dt} \qquad ...(14)$$

From Newton's Second Law, the rate of change of momentum is equal to the force. Hence on combining eqns (12) and (14), we get

$$\frac{d\Phi}{dt} = 2\pi R^2 \frac{dB}{dt} = 2\frac{d}{dt}(\pi R^2 B)$$

This relation shows that the rate of change flux Φ within the orbit of radius R is always twice what it would have been if the magnetic field intensity were uniform throughout the orbit. This relation is known as the betatron condition or flux condition. It is also called as "2-1" rule of the betatron, this can be written as

$$dB_C/dt = 2\ dB_G/dt, \qquad ...(15)$$

where B_C is the space-averaged field strength in the core and B_G is the guiding field at the orbit. Since the induced potential is determined by the rate of change of flux, the iron core is laminated as in a transformer and alternating potential at 60 or 180 cycles is used to produce varying magnetic field. The electrons are injected at an instant when the magnetic field is just rising from its zero value in the first quarter cycle. The increasing magnetic field induces a potential within the doughnut which increases the energy of the electrons. When the field strength passes its maximum value and starts to decrease, the direction of the induced electromagnetic force is changed and the electrons start to slow down. This effect is avoided by removing the electrons from their stable orbit by passing a pulse of current through an auxiliary coil when the field reaches its positive maximum. These high energy electrons leave the path tangentially to strike a tared which then emits X-rays. The emission from the machine consists, therefore, of a succession of short pulses of X-radiations, each occurring at the peek of the a.c. maximum. Only the first quarter of each a.c. cycle is usefully employed in acceleration.

The energy of electrons can be estimated from the average Induced emf and the total number of revolutions made by the electrons. If be flux variation is given by the relation $\Phi = F_0 \sin \omega t$.

The time during which acceleration takes place will be $\pi/2\omega$. The energy gained by the electron per turn when the flux changes will be

$$eV = e\ d\Phi/dt = ew\Phi_0 \sin \omega t \qquad ...(16)$$

Thus, during the acceleration period, its average value will be $(2/\pi)\ e\omega\phi_0$. For most of the time electrons travel with a velocity close to the velocity of light and, therefore, the total distance travelled during the acceleration process will be $c\pi/2\omega$. If R is the radius of the orbit, the number of revolutions N will be $c/4\omega\ R$, hence the total energy gained by electrons will be the product of the number of revolutions and the mean energy per turn.

$$\therefore \qquad E = e\ c\phi_0/2\pi R. \qquad ...(17)$$

The energy of electrons can also be estimated by the help of relativistic equation for energy

$$K.E. = pc = B_0eRc, \quad ...(18)$$

where B_0 is the magnetic field at the end of the acceleration cycle at the orbit. The energy obtainable is, therefore, limited by the radius and the peak strength of the magnet at the orbit. Using rest energy W_0 (= m_0c^2) of the electron. The kinetic energy T is given by

$$B_0 = \frac{mv}{eR} = \frac{(T^2 + 2TW_0)^{1/2}}{ceR} \quad ...(19)$$

This relation shows that the high energy electrons can be retained in reasonably small orbits by readily available magnetic field.

Orbital Stability: Stability of the betatron orbit is achieved by making the para-axial field B: at the orbit decreasing with radius according to the relation

$$B_z = B_0\ (R/r)^n = B_0\ (r/R)^{-n}, \quad ...(20)$$

where n is a constant, R the radius of *equilibrium orbit* and By is the value of Bz at equilibrium orbit. Since r makes small variation from R, hence can be written as r = R + x, Maxwell's equation for this case curl B = 0 gives

$$\partial Bz/\partial r = \partial B_r/\partial z$$

$$\therefore \quad B_r = -\,nzB_0/R \quad ...(21)$$

The *radial equation of motion* is

$$\frac{d}{dt}(mr) = \frac{mv^2}{r} - B_zev \quad ...(22)$$

For radius r, eqn. (34) gives

$$B_z = B_0\left(\frac{R+x}{R}\right)^{-n} = B_0\left(1 - \frac{nx}{R}\right) \quad ...(23)$$

Hence eqn. (22) becomes

$$\frac{d}{dt}(m\dot{x}) = \frac{mv^2}{R+x} - B_0\left(1 - \frac{nx}{R}\right)ev$$

or

$$m\ddot{x} = \frac{mv^2}{R}\left(1 - \frac{x}{R}\right) - ev\ B_0\left(1 - \frac{nx}{R}\right) \quad ...(24)$$

For equilibrium orbit x and their derivatives are zero, hence we get $mv^2/R = evB_0$ and thus eqn. (24) becomes

$$m\ddot{x} = \frac{mv^2}{R}\left(1 - \frac{x}{R} - 1 + \frac{nx}{R}\right) - \frac{mv^2}{R^2}\, x\,(-1 + n)$$

or $$\ddot{x} + \omega^2 (1 - n)\, x = 0, \qquad ...(25)$$

where $\omega = v/R$ is the angular velocity in the equilibrium orbit. Above relation shows that the frequency of radial oscillations

$$f_{r}, = \omega\,(1 - n)^{1/2}/2\pi \qquad (n < 1) \qquad ...(26)$$

The *vertical equation of motion* for a small axial displacement z out of the median plane is written as

$$\frac{d}{dt}(mz) = B_r\, ev = ev\left(\frac{-nB_0 z}{R}\right)$$

or $$\ddot{z} + \omega^2 nz = 0 \qquad ...(27)$$

Thus, the frequency of vertical oscillations is

$$fz = \omega n^{1/2}/2\pi \qquad ...(28)$$

For equations (26) and (41) to describe oscillations, the coefficient of x and z must be positive, *i.e.*, $1 > n > 0$. It shows that the field must decrease from the centre to the periphery, or in other words that the lines of force must be concave towards the axis of the machine. This decrease must not be too rapid, as $n < 1$. These oscillations are named as *betatron oscillations*, and are naturally damped as the field grows, the oscillation amplitude is proportional to $B_0^{-1/2}$. These principles are utilized in the designing of the magnet. The same magnet structure is used to provide the accelerating flux across the short central gap and the guiding field in the wider gap which surrounds the flux gap. The gap lengths are chosen to satisfy relation (29) and the guide field gap is tapered to given an n value between 0 and 1.

Most of the modern betatrons arc operated from a 60 cycle/sec a.c. source. If the length of the circular path within the doughnut of a betatron is supposed to be 3 metre and the energy gained by the electron in each revolution 400 eV, the electrons will then have a total energy of 167 MeV. In practice the energy of electrons can be varied from 10 to 100 MeV by applying the orbit shifting magnetic field at different times during the quarter cycle in which the field is increasing. The betatron, constructed at the General Electric Research Laboratories, produces 100 MeV electrons. The diameter of the pole face is 76 inches and that of stable orbit is 66 inches. The magnet weighs 130 tons and the maximum magnetic field at the orbit is 4000 gauss. Electrons, injected at energies of 30-70 keV, travel around the doughnut about 2.5 × 106 times, gaining about 400 eV of energy at per revolution. University of Illinois has a large machine with doughnut diameter 97 inches and magnet weighing 350 tons giving 340 MeV electrons.

Since the operation of the betatron is unaffected by the increasing mass of the electron, as it gains energy it might be thought that there would be no upper limit of energy obtainable by means of this device. In fact, there is, not because of mass increase but because of radiation loss, a limit of energy where the radiation off-sets the gain. Because of the centripetal force, the circulating electrons are expected to emit radiations. Using classical arguments, Schwinger in 1949 gave a relation for the total energy radiated by an electron. According to him the total energy radiated by an electron of energy W in one revolution is

$$U = \frac{1}{3\in_0}\left(\frac{e^2}{R}\right)\left(\frac{W}{m_0c^2}\right)^4 = 88.5\,\frac{W^4}{R} \qquad \text{...(29)}$$

where U is in keV, W in geV (giga electron volt, in Europe which is known as beva electron volt in U.S.A.) and R in metres. This relation indicates that the energy radiated per revolution increases with the fourth power of the electron energy and thus becomes very serious at very high energies. For the 300 MeV betatron the total loss is about 13% of the total energy. If no compensating means are included the orbit will shrink about 20% and the beam will loss by collision with the inside wall of the vacuum chamber. Electrons can however be accelerated to higher energies without appreciable radiation losses by means of a linear accelerator.

PRINCIPLE OF PHASE STABILITY

In a phase stable accelerator, particles will be accelerated at a series of gaps by an alternating electric field. *The gap separation and the frequency and strength of the field are adjusted so that a particle of a specified energy arriving at a particular gap at a specified equilibrium phase of the accelerating field will arrive at the next gap at the same phase.* This principle is most important in some accelerators, known as *synchronous or phase stable accelerators*. These are most easily understood by analyzing the electric field patterns as travelling waves. If x represents distance travelled by the particles, the significant field component can be represented as

$$E_x = E \sin\left[\int \omega \, dt - \int \frac{\omega}{v} dx + \phi_0\right] \quad ...(1)$$

where ϕ_0 is the equilibrium phase, ν the wave velocity and $E = V/x_0$. Here V is the voltage across the accelerating gap and x_0 the distance between two consecutive gaps. If the particle is travelling on the equilibrium orbit at the equilibrium phase ϕ_0, we have to adjust ω and ν such that bracket term of eqn. (1) reduces to ϕ_0. If the particle is not at the equilibrium phase but as at a phase $\phi_0 + \Delta\phi$. Associated with the phase error will be a momentum error Δp from the equilibrium momentum p_0 and position error Δx from the position of the equilibrium particle. Since a phase change of 2π results in a position change Δx of one wavelength, hence

$$\Delta\phi = -\omega\Delta\, x/v \quad ...(2)$$

and $$\frac{\Delta p}{p_0} = \frac{1}{p_0}\left[\frac{m_0\dot{x}}{(1-\dot{x}^2/c^2)} - \frac{m_0 v}{(1-v^2/c^2)^{1/2}}\right] = \frac{\Delta\dot{x}}{v(1-v^2/c^2)} \quad ...(3)$$

Δp also appears in the equation of motion

$$\frac{d}{dt}(p_0 + \Delta p) = qE \sin(\phi_0 + \Delta\phi)$$

$$\therefore \quad \Delta\dot{p} = qE\, \Delta\phi \cos\phi_0 \text{ (for small } \Delta\phi). \quad ...(4)$$

The combination of equations (2), (3) and (4) gives

$$\frac{d}{dt}\left[\frac{p_0}{v(1 - v^2/c^2} \frac{d}{dt}\left(\frac{v\Delta\phi}{\omega}\right)\right] + qE\, \Delta\phi \cos\phi_0 = 0 \quad ...(5)$$

For a *linear accelerator ω* is kept constant and we get

$$\frac{d}{dt}\left[\frac{1}{(1 - v^2/c^2)^{3/2}} \frac{d}{dt}\left(\frac{\Delta\phi v}{c}\right)\right] + \frac{\omega qE}{m_0 c}\Delta\phi \cos\phi_0 = 0 \quad ...(6)$$

In the case of particles travelling on a *circular or spiral orbit,* the momentum error results not only in an azimuthal position error but also in a radius error Δr from the equilibrium radius R. We shall consider acceleration by the hth harmonic of the revolution frequency. Thus, eqn. (2) becomes

$$\Delta\phi = h\omega\, \Delta x/v = h\omega r\, \Delta\theta/v = -h\Delta \quad ...(7)$$

$$\therefore \quad \Delta\phi = h\Delta\dot{\theta} = -h\, [\Delta(r\dot{\theta}) - \dot{\theta}\Delta r]/r \quad ...(8)$$

or $$\Delta\dot{\phi} \simeq \frac{hv}{R}\left[\frac{Dr}{R} - \frac{D\dot{x}}{v}\right] \qquad ...(9)$$

Since p = Bqr and B $\propto r^{-1}$ for stability of betatron oscillation

$\therefore$ $$\Delta p/p_0 = (l - n)\ \Delta r/R \qquad ...(10)$$

Substituting in eqn. (9) for Δr from eqn. (10) and $\Delta\dot{x}$ from eqn. (3), we get

$$\Delta\dot{\phi} = \frac{hv}{R}\frac{\Delta p}{p_0}\left[\frac{1}{1-n} - \left(1 - \frac{v^2}{c^2}\right)\right] \qquad ...(11)$$

In a circular accelerator the equation of motion is

$$\frac{1}{r}\frac{d}{dt}(rp) = \dot{p} + \frac{\dot{r}\,p}{r} = qE\sin(\phi_0 + \Delta\phi) + qE' + qr\ B_z$$

As $q\dot{r}\ B_z = rp/r$, hence

$$\dot{p} = qE\ \sin\ (\phi_0 = \Delta\phi) + qE'. \qquad ...(12)$$

In this eqn. E' represents other accelerating or decelerating forces such as betatron acceleration term

$$E_b' = \frac{1}{2\pi r}\int\frac{\partial B}{\partial t}\,2\pi r\ dr \qquad ...(13)$$

and radiation deceleration term

$$E_r' = 9.60 \times 10^{-10}\ (1 - v^2/c^2)^{-2}\ r^{-2}\ \text{volt/metre}.$$

E in eqn. (12) is l/2πr times the voltage V, which is the sum of the peak voltages applied across all accelerating gaps. Thus eqn. (12) becomes

$$rp = \frac{qV}{2\pi}\sin\ (\phi_0 + \Delta\phi) + q\int\frac{\partial B}{\partial t}r\ dr \qquad ...(14)$$

In terms of momentum, radius and phase errors thus becomes

$$\dot{p}_0\,\Delta r + R\Delta\dot{p} = (qV/2\pi)\ \Delta\phi\ \cos\ \phi_0 + q(\partial B/\partial t)\ R\ \Delta r \qquad ...(15)$$

Since p_0 = BqR hence first and last terms of this eqn cancel each other, and we are left with

$$\Delta\dot{p} = (qV/2\pi R)\ \Delta\phi\ \cos\ \phi_0 \qquad ...(16)$$

Combining this eqn with eqn. (11), we get the phase equation for the *synchrotron*. Using $\gamma = \left[\frac{1-v^2}{c^2}\right]^{1-1/2}$ we have

$$\frac{d}{dt}\left[\frac{\gamma\Delta\dot{\phi}}{1/(1-n) - 1/\gamma^2}\right] = \frac{hqV}{2\pi}\frac{\cos\phi_0}{R^2 m_0}\,\delta\phi \qquad ...(17)$$

In the *synchrocyclotron* B is constant and radiation losses are negligible, hence E' in eqn (12) is zero. The harmonic number h is usually unity. n is very small can be neglected and R is not constant, hence eqn. (11) becomes

$$\Delta\dot{\phi} = \frac{v}{R}\frac{\Delta p}{p_0}\left(1 - \frac{1}{\gamma^2}\right) = \frac{qB}{m_0^2 c}\frac{\beta}{\gamma^2}\Delta p \qquad ...(18)$$

Equation (14) takes the form

$$r\dot{p} = p\dot{p}/Bq = (qV/\pi)\sin\phi \qquad ...(19)$$

$$\therefore \quad \frac{1}{m_0 c}\frac{d}{dt}(p\Delta p) = \frac{d}{dt}(\beta\gamma\,\Delta p) = \frac{q^2 BV}{\pi m_0 c}\Delta\phi\cos\phi_0 \qquad ...(20)$$

Combining eqns (18) and (20) we get the second order differential eqn in $\Delta\phi$ and is known as the *phase equation or the synchrocyclotron.*

$$\frac{d}{dt}(\gamma^2\,\Delta\phi) = \frac{(q/m_0)^3\,VB^2}{\pi c^2}\Delta\phi\cos\phi_0 \qquad ...(21)$$

This again describes a stable damped oscillation provided only that $\cos\phi_0$ is negative. Since $\sin\phi_0$ must be positive to permit acceleration, hence ϕ_0 must lie between $\pi/2$ and π.

HIGH ENERGY CIRCULAR ACCELERATORS

The acceleration of nuclear particles to relativistic energies was only made possible by the advent of the *principle of phase stability*. When applied in conjunction with a magnetic guiding field, this principle has resulted in accelerators whose cost per MeV is even lower than that of smaller machines. According to the way in which this phase stability is achieved, we shall have to study the *synchrocyclotron,* the electron *synchrotron* and the *proton synchrotron*. All these machines are also called *synchrotrons*.

1. The Synchrocyclotron (Frequency Modulated Cyclotron)

The relativistic limitation on energy for fixed frequency cyclotrons has restricted the useful size of the magnets. This limitation can be removed and the ions can be accelerated indefinitely if the applied frequency is varied to match exactly the ion revolution frequency. The instantaneous augular frequency of the phase oscillation is $(q/m_0)^3\ VB^2 \cos\phi_0/\pi c^2\gamma^3$.

If V = 10 kV and the particle energy is 300 MeV, the frequency of phase oscillation is about 1/1000 of the rotation frequency of the ions. The oscillation is damped as energy increases so that the acceleration can be carried on to whatever limit is set by the greatest radius at which magnetic field is usable.

This upper limit is set by the fact that the field falls of rapidly with radius towards the edge of the magnet pole. The limit is set in practice by the passage of n through a value of 0.2 at which the radial betatron oscillation frequency is exactly twice the vertical betatron oscillation frequency

$$\frac{f_r}{f_z} = \frac{(1-n)^{1/2}\ f_0}{n^{1/2}\ f_0} = \frac{(1-0.2)^{1/2}}{0.2^{1/2}} = \frac{2}{1} \qquad ...(1)$$

At this point a resonance takes place between two oscillations and the vertical oscillations amplitude increases until the beam is lost. It is found generally about one magnet gap width in from the edge of the magnet pole. In synchrocyclotron light +ve ions (p, d, α) are accelerated to energies significant relative to the rest energy of the particle. The ions traverse circular orbits with increasing radii as energy increases. The ions pass many times (2 times in an one revolution) through the rf electric field of a large D-shaped hollow electrode and experience an acceleration on each traversal of the accelerating gap. For each value of panicle energy in the magnetic field there is a particular orbit radius and a specific frequency of revolution.

$$r = \frac{mv}{qB} = \frac{[T(T + 2\ m_0c^2]^{1/2}}{cqB} \qquad ...(2)$$

At every instant the synchronous ion and the oscillator have the same frequency, given by

$$f = \frac{v}{2\pi r} = \frac{qB}{2\pi m} = \frac{qB}{2\pi m_0}\ \frac{m_0c^2}{(m_0n_0 + T)} = \frac{f_0}{1 + T/m_0c^2}$$

where $f_0 = qB/2\pi m_0$ is the non-relativistic cyclotron frequency.

The decrease in ion revolution frequency is caused primarily by the increasing T but is also affected by the slight decrease in B with increasing radius required for orbit stability. Above relation also shows that the fractional change in oscillator frequency is proportional to $\frac{T}{M_0c^2}$ electrons, because of the small rest mass, cannot be readily accelerated in the synchrocyclotron. The change in frequency is ordinarily made with a rotating multibladed capacitor.

The *principle of phase stable orbits* can be explained by reference of Fig. 10, which represents the variation of the radio-frequency oscillating potential with time. Particles which arrive at each gap at the instant corresponding to the point indicated by P in the figure will always receive the appropriate impulse. A particle which has more energy than that culled for at a particular gap will necessarily traverse an orbit of larger radius which requires a time longer than the r-f-period and will arrive later as indicated by the point L.

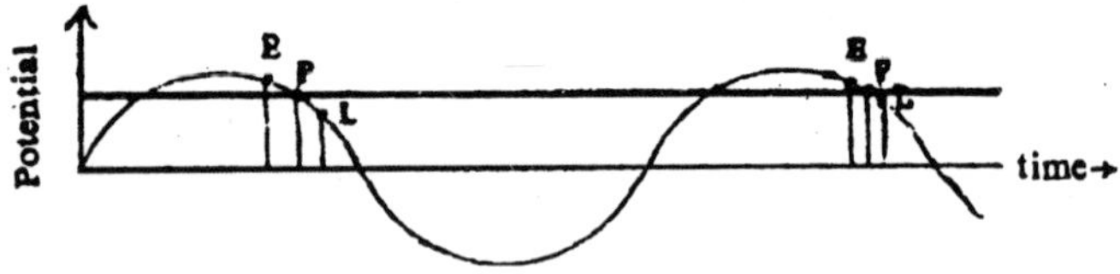

Fig. 10 : Principle of phase stability.

The potential will then have decreased and this particle will receive a smaller energy impulse. The increase in mass is hence small. It is clear that the time T required to traverse the path will be less and its next arrival will be closer to the point indicated by P (prompt). After traversing several times, the particle will be brought to the stable phase and the resonant energy. On the other hand, if particle has smaller energy, it will arrive earlier as indicated by the point E. This particle will receive greater energy and mass than normal, since the potential is now higher than at P- It will thus take a slightly longer time to return to the point where it again receives impulse.

This point will be nearer to the point P. In this way after several revolutions the particle will be brought to the stable phase. *The revolution of the charged particles is thus automatically synchronized with the changing frequency of the accelerating potential.* It should be noted that phase oscillations occur if phase lies between $\pi/2$ and π.

Dee Voltage

If the final energy is 500 MeV, obtained in time 0.01 sec by a source of frequency 20 Me/sec. The average energy gained per revolution = $(500 \times 10^6/0.01)/20 \times 10^6$ = 2500eV/rev. If the synchronous phase angle is ϕ_0 = 150°, so that sin ϕ_0 = 0.5.

As there are two accelarations per turn, a single dee with a peak potential of 2500volts is adequate, which is very small in comparison of 100 kilovolts required for cyclotrons. The *D-shaped electrode* is similar to the electrodes used in the standard cyclotron. The *ion source* is quite -similar to that used for the standard cyclotron.

The *magnetic field* is produced by a solid core magnet or large pole-face area. It must be steady and approximately uniform over the pole face, decreasing slightly with increasing, radius to provide focusing. The weight of the magnet varies as T^3.

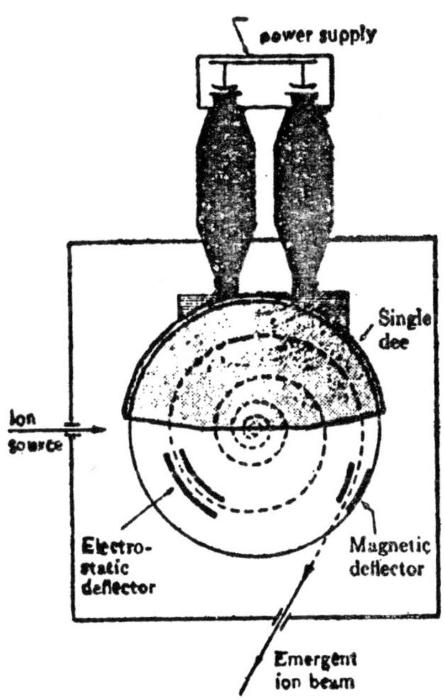

Fig. 11 : Schematic diagram of the 184 inch synchrocyclotron.

Exciting coils for the magnet have also been developed. Some designers use oil-bath cooling and others prefer dry-wound coils. The *radio frequency oscillator* for giving power to the D depends for its frequency variation on the mechanical modulation of the resonant D circuit by a variable capacitor.

In half wave and three quarter wave circuits, the capacitor is located at the outer end of the D stem and can have its own separate vacuum system. For maximum capacitance, it must have close spacing of the blades, smooth rounded corners, highly polished surfaces and fast pumps for a good vacuum. In quarter wave circuits, the capacitor operates in the fringing magnetic field at the back of the D and is of multiple parallel plate design. *The deflection* of resonant ions to produce an emergent beam is complicated as the change in orbit radius per turn is very small. The deflector must pull the beam through a large radial distance. The magnetic deflector known as regenerator is used as it acts on all particles of the beam. The emergent beam is focused to small diameter and passed through channels in a shield.

As soon as the phase stability was recognized, frequency modulation was applied to cyclotrons, first in Berkeley, then in many other laboratories. The first fre-quency modulated cyclotron was the 184 inch machine at the university of California at Berkeley which produced 190MeV deuterons and 380 M eV α-particles. It consists of only a single dee placed inside a vacuum chamber between the poles of a very large electromagnet. The pole faces are specially shaped to provide a field which de-creases almost linearly from the centre (15 kilo gauss) out to the position of maximum orbital radius.

The field thus produced will focus the particles in the median plane. The oscillating potential is applied between the dee and the ground plane. The radio-frequency (peak voltage 15 kV) is supplied by a tube oscillator circuit modulated by being coupled to a variable capacitor. By means of this capacitor the frequency of the oscillating potential is decreased to compensate for the gain in the effective mass of the particle as its speed increases. For deuterons and α-particles the frequency was modulated from 11.5 mega cycles/sec at the instant of injection to 9.8 mega-cycles/sec when the particles reached the periphery of the dee Protons with energies of about 350 MeV were obtained with an oscillator frequency modulated from 23 to 15.6 mega cycles/sec. In 1957, protons of 720 MeV energy were produced by increasing the strength of the magnetic field from 15-28kilo gauss. Another large synchrocyclotron with a magnet 236 inches in diameter weighing 7200 tons was constructed at Dubna in U.S.S.R. in 1954. This produces protons of energy 680 MeV. A 600 MeV synchrocyclotron with magnet 196 inches in diameter weighing 2500 tons was completed in 1958 at the CERN laboratory in Geneva, Switzerland. Designs for even larger machines seen practical. But the excessive cost of solid core magnets for higher energies (weight proportional to T^3) has transferred interest to the proton synchrotron.

The main differences between the operation of the ordinary cyclotron and synchrocyclotron may be summarized as: In the latter case;

1. The magnetic field decreases towards the edge of the magnet pole.
2. The frequency is modulated.
3. The particles may turn around 10^5 times before reaching the maximum energy, instead of 10^2 times in the former case.
4. The ions from slugs that circulate in the machine and come out in spurts lasting about 50msec and repeating roughly 100 times per sec. The synchrocyclotron has been most valuable in providing information about the production and properties of mesons.

2. The Electron Synchrotron

When studying the betatron we found that maximum energy attainable was limited mainly because of the very low energy gain per turn. The gain was compensated by the losses due to electromagnetic radiation at high energies. We have seen in the last sub-article that electrons cannot be accelerated by synchrocyclotron because of their small mass. By adapting the betatron to use the synchrotron principle, Goward and Barnes, in 1946, in England, gave a machine which is known as electron synchrotron. In the synchrocyclotron the nature of particles was such that their acceleration entailed a decrease of their frequency of revolution in the presence of a particularly constant magnetic field. Conversely, since relativistic electrons (energy $\geq$ 2 MeV) revolve on a fixed orbit with constant frequency (as they have reached the velocity of light v = 0.98c at 2 MeV), the synchronous energy can only increase with time in the electron synchrotron if the magnetic field is also made to increase so as to satisfy equation.

$$\omega_s = eBc^2/W_s, \qquad ...(1)$$

where the subscript A' denotes synchronous particles.

The electron synchrotron has, like the betatron, the doughnut in an ac-magnetic field. The magnet can be a ring of C-shaped units, not filling the hole in the doughnut as the case with the betatron. The weight of the magnet is decreased in this way. In some synchrotrons the magnetic field is produced by coils only, no iron being present. In the central gap some flax bars serve as the central core of the magnet to start up the machine us a betatron.

These bars are made of high permeability metal and do not have to be large; they short the magnetic field at low inductions but become saturated at high inductions and the transition from betatron action to synchrotron action can be made smoothly. Part of the interior of the doughnut is coated with copper or silver to give a resonance cavity. A small break in the coating separates it into two parts. A high frequency electric field from a radio-frequency oscillator is applied across this gap at the proper time in the magnetic cycle. When the accelerator is on, the electron is accelerated each time it crosses through the resonator.

The electron synchrotron accelerates electron in an orbit of constant radius by means of a radiofrequency electric field applied across a gap. A ring shaped magnet provides the magnetic field over the doughnut shaped vacuum chamber. The pole faces are accurately shaped to provide a field which decreases with increasing radius, with n~0.6, to supply focusing forces for the electrons, similar to the betron cash. The maximum energy of the electrons depends on orbit radius and on to maximum magnetic field.

$$T = ceBR = 300 \text{ BR MeV}. \qquad ...(2)$$

Synchronous acceleration starts normally when the electrons reach the velocity of light. At this constant velocity frequency

$$f = c/2\pi R = 47.8 \times 10^6 \text{ cps} \qquad ...(3)$$

The rf-accelerating field must have a frequency equal to this electron frequency or to some harmonic of it. The volts per turn requirement comes from the rate of magnetic field and other constants of motion. We can obtain it by the relation

$$V_e = 2\pi R^2 \, dB/dt. \qquad ...(4)$$

During acceleration the electrons are kept in step with the accelerating field by the phase stability. The equilibrium electrons continue to return to accelerating point at the correct phase; other electrons in the phase stable range oscillate in phase around this equilibrium phase. The phase oscillations are accompanied by oscillations in energy and so in orbit radius. Hence, the electrons are accelerated in a bunch as are the protons in synchrocyclotron.

By a little manipulation it can be shown that for electron energies higher than 2 or 3 Mev, we have

$$\frac{\text{Frequency of phase oscillation}}{\text{Frequency of revolution}} = \left[\frac{\text{he V}\cos\phi_0}{2\pi m_0 c^2(1-n)}\right] \quad ...(5)$$

Thus, for V = 1500 volts, h = 1, n = 0.6 and energy 3MeV, the ratio is about 0.014.

Electrons are injected into the doughnut after a preliminary acceleration in an electrostatic field to 50-100 keV. When these are injected at the beginning of the magnetic cycle, the device operates as a betatron.

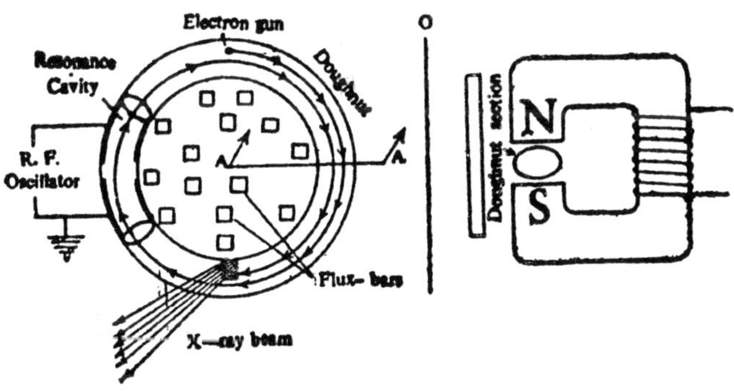

Fig 12 : Electron Synchrotron, (left) Plan with magnet removed (right) vertical section showing annular magnet.

The electrons travel in a circular paths and increase their energy as the field increases, until the electrons reach some 2 MeV at which point the bars are magnetically saturated and are no longer able to induce an e.m.f. This saturation causes the betatron mechanism to stop. At this point the synchrotron mechanism begins to operate. If the potential applied to the resonator operates at the proper frequency, the electrons are all kept in phase and receive increments of energy at each revolution, as they pass through the cavity. The high frequency field remains on while the magnetic field is increasing and is automatically cut off when the electrons have acquired their maximum energy or if desired at an earlier time in the cycle. As the magnetic field still has a little increase to go before reaching its maximum, the radius of the beam of constant velocity (~c) diminishes a little so that the beam spirals in slightly to strike a target projecting from the inner edge of the doughnut which gives off short wavelength X-rays. The rays emerge in pulses as in the betatron.

Several methods of pre-acceleration have been used in the electron synchrotron. Van de Graaff generator or a linear accelerator used as an injector in many of the recently constructed electron synchrotrons. This type of injector has the advantage of providing a well collimated highly mono-energetic beam of electrons. Electrons are injected at 80 keV, accelerated to 7 MeV by betatron action and finally reach energies of 330 MeV in the machine which is at the Massachusetts Institute of Technology. This machine has a 50 ton magnet and radio frequency of 46.5 mega cycles. The orbit radius is 40 inches.

Higher-energy (1500 MeV) electron synchrotron consists principally of four quadrants in each of which a magnetic field is created by electromagnets with C-shaped cross-sections. The quadrants are separated by short straight sections where there is no magnetic field. All these sections are exploited to connect pumps and instruments to the accelerator tube and two of them are used for injection and acceleration of the electron beam.

3. The Proton Synchrotron

We know that 1500 MeV is an upper limit for betatrons and electron synchrotrons because of the energy loss by radiation of the high energy electrons revolving in an orbit. With protons, however, the situation is different as the rate of loss of energy by a charged particle revolving in an orbit is not only proportional to E^4/R but also to $1/M_0^4$. Since the rest mass of the proton is nearly two thousand times that of an electron, the proton energy can hence be raised to extremely high values.

The principles on which the proton synchrotron is designed and operated are basically the same as those of the electron synchrotron. Its name does not imply that its principles of operation apply exclusively to protons (they are equally valid for deuterons, α-particles or other ions), but results from the fact that protons, being elementary particles, are preferred for investigations in elementary particles physics. Protons revolve in an orbit of constant radius in a doughnut shaped vacuum chamber. A ring shaped magnet surrounding the doughnut is required. In the synchrocyclotron, however, the protons follow a spiral path and the magnetic field must cover the whole area in which the particles spiral around and solid core magnet is required. Because of the large reduction in the weight of the magnet a proton synchrotron is much cheaper to construct than a synchrocyclotron.

Protons are accelerated to a few MeV and then injected into the doughnut, tube, when the magnetic field is small. A magnetic field which increases with time as the proton gains energy is applied normal to the chamber, as in the betatron. Unlike the electrons, which approach the velocity of light at relatively low energies ($v = 0.98$ c at 2 MeV) and so have an essentially constant frequency of rotation during acceleration, protons do not reach a speed of 0.98c until their energy is 4 BeV. A velocity and frequency of rotation of protons in an orbit of constant radius increase during the acceleration. The frequency increase is rapid at first but slows down as the proton velocity approaches c.

The frequency of the oscillating potential must be increased, as in the synchrocyclotron to synchronize with the rotation frequency of protons and to keep the protons in a constant average orbit radius. This is the main difference between electron synchrotron and proton synchrotron. In most of the present machines the frequency modulation is accomplished electronically rather than by rotating condensers, as in the frequency modulated cyclotrons. The same type of phase focusing exists to bunch the particles about an equilibrium phase of the accelerating field as for the electron synchrotron.

The largest and most costly component of a proton synchrotron is its magnet. The first step in design is a decision on magnet dimensions and properties. It determines characteristics of the other components. The choice of the magnet structure and maximum field determines the orbit radius for a chosen particle energy. The K.E. of the particle T is given by

$$T(T + 2M_0c^2) = q^2c^2B^2R^2 \qquad ...(6)$$

When T and M_0c^2 are expressed in BeV units, B in weber/m^2 and $q = 2e$. Then

$$R = 3.33[T(T + 2M_0c^2)]^{1/2}/Bz \text{ metres.} \qquad ...(7)$$

The design involved the use of parallel bundles of iron laminations arranged in a radial array around a circle, leaving wedge shaped spaces between the bundles at the orbit. To focus particles in orbit, the magnetic field must decrease with increasing radius across the pole face. In both the *cosmotron* and *bevatron* the n value was chosen as 0.6, and was obtained by doping the pole faces so the gap on the outside of the orbit is longer than on the inside.

Magnetic field increases with time during the pulse, the rate of increase is determined by the characteristics of the power supply and the magnet windings. The average value of dB/dt is given by the maximum field and the rise time. The rate of increase of particle energy with lime is directly related to the rate of magnetic field increase. By differentiating eqn. (6), we get

$$\frac{dT}{dt} = \frac{q^2c^2R^2B}{T+M_0c^2}\frac{dB}{dt} \quad ...(8)$$

The ion revolution frequencies at injection and at high energy determine the range of frequencies required in the radiofrequency system. The relativistic relation for the frequency of revolution in circular orbit is

$$f_c = \frac{c^2qB}{2\pi(T+M_0c^2)} \quad ...(9)$$

In the orbit with for quadrants of radius R and straight sections of length L the orbital frequency is decreased by the path length ratio.

$$\frac{f_0}{f_c} = \frac{2\pi R}{[2\pi R+4L]} \quad ...(10)$$

The energy increase per turn can be obtained by dividing relation (8) by f_0 given in eqns (9) and (10).

$$\Delta T/\text{turn} = qR(2\pi R + 4L)dB/dt. \quad ...(11)$$

The variation of frequency of revolution fo for cosmotron with the magnetic field shows a range of frequencies from 0.4 megacycle/sec. at 4 MeV injection (B = 3.11 gauss) to 4.2 mega-cycles/see. at the maximum energy of 3BeV. f_0f_c = 0.824 and for the average value of dB/dt = 14 kilo gauss/sec. ΔT/turn = 880eV/turn.

Injecting particles into a synchrotron is not very easy. A high intensity, pulsed beam of panicles is needed, with small energy spread and delivered in a well-collimated beam of small cross- section. It is pulsed at the start of each magnet cycle for a short time during which the particles come in stable orbits. A particle directed into the synchrotron field on a path 11 to its equilibrium orbit but r > R, will oscillate in the radial plane about the equilibrium orbit with a frequency

$$f_r = \left(1+\frac{L}{\pi R}\right)(1-n)^{1/2} f_0 \qquad ...(12)$$

Angular divergence also introduces vertical oscillations about the median plane with a frequency

$$f_z = \left(1+\frac{L}{\pi R}\right)(1-n)^{1/2} f_0 \qquad ...(13)$$

To obtain a fine *vacuum*, pump stations are distributed around the periphery. Cosmotron chamber requires 12 pumping stations, 3 in each quadrant. A target is permanently mounted just inside the inner wall of one of the straight section vacuum chambers.

The first proposal of a proton accelerator using a ring magnet, in which both magnetic field and frequency of the accelerating electric field were varied, was made in 1943 by Prof. M.L. Oliphant of Birmingham University. Design studies of proton synchrotrons in the U.S.A. were started in 1947 and the first machine was in operation in 1952 at the Brookhavan National Laboratory, This machine produced protons of energy 2.3 BeV, although weight of the magnet was 1650 tons. The energy was later increased to 3 BeV. The Brookhaven machine is called *cosmotron* because it makes possible the study of some of the nuclear reactions that occur with particles with energies comparable to those of the primary cosmic ray particles.

In 1954, 6.4 BeV protons were obtained in the machine at the Radiation Laboratory of the University of California. This machine is called *bevatron* (billion-electron-volt accelerator). A 10BeV machine was completed at Dubna near Moscow in 1957 and is known as *synchrophasotron*. Several other conventional proton synchrotrons nave been built in Saclay, France (Max. energy 2.5 BeV, magnet weight 1080 tons); Harwell, U. K. (Max. energy 7.0 BeV, magnet weight 7000 tons); Princeton, U.S.A. (Max. energy 3.0 BeV, magnet weight 350 tons) ; Argonne U. S. A. (Max. energy 12.5 BeV, magnet weight 4000 tons); Canberra, Australia (Max. energy 10.6 BeV, magnet weight 0 tons due to air core).

The Brookhaven cosmotron is not circular but in four quad- rants each of 30ft. radius and are joined by four straight sections each of 10ft. long. The magnetic field is applied over four quadrant and straight sections are kept free from magnetic field. The magnet, with a maximum diameter of 75 ft is a steel ring about 88 ft in cross-section and weighs about 2000 tons. Protons of about 4 MeV energy, derived from a 4 MeV Van de Graaff machine, an injected into one of the straight portions of the tube when the magnetic field has reached a value of about 300 gauss. The maximum magnetic field is 14000 gauss when the peak current through the magnet is 7000 amps. The electrostatic inflector plates, located in this straight section, deflect the protons entering at a 30° angle to the equilibrium orbit, by this angle.

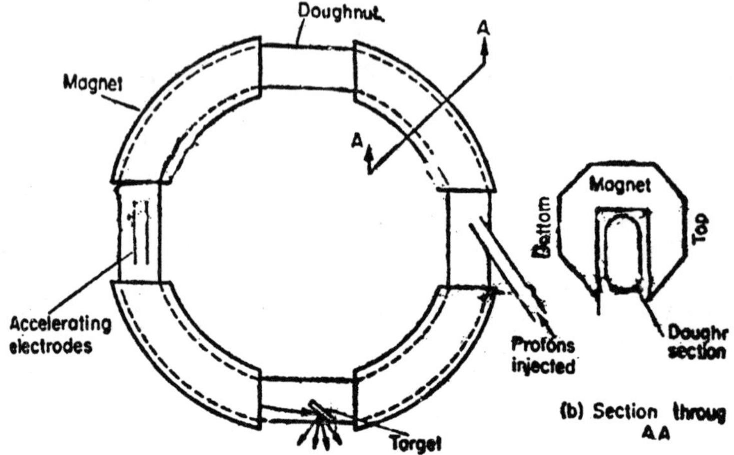

Fig. 13 : Schematic diagram of cosmotron.

The magnetic field is increased in proportion to the momentum to keep the protons in the synchrotron moving in a circular path of constant radius. As a result of the increasing speed of the particles as they gain energy in successive revolutions, the time required for the protons to make a turn of constant radius decreases, hence the frequency of revolutions increases correspondingly.

To maintain the condition of phase stability, the frequency of the radio frequency potential is increased in the same manner. It is applied to one of the straight sections.

This straight section is made of a non-metallic ferromagnetic material, called *ferrite*. This ferrite tube consists of one turn of wire supplied with high frequency current from an oscillator, the secondary is the proton beam itself and tube forms the core of a high frequency transformer. The proton receives an average energy of about 1000eV per revolution and thus makes about 2.3×10^6 revolutions per sec.

The frequency of the oscillations is increased fairly rapidly at first, as the protons gain speed and then slowly. The frequency becomes almost constant when the speed attained is within a few per cent of the velocity of light. During the acceleration period the radio frequency changes from about 0.37 mega-cycles/sec to about 4 megacycles/sec.

In the fourth straight section are placed the targets and arrangements for extracting the beams. The energy of the protons striking the target can be varied at will upto the maximum energy limit. The whole machine is surrounded with 8 ft. thick concrete shields. Studies of the nuclear reactions, possible with cosmotron, are essential to an understanding of nuclear structure. The Bevatron has been used to create proton-anti-proton pairs as well as neutron anti-neutron pairs.

MICROTRONS

The microtron, also called an *electron cyclotron*, is an electron accelerator, originally proposed by Veksler, since relativistic electrons cannot be accelerated by fixed frequency cyclotrons. In a microtron a small radio frequency accelerating cavity is located near the edge of a uniform magnetic field. The source of electrons is a hot filament in the entrance hole.

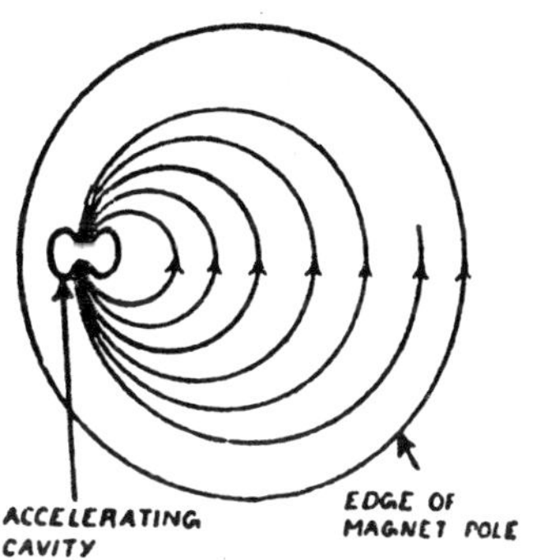

Fig. 14 : Orbits in a microtorn.

The electrons are drawn from it by the intense electric field. They then travel on a circular orbit in the magnetic field and return to the resonator in approximately an integral number of radio frequency cycles so that they are again accelerated. Thus, the electron-path is a sequence of circles of ever-increasing radius, all tangent at the resonator. The time period of electron of energy W is

$$\tau = \frac{2\pi}{\omega} = \frac{2\pi}{\frac{Be}{m}} = \frac{2\pi mc^2}{Bec^2} = \frac{2\pi W}{Bec^2} \qquad ...(1)$$

$\therefore$ Change in period $$\Delta\tau = \left(\frac{2\pi}{Bec^2}\right)\Delta W \qquad ...(2)$$

Thus, we see that the change in period depends on the change in energy, quite independent of total energy. The microtron operates on this principle.

Resonance Conditions

If the energy gained per turn is expressed as a fraction $\in$ of the rest mass energy W_0, *i.e.*, $\Delta W = \in W_0$. Electron started at rest gains $\in W_0$ in crossing the gap and hence has total energy $W_1 = W_0 = \in W_0$. Thus, eqn. (1) becomes

$$\tau = 2\pi W_0 (1 + \in) Bec^2$$

It must be equal to some integral multiple μ of the period τ_{rf} of rf oscillator, *i.e.*,

$$2\pi W_0 (1 + 2\in)/Bec^2 = \mu\tau_{rf} \qquad ...(3)$$

After a second acceleration electron acquires same energy increment $\in W_0$, hence its time period

$$\tau_2 = 2\pi W_0 (1 + 2\in)/Bec^2 \qquad ...(4)$$

$$\therefore \qquad \tau_2 - \tau_1 = (2\pi W_0/Bec^2)\in \qquad ...(5)$$

This difference in periods must be integral multiple of oscillator frequency τ, *i.e.*, $\tau_2 - \tau_1 = v\tau rf$. Hence

$$(2\pi W_0/Bec^2) \in = v\tau rf \qquad ...(6)$$

After the second rotation, the electron reaches the gap at the original phase and receives the same energy increment. Eqns. (3) and (6) give

$$\in = v/(\mu - v) \qquad ...(7)$$

$$\therefore \qquad \tau rf = \lambda/c = 2\pi W_0/Bec^2 /(\mu - v)$$

$$\text{or} \qquad B\lambda = 2\pi W_0/ec (\mu - v) \qquad ...(8)$$

Above relations show that m must exceed v. The difference in diameters between two consecutive orbits is

$$D_{n+1} - D_n = v\lambda/\pi \qquad ...(9)$$

The output kinetic energy in N turns $T = N\epsilon W_0$ ...(10)

If the electrons are injected into the resonator with an initial energy kW_0, above relations can be replaced as

$$\tau = 2\pi W_0 (1 + k + \varepsilon)/Bec^2 = \mu\rho rf$$

$$\epsilon = (1 + k)v/(\mu - v)$$

$$B\lambda = 2\pi W_0 (1 + k)/ec\,(\mu - v) \qquad ...(11)$$

$$T = kW_0 + NeW_0.$$

The electrons in the microtron experience a variety of phase stability as found in the synchrotron. The microtron has the advantage over the betatron that its accelerated beam is rather easily extracted for use outside the machine.

THE ALTERNATING GRADIENT (STRONG FOCUSING) ACCELERATOR

Although the energies which could be reached in the proton synchrotron are much greater, in theory, than several BeV, practical limits are set by the size of the machine and by its cost. By the use of the *strong focusing principle*, the straying of the accelerated particles from the ideal circular path can be markedly reduced in both radial and axial directions. The size of the doughnut can be decreased correspondingly and high particle energies can be attained without the need for large and heavy magnets. A suggested method for achieving strong focusing was published in 1952 by E. D. Courant, MS. Livingston and H.S. Snyder of Brookhaven National Laboratory, based on the use of magnetic fields with alternating gradients. In this system a synchrotron is modified such that the particles in going round their orbiting vacuum chamber pass alternately through strong focusing and defocusing magnetic lenses. The result of this is a very strong net focusing.

The principle of alternate-gradient focusing can best be understood in terms of a very simple optical analogue. If two optical lenses, one convex and one concave, both with the same numerical focal length, are placed as shown in Fig. 15. In any lens, whether optical or ion optical, the angle of deflection experienced by a given ray is proportional to the distance from the central ray.

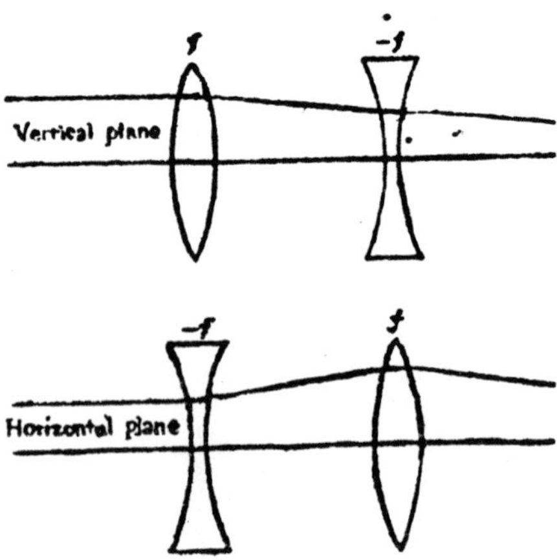

Fig. 15 : Optical analogue of both sections for each plane.

The angle of deflection is, therefore, larger in the defocusing lens than in the focusing lens, and the net focusing effect always results. It can also be proved by using lens formula. If the lenses are separated by a distance d, the overall focal length F is

$$\frac{1}{F} = \frac{1}{f_1} + \frac{1}{f_2} - \frac{d}{f_1 f_2} \qquad ...(1)$$

If $f_1 = f$, $f_2 = -f$, then $F = f^2/d$. which is a +ve quantity. Thus, we see that *the combination of two lenses one covergent and other divergent is always he a convergent lens and is independent of the order of the lenses provided separation remains same.* The beam of particles is similar to a pencil of light, for which the sector with field index $n\left(= \frac{-dB}{dr}\frac{r}{B}\right) > 0$ behaves as a convergent lens in the vertical plane and as a divergent lens in the horizontal plane. The sector with field index $n < 0$ behaves as a convergent lens in the horizontal plane and divergent lens in the vertical plane. In this way there is a succession of convergent and divergent lenses in both planes thus, providing the vertical and radial focusing. The action of the alternating gradient magnets on the particles in the proton synchrotron is the same as the action on the light ray by a series of lenses which are alternatively focusing and defocusing.

If a beam of charged particles is moving along the z-axis across a region of transverse field E_x, given as $E_x = -G_x$, where G is a constant, the beam will be focused in the xz plane. To have div E = 0, E_y must have the form $E_y = Gy$ and the beam will be equally strongly defocused in the y-plane. If the beam now passes) through a second region of reversed gradients, the net effect will be focusing in both planes. In the case of magnetic fields G = particle velocity v × magnetic field gradient B'.

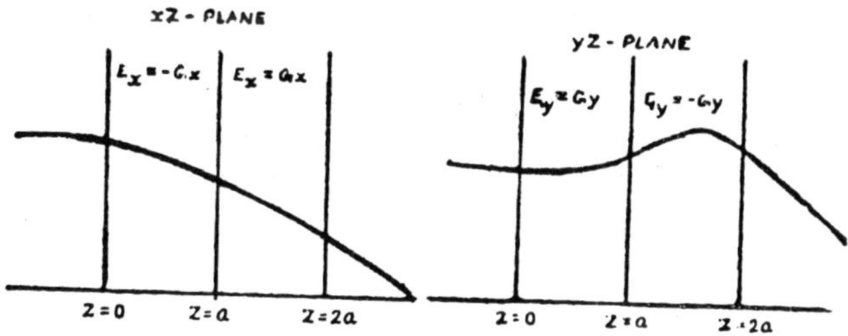

Fig. 16 : Focusing in a two-element AG lens.

In the xz-plane the eqn. of motion in the first region is

$$m\ddot{x} = -qGx \qquad ...(2)$$

Its solution is

$$X = X_0 \cos (qGt^2/m)^{1/2} + \dot{x}_0 (qG/m)^{-1/2} \sin (qGt^2/m)^{1/2}$$

$$= X_0 \cos (qGz^2/mv^2)^{1/2} + \dot{x}_0 (qG/m)^{-1/2} \sin (qGz^2/mv^2)^{1/2}$$

At z = a, we have

$$x = x_0 \cos ka + x_0 (1/kv) \sin k a$$

$$x/v = -kx_0 \sin ka + x_0 (1/v) \cos ka, \qquad ...(3)$$

where $k = (qG/mv^2)^{1/2}$

For the second region eqn. (3) becomes

$$x = x_0 \cosh ka + x_0 (1/kv) \sinh ka$$

$$x/v = k x_0 \sinh k a + \dot{x}_0 (1/v) \cosh ka \qquad ...(4)$$

Using matrix method, the displacement and velocity of the particle after passing through these two field regions can be written as

$$\begin{vmatrix} x \\ \dot{x}/v \end{vmatrix} = \begin{vmatrix} \beta_2 & \beta_1/k \\ k\beta_1 & \beta_2 \end{vmatrix} \begin{vmatrix} \alpha_2 & \alpha_1/k \\ -k\alpha_1 & \alpha_2 \end{vmatrix} \begin{vmatrix} x_0 \\ \dot{x}_0/v \end{vmatrix}$$

$$= \begin{vmatrix} (\beta_2\alpha_2 - \beta_1\alpha_1) & +(\beta_2\alpha_1 - \beta_1\alpha_2)/k \\ k(\beta_1\alpha_2) - \beta_2\alpha_1) + & (\beta_2\alpha_2 - \beta_1\alpha_1) \end{vmatrix} \begin{vmatrix} x_0 \\ \dot{x}_0/v \end{vmatrix} \qquad ...(5)$$

Similarly in the y-z plane or the defocusing plane

$$\begin{vmatrix} y \\ \dot{y}/v \end{vmatrix} = \begin{vmatrix} (\beta_2\alpha_2 + \beta_1\alpha_1) & +(\beta_2\alpha_1 + \beta_1\alpha_2)/k \\ k(\beta_1\alpha_2) - \beta_2\alpha_1) + & (\beta_2\alpha_2 - \beta_1\alpha_1) \end{vmatrix} \begin{vmatrix} y_0 \\ \dot{y}_0/v \end{vmatrix} \qquad ...(6)$$

where we have assumed α_1 = sin ka, α_2 = cos ka, β_1 = sinh ka and β_2 cosh ka. After simplifying above equations, assuming ka very small, we see that two velocities ($\dot{x}$ and y) are equal for the same initial displacements and both are directed toward the axis.

Above treatment is modified to the lens element of an AG focusing system, consisting of the half focusing, full defocusing and the half focusing magnets. The transformation matrix is thus given by

$$\begin{vmatrix} \cos\frac{ka}{2} & \frac{1}{k}\sin\frac{ka}{2} \\ -k\sin\frac{ka}{2} & \cos\frac{ka}{2} \end{vmatrix} \begin{vmatrix} \cosh ka & \frac{1}{k}\sinh ka \\ k\sinh ka & \cosh ka \end{vmatrix} \begin{vmatrix} \cos\frac{ka}{2} & \frac{1}{k}\sin\frac{ka}{2} \\ -k\sin\frac{ka}{2} & \cos\frac{ka}{2} \end{vmatrix}$$

$$\begin{vmatrix} \cosh ka \cos ka & (\sinh ka + \cosh ka \sin ka)/k \\ k(\sinh ka - \cosh ka \sin ka) & \cosh ka \cos ka \end{vmatrix} \qquad ...(7)$$

This matrix looks like the matrix, if we put

$$\cos\mu = \cosh ka \cos ka. \qquad ...(8)$$

For a circular synchrotron composed of 2N magnets, which have the same strength of field at the midpoints of their radial apertures, the radial gradients are alternate in direction and having field indices n and –n. The shift in phase μ_z, of the axial betatron oscillation and μ_x for radial motion are thus given as

$$\cos\mu_z = \cos\mu_z = \cos(\pi n^{1/2}/N)\cosh(\pi n^{1/2}/N) \qquad ...(9)$$

$$\left[\because ka = \left(\frac{qdB/dr}{mv}\right)^{1/2}\frac{2\pi r}{2N} = \left(\frac{qnB/r}{mv}\right)^{1/2}\frac{\pi r}{N} = \frac{\pi n^{1/2}}{N}\right]$$

cos μ_z and cos μ_x both lie in between +1 and–1. As cosh ($\pi n^{i/2}$/N) never vanishes, hence both the above quantities will became zero if

$$n/N^2 = 1/4 \qquad ...(10)$$

Large values of n mean powerful forces and reduced amplitudes of betatron oscillations, so that magnets with apertures only 1 and 2 inches on a side appeared possible eqn (9) can represent the stability diagram for the AG synchrotron.

Phase Stability

The charged particle can acquire the synchronous energy increment if passes the gap at the appropriate phase. The ions which do not reach the gap at exactly the correct moment will gain too much or too little energy and their momenta will differ from the ideal value. The momentum compactness α is defined as: consider two particles, one with momentum p and orbit length L and the other with momentum p + dp and orbit length L + dL. The momentum compactness

$$\alpha = \frac{\frac{dp}{p}}{\frac{dL}{L}} = \frac{\frac{dp}{p}}{\frac{dR}{R}} \quad (\text{As } L = 2\pi R) \qquad \text{...(11)}$$

It depends on the machine parameters. For an AG accelerator it can be given as

$$\alpha = \frac{n^{3/2}\pi}{4N}\left(1+\frac{L_s}{L_m}\right)\left(\coth\frac{n^{1/2}\pi}{2N} - \cot\frac{n^{1/2}\pi}{2N} + \frac{L_s n^{1/2}\pi}{L_m N}\right),$$

where L_s is the length of the straight sections and L_m that of magnets.

As a special case, consider that there are no straight sections ($L_s = 0$) and $\cos\mu_z = \cos\mu_x = 0$ for which $n/N^2 = 1/4$. Under these conditions

$$\alpha = \frac{n\pi}{8}\left[\coth\left(\frac{\pi}{4}\right) - \cot\left(\frac{\pi}{4}\right)\right] = \frac{n\pi}{8}[1.525 - 1] = \frac{n}{4.85} \qquad \text{...(12)}$$

The first two AG proton synchrotrons to go into operation (in 1961) were the 28 BeV machine at CERN (European Council for Nuclear Research, Geneva, Switzerland) and the 33 BeV machine at Brookhaven National Laboratory. The Brookhaven AGS. is essentially circular, with a diameter of 842.9' (about one half mile in circumference). The magnet, weighing about 4000 tons, consists of 240 sectors, 36" × 39" in cross-section, and has a maximum field strength of 13000 gauss. The magnet bends the protons into a path of just this diameter and applies strong focusing forces which always tend to bring the protons back toward their orbit within the vacuum chamber. The vacuum pipe is only 7" wide and $2\frac{3}{4}$" high and is maintained at a pressure of less than 10^{-5} mm of Hg.

The pole pieces of each successive pair of magnets are shaped so that the magnetic field alternatively increases and decreases in the radial direction. This alternating magnetic gradient causes the circulating proton beam alternately to focus and defocus in vertical and horizontal plane and yields a tightly focused beam after many traversals of the magnets.

The magnets are arranged in 12 super-periods or identical groups of 20 magnets each. Between super periods and in the middle of each superperiod is a 10' straight section, that can be used either for one of the 12 radio frequency accelerating stations, for injection, for targets or for other beam ejection means. The variable frequency increases from 1.4-4.5 mega-cycles per sec in each pulse. Protons of 50 MeV energy arc injected into one of the straight sections.

These protons are cbtained from a hydrogen gas by stripping the negatively charged electrons from the gas molecules, leaving as positively charged protons in the ion source, and are accelerated by Cockcroft-Walton generator to 750 keV and then directed into the linear accelerator, of a long cylindrical tank about 3' in diameter and 110' long containing 124 drift tubes along its axis. These protons, circulating around the ring, are accelerated by electric fields produced in acceleration stations. A high frequency voltage is impressed across two gaps at these stations. If the frequency applied is correct, the same proton will always be accelerated at each gap. The protons gain about 90 keV of energy per revolution. To reach about 30 BeV 325000 revolutions around the ring are required. The proton beam thus obtained can be directed at appropriate target substances and the resulting reactions can be studied by means of the emitted radiations.

The 70 BeV proton AGS designed at Serpukhov near Moscow, U.S.S.R. in 1967 has 120 magnet units separated by straight sections of three different lengths, namely, about 15, 8.5 and 4.25ft. The maximum field strength used is 12000 gauss. The doughnut chamber is 6.5" wide and 4.4" high and is of diameter 1500'. The 100 MeV energy protons, injected from a linear accelerator, are accelerated by means of 53 accelerating stations distributed around the chamber. The energy gain is 350 keV per revolution. The maximum beam intensity is of the order of 1012 protons per pulse and the repetition rate is 5-10 pulses per min.

One of the World's most powerful accelerators is the 200 BeV A.G. synchrotron at the Lawrence Radiation Laboratory. It was ready in 1974.

The machine is a system of four accelerators, feeding one another in series. The first, a Cockcroft-Walton generator injects 750 keV protons into a 200 MeV linear accelerator of length 160 metres and thus serves as a pre-injector into an 8 BeV injector alternating gradient synchrotron, which in turn delivers high energy protons into the main synchrotron. The latter is a strong focusing machine with 528 magnets in the ring about 1500 metres in diameter. The machine operates at a pulse rate of 30 per min producing 3×10^{13} protons per pulse at 200 BeV energy. A 300 BeV alternating gradient proton synchrotron is now available at CERN. It would use the same method of injection but the main accelerator would be larger. It is improbable that construction of the 600-1000 BeV accelerator started in 1972 in the United States will be completed upto the end of 1982.

AVF CYCLOTRONS

In the synchrocyclotron the frequency is varied with the radius in the same way as the mass increases with the radius. Another way is to increase the magnetic field with the radius as particle mass increases, keeping the frequency of the accelerating voltage as fixed. This requirement conflicts with the condition of axial stability of the particle orbit. For this Thomson proposed the use of alternately high and low regions of magnetic field around the orbit. Machines using such magnets arc, therefore, called AVF (*Azimuthally Varying Field*). As azimuthally varying fields are produced by *sector or spiral magnets*. On account of focusing process these cyclotrons are named as *Sector Focused (SF)* or *Spiral Focused Cyclotrons*.

Principle

The magnetic field varies with azimuth, regions where it is high are called *hills* (H) and where it is low as *valleys* (V). The radial velocity Vr interacts with the field Be to produce an axial force known as Thomas force. In the spiral shaped magnets, the convex side works as a focusing lens and concave side as defocusing lens. The net effect is focusing and the force as Kerst force. Laslett suggested that the particle entering the focusing surface of the hill would reach the defocusing surface normally or very close to the normal, hence the net effect would be focusing. When the focusing always takes place at each hill, the focusing is known Kerst and Lasslet focusing.

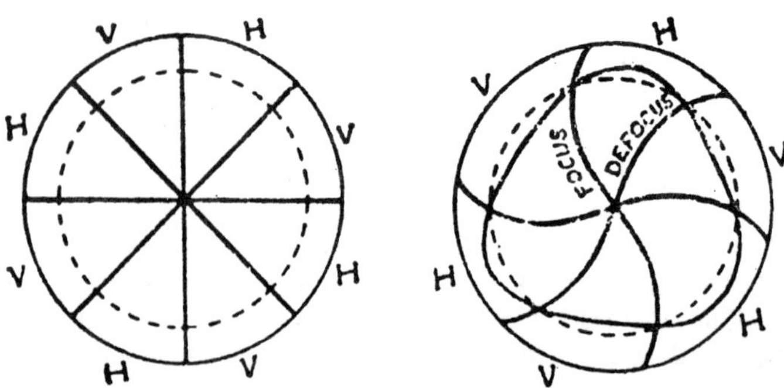

Fig. 17 : Thomas three sector field (left), Kerst and Laslett spiral shaped field (right).

Flutter

The change in field strength between the hills and valleys is defined in terms of a quantity known as flutter amplitude

$$B = B_{av} (1 - f \sin N\theta) \quad ...(1)$$

$$\therefore \quad B_H = B_{av} (1 + f), N\theta = 3\pi/2 \quad \text{or} \quad \theta = 3\pi/2N$$

$$B_V = B_{av} (1 - f), N\theta = \pi/2 \quad \text{or} \quad \theta = \pi/2N.$$

Thus for first harmonic (N = 1). B_H corresponds to $3\pi/2$ and B_V to $\pi/2$. Another quantity known as *flutter function*

$$F = [(B^2)_{av} - (B_{av})^2]/(B_{av})^2 \quad ...(2)$$

Eqn (2) gives $B^2 = (B_{av})^2 (1 + f^2 \sin^2 N\theta - 2f \sin N\theta)$.

$$\therefore \quad (B^2)_{av} = (B_{av})^2 [1 + 1/2\ f^2],$$

$$\text{or} \quad F = 1/2\ f^2 \quad ...(3)$$

$$\text{Flutter amplitude } f = \frac{B_H - B_v}{2B_{av}}. \quad ...(4)$$

Thomas Focusing

For the arrangement of n hills or valleys, where the z-componetn of the magnetic field vary as

$$B_z = (B_z)_{av} (1 - f \sin B\theta) \quad ...(5)$$

For an assumed constant momentum of a particle of mass M moving with a velocity v, we get

$$Mv = qB_z,\ \rho = (B_z)_{av}\ r_0$$

$\therefore$ radius of curvature $\rho = r_0\ (B_z)_{av}/B_z$

or
$$\frac{1}{\rho} = \frac{1 - f\sin N\theta}{r_0} \quad ...(6)$$

For a particular curvature

$$\frac{1}{\rho} = \frac{r^2 + 2r'^2 - rr''}{(r^2 + r'^2)^{3/2}}$$

Since r changes very slowly with θ, hence $r' = dr/d\theta$ is small and r'^2 can be neglected.

$$\therefore \quad \frac{1}{\rho} \simeq \frac{r^2 - rr''}{r^3} \simeq \frac{r_0 - r''}{r_0^2} \quad ...(7)$$

Solving eqns. (6) and (7), we get

$$r'' = r_0\ f \sin N\theta,\ r' = (r_0\ f/N) \cos N\theta$$

Here constant of integration is zero as $r' = 0$ when $\theta = \pi/2$

$$\therefore \quad v_r = \frac{dr}{dt} = \frac{dr}{d\theta}.\frac{d\theta}{dt} = -\omega\frac{dr}{d\theta} = \frac{r_0}{N}\omega f \cos N\theta$$

Let us find B_θ, which can be obtained by the first term of Taylor's Expansion.

$$B_\theta = \left[B_\theta\right]_{z=0} + z\left[\frac{\partial B_\theta}{\partial z}\right]_{z=0} + ... = z\left[\frac{\partial B_\theta}{\partial z}\right]_{z=0}$$

$$= z\left[\frac{\partial B_z}{r_0 \partial\theta}\right] \quad \text{(As curl B = 0)}$$

$$= \frac{z}{r_0} fN\ .\ (B_z)_{av} \cos N\theta$$

$\therefore$ Thomas Force $F_z = qv_rB_\theta = -zN\omega^2f^2 \cos^2 N\theta$

Average vertical force $(F_z)_{av} = -1/2\ M\omega^2f^2z$

$$\therefore \quad M\frac{d^2z}{dt^2} + \frac{M\omega^2f^2}{2} z = 0 \quad ...(8)$$

and Thomas Frequency $\omega_{ih} = \omega f/\sqrt{2}$

The ratio $v = \frac{\omega_{ih}}{\omega} = \frac{f}{\sqrt{2}}$ or $v^2 = \frac{f^2}{2} = F$ ...(9)

As in axial field $v^2 = n$, hence flatter F plays the same role in azimuthally varying fields as n does in axial fields. By using the flatter we could adjust cu in resonance due to the variation of B.

Spiral Focusing

In the region where the field lines are bowed there is a horizontal component B_h at points off the mid plane ⊥ to the spiral locus of constant phase of field. If ϕ is the spiral angle, $B_r = -B_\theta \tan\phi$ and the axial force $F_z = -qv_\theta Br = -\omega^2 zf\, NM \cos N\theta \tan\phi$. This is focusing and defocusing depending on the field sign of $\tan\phi$ and changing sign of $\cos N\theta$. Let us take account of the different path lengths in the focusing and defocusing fields. Now F_z becomes

$$F_z = \omega^2 zfNM \cos N\theta \tan\phi[1 - (f/N)\cos N\theta \tan\phi]$$

$$d^2z/d\theta^2 = zfN \cos N\theta \tan\phi - zf^2 \cos^2 N\theta \tan^2\phi \qquad ...(10)$$

or

$$< d^2z/d\theta^2 > = 1/2\ zf^2 \tan^2\phi$$

Its solution is $\bar{z} = \bar{z}_m \sin(\omega_{a\upsilon}\theta)$ ...(11)

where radial frequency $\omega_{av} = (f/\sqrt{2}) \tan\phi$

The larger the value of $\bar{z}$, the stronger are the in and out radial components of field, so the ion is forced into rapid axial oscillations about the path $\bar{z}$. Using $z = \bar{z} + \Delta x$, and solving, we get

$$\Delta z = -\bar{z}\ (f/N) \cos N\theta \tan\phi \qquad ...(12)$$

and $z = \bar{z} - \bar{z}\ (f/N)\cos N\theta \tan\phi$

Using this value of z in eqn. (10), and neglecting term in f^3 as f is very small (~ 1/20), we get

$$d^2\bar{z}/dt^2 = \omega^2 d^2\bar{z}/d\theta^2 = -\bar{z}\ 2\omega^2 f^2 \cos^2 N\theta \tan^2\phi$$

Using its average value as $\bar{z}\omega 2\ f^2 \tan^2\phi$, hence

$$d^2\bar{z}/dt^2 + \bar{z}\omega^2 f^2 \tan^2\phi = 0 \qquad ...(13)$$

But the Thomas force is also present, hence the proper eqn of motion in any azimuthally variation of field is

$$\frac{d^2z}{dt^2} + \omega^2 \left(\frac{f^2}{2} + f^2 \tan^2 \phi \right) z = 0 \qquad ...(14)$$

Hence the axial betatron frequency

$$\omega_z^2 = \frac{1}{2} \omega^2 f^2 (1 + 2 \tan^2 \phi)$$

and $v_z^2 = \omega_z^2/\omega^2 = \frac{1}{2} f^2 (1 + 2 \tan^2 \phi) = F(1 + 2 \tan^2 \phi)$

The *average field index k* is defined as

$$k = \frac{d < B >/< B >}{dR/R} = \frac{dp/p}{dR/R} - 1 = \alpha - 1$$

Above method of finding v_z is an approximate. The complicated mathematics gives

$$v_x^2 = 1 + k \text{ ....}$$

$$v_z^2 = -k + \frac{N^2}{N^2 - 1} F(1 + 2 \tan^2 \phi) \qquad ...(15)$$

The first SF cyclotron to accelerate protons was operated at Delft in 1958 and it gave 12 MeV protons. The first SF cyclotron in the relativistic energy region came into operation in 1961 at UCLA (California)—protons of 50 -75 MeV range were successful accelerated. The largest of 850 MeV with eight spiral sectors is in progress at Oak Ridge.

INDIAN ACCELERATION

The accelerators in operation in India are:

(a) Cock Croft-Walton Accelerators

(a) Physics University Andhra, University Waltair, maximum proton energy 400 keV, in operation since 1970.

(b) B.A.R.C. Bombay (400 keV. 1970) used for ion-implantation,

(c) Bose Institute of Science & Technology, Calcutta (250 keV, 1959),

(d) Saha Institute of Nuclear Physics, Calcutta (250 keV, 1956),

(e) B.A.R.C, Bombay (150-250 keV, 1956, (f) Physics Deptt., A.M.U. Aligarh (150 keV, 195K).

(b) Van de Graaff Accelerators

(a) Physics Deptt, B.H.U. Varansi (400 keV, 1976),

(b) Physics Deptt, Punjabi University, Patiala (400 keV, 1977),

(c) I.I.T. Kanpur (2 MeV, 1971),

(d) B.A.R.C. Bombay (5 MeV, 1961). Accelerators at Kanpur and Bombay are vertical type, the beam is bent to 90° by electro-magnetic focusing.

(c) Fixed Energy Cyclotron

It is in operation since 1965 at Saha Institute of Nuclear Physics, Calcutta. The maximum proton energy available is 3.7 MeV.

(d) Variable Energy Cyclotron

(Weak Focusing or Classical type). It is in operation since 1975 at Physics Deptt, Panjab University Chandigarh. The maximum possible energy is 8 MeV for protons. The deuterons, He^3 and alphas are accelerated and used in research. The machine has following characteristics:

(i) maximum magnetic field 14 kilo gauss,

(ii) weight of the magnet 20 tonnes,

(iii) size of pole pieces 26",

(iv) frequency of the oscillator 10-20 MHz,

(v) dee voltage 30-40 kilo volts,

(vi) ion source-hooded arc type,

(vii) vacuum 5×10^{-6} mm of Hg,

(viii) total power consumption 100 KVA.

(e) Variable Energy Cyclotron (AVP Type)

It is in operation since 1977 at Calcutta and is capable of delivering a beam of protons of energies between 6-60 MeV, deuterons between 12-65 MeV and alpha particles between 25-130 MeV. The machine has following characteristics:

(i) Electro-magnets of weight 262 tonnes, 610 cm length, 224 cm width and 290 cm height. 3 spiral shaped sector pole tips are placed 120°, apart. Valley coils are five sets of coils, each set having three pairs. The seventeen trim coils have 34 leads.

(ii) A single RCA 6949 is used as the main oscillator tube. The frequency range is 5.5 to 16.5 MHz.

(iii) Ion source is a hot cathode P.I.G type source.

HIGH ENERGY n-p AND p-p SCATTERING

High energy scattering experiment can be used to investigate details of the potential shape and to learn about the two nucleon forces in states other than the S-state. The development of the high energy accelerators has made it possible to extend the energy range of n-p and p-p scattering to energies at which the de Broglie wavelength of the relative motion is considerably smaller than the range of nuclear forces.

At high energy, the presence of higher angular momentum states results in a large number of parameters, making the data much more difficult to interpret. In the words of Blatt and Weisskopf, *the data available are strange and unexpected.* The analysis is still more complicated for neutrons since a mono-energetic neutron beam is not available. The cross-section, experimentally observed is an average over a rather large energy region and may consequently fail to show rapid energy dependence.

The high energy p-p scattering data are easier to interpret than the n-p scattering data, because the Pauli exclusion principle cuts the number of phase shifts in half and the experimental cross-section can be obtained with mono-energetic protons.

1. n-p Scattering Above 10 MeV

Incident neutrons upto 20MeV laboratory kinetic energy are scattered by protons isotropically. Deviations from spherically symmetric scattering are clearly measurable at energies of 27 MeV. At still higher energies the angular distribution of the scattered neutrons takes on a characteristic valley shape. The differential cross-section for neutrons of 40, 90 and 260 MeV lab kinetic energy, as a function of neutron scattering angle φ(C.M. system) is shown in Fig. 18.

The most striking fact about these distributions is the forward (ϕ = 0) and backward (ϕ = 180°) maxima. The peak at 0° can be understood in terms of absorption and diffraction of neutrons in the scattering medium. The backward peaking at 180° is the result of the exchange force which interchanges neutron and proton position co-ordinates, the Majorana.

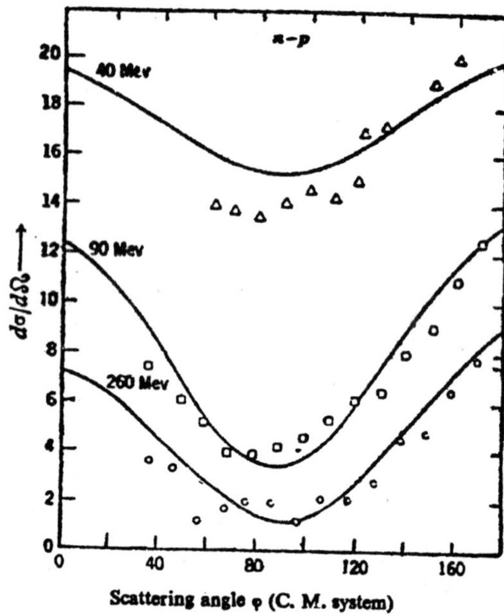

Fig. 18 : Angular distribution of neutrons scattered by protons.

Serber Force

The observed neutron angular distribution has been explained by assuming a half and half mixture of ordinary and exchange force. This mixture is known as a Serber force and is represented by the operator $1/2(1 + P^M)$. The Majorana exchange operator P^M (has a value +1 for even values of l and –1 for odd values of l. Thus, the Serber force is zero for odd states (P, F,...), and hence cannot account for the saturation. In contrast with n-p scattering below 10MeV, the effective range for high energy scattering is dependent on the assumed shape of the potential well. Christian and Hart have used several potential shapes and found that the distribution near 90° was flat with pure central force. The addition of some tensor force makes the theoretical curve less flat near 90°.

Born Approximation

The influence of exchange force and of ordinary forces on high energy scattering can be understood qualitatively by employing the Born approximation, inspite of the fact that the nucleon interactions are not weak. Born Approximation can be expected to give some indication of the true behaviour at least of the less strongly interacting partial waves of higher *l*. The scattering amplitude as defined is given in the Born approximation by

$$f(\theta) = -\frac{M}{4\pi\hbar^2}\int dr\ V(r)e^{i}q.r \qquad ...(1)$$

where V(r) is the interaction potential and q the momentum transferred in scattering, having a value $(k_{final} - k_{initial})$. For elastic collisions with $E_f = E_i$, $|q| = 2k \sin \theta/2$. Here θ is the angle of deflection in the C.M. system and $2\hbar^2 k^2/M = E_i$. Calculation of the high energy cross-section involves the shape of the potential, and no simple general treatment of the problem can be given except at such high energies that the Born approximation is applicable. The approximation is not qualitatively wrong for energies well above the depth of the nuclear well (say 30 MeV or so) and is not satisfactory for quantitative work at any energy.

If the potential V(r) vanishes outside the range R, the Born approximation predicts isotropic scattering, more or less independent of energy if $qR < 1$. The oscillations of exp. (iq.r.) cause the scattering to fall off rapidly if $qR > 1$. Near $q = 0$ (forward scattering) the cross section should stay rather large even as the energy increases. For $q = 2k$ (backward scattering) the cross-section will fall rapidly with increasing energy. The general behaviour of the above eqn. is shown in Fig. 19. Eqn. (1) indicates that neutrons scattered at energies in the 100 MeV range mostly go forward and recoil protons mostly have low energies in the neighbourhood of 10 MeV.

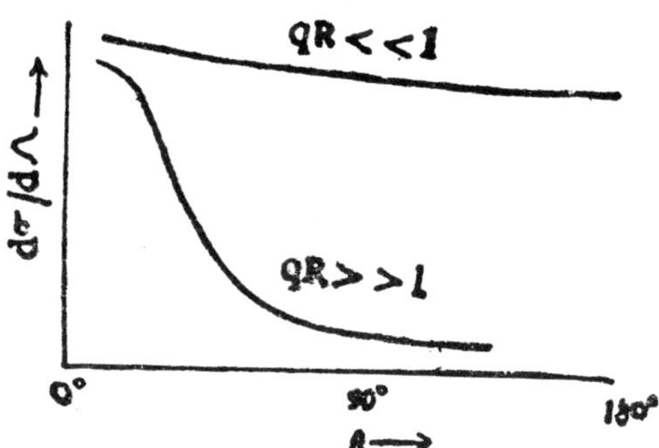

Fig. 19 : Born approximation result for n-p scattering.

But experimental results showed that the majority of the emergent particles were fast protons moving with nearly the energy and the direction of the incident neutrons. This can be explained if the neutron and proton change roles, or is exactly the consequence of Majorana force. As the S wave scattering amplitude is independent of the exchange character of the nuclear force, it can be shown that, for the same radial dependence of the force and to the extent that the Born approximation tor scattering in states of $l \neq 0$ is valied. A pure Majorana force given the same scattering in the direction θ as a Wigner force does in the direction π – θ.

In other words we can say that the cross-section shows a strong maximum in the forward direction for ordinary forces and in the backward direction for Majorana exchange force. Since the recoiling proton goes forward in the latter case, the exchange potential can be said to cause an exchange of the electric charge from the proton to the neutron which then continues on its way as proton.

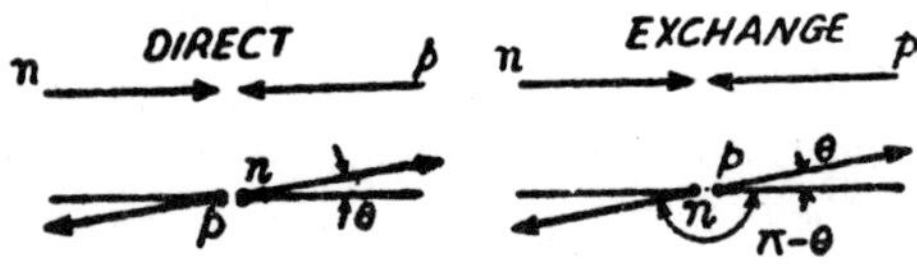

Fig. 20 : Direct and exchange n-p scattering.

2. p-p Scattering Above 10 MeV

Just as in the case of n-p scattering, it is expected that more information concerning the shape and exchange character of the potential can be obtained from high energy p-p scattering. The Coulomb potential plays a smaller and smaller part in the scattering as the energy is increased and a direct comparison of the n-p and p-p interactions should be possible at very high energies. In the p-p scattering the identity of the two protons reduces considerably the number of parameters necessary to describe the scattering. As we have already noted, two protons in a singlet state have even parity while those in triplet states have odd parity. If the Serber exchange, suggested by the high energy neutron-proton data, is assumed to act between protons, no scattering occurs in states of odd parity. This result indicates that there would be no triplet scattering of one proton by another. An estimate of high energy scattering can be again made by the Born approximation. If the Coulomb scattering is ignored the differential cross-section calculated in this way for any potential with *Serber Exchange* is just twice that for the n-p scattering calculated in the same approximation with the same potential. A central potential with the range for singlet and triplet states leads, in Born, approximation, to the same relative angular distribution in both states. Therefore, the p-p angular distribution calculated in this case is the same as in the n-p.

The experimental differential cross-sections for high energy p-p scattering are shown in Fig. 21. At low angles, cross-section is large because it is dominated by Coulomb scattering. For angles above this small angle region, the cross-section is nearly independent of angle (or *isotropic*). The accuracy is not high at all energies, but around 150 MeV the departure from isotropy is about 3%. An isotropic angular distribution generally implies pure S-wave scattering. This is highly unlikely at incident proton energies of 100 MeV or more, as the total cross-section is almost twice as large as the maximum possible for pure S-wave scattering. At one particular energy, it might happen that interference between P and D wave scattering results in a nearly isotropic scattering intensity, but this would not occur over a range of energies. Another puzzling result is the energy dependence of the total cross-section. The total cross-section is nearly independent of energy for energies from 150 to 400 MeV, and has a value 3.4 ± 0.4 millibarns/steradian. Above 400 MeV the observed cross-section begins to rise significantly as the creation of pions becomes important.

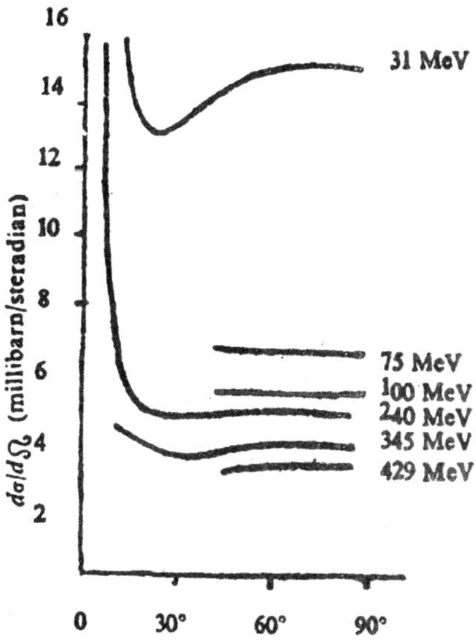

Fig. 21 : Experimental differential cross-section for high energy p-p scattering.

Nuclear Core

An interesting explanation of these strange results was suggested by Jastrow. For any conventional potential the S-phase decreases to zero with increasing energy and approximate isotropy is not possible to obtain when phases other then S-phases are involved. It is possible if the S-wave phase shift is assumed-negative which corresponds to a repulsive potential. Since at high energies the S-wave is strongly influenced by the behaviour of the potential at very short distances hence we require a potential that attractive at larger distances but has a strongly repulsive core.

The S-wave will have a phase shift which will depend on this central repulsive core and become repulsive over all the high enough energies. The higher l-values are kept away from the core by centrifugal forces. The core radius is of the order of 0.5 fermi. Another useful feature of the repulsive core hypothesis is that it helps to explain the saturation of nuclear forces.

At high energies in n-p and p-p interactions, tensor forces are important. In 1958. Marshak has introduced another term V(r) L.S into the two body potential, where L is the orbital and S the spin, angular momentum of the system. Recently a very promising technique involving the scattering of polarized protons has been developed for the more detailed informations about the nucleon-nucleon interaction.

POLARIZATION, HIGH-ENERGY NUCLEAR SCATTERING

The spin orientations of particles in a beam from an accelerator are normally randomly distributed. If some how the spins are oriented in some preferred direction, the beam is then said to be polarized in that direction. If the number of particles in the beam with spin component parallel to this preferred direction is N_+ and the number with anti-parallel spin component is N_-, then the polarization P, often called the polarizing power, is defined as

$$P = (N_+ - N_-)/(N_+ + N_-). \qquad ...(1)$$

Nucleons may be polarized by elastic scattering on a spinless centre provided that the interaction producing the scattering is spin dependent. We may scatter a beam of protons on a target and select the beam scattered under angle θ_1. By scattering this beam a second time on a target under angle θ_2, we find that the scattered intensity depends only on θ_2, but also on angle ϕ, between the two planes of scattered and can be written as

$$I = A(\theta_2) + B(\theta_2) \cos \phi \qquad ...(2)$$

Let n be the number of particles in the initial unpolarised beam. We can assume that in this beam there are 1/2n particles with spins of each sign (+ or –). After scattering by target T_1 to the left, the number of particles reaching the target T_2 with polarization P_1 and spin up is $1/2nf_1(1 + P_1)$ and with spin down is $1/2nf_1(1 - P_1)$, where f_1 is th fraction of all particles reaching T_2 Similarly those particles scattered to the right by T_1. $1/2nf_1(1 - P_1)$ will reach T_2 with spin up and $1/2nf_1(1 + P_1)$ will reach T_2 with spin down. Let P_2 and f_2 be respectively to polarization with spin up and the fraction of all particles scattered by the target T_2.

The number scattered twice to the left with spin up = $1/2nf_1f_2(1 + P_1)(1 + P_2)$ and with spin down = $1/2nf_1f_2(1 - P_1)(1 - P_2)$. Hence, the number of particles scattered twice to the left (LL) is $1/2nf_1f_2(2 + 2P_1P_2)$. The number of particles scattered twice, first to the left by T_1 and then to the right by T_2, with spin up is $1/2nf_1f_2(1 + P_1)(1 - P_2)$. Hence the number of particles scattered twice, first to the left and then to the right (LR) = $1/2nf_1f_2(2 - 2P_1P_2)$.

$$\therefore \text{ Left right symmetry } \epsilon = \frac{(LL) - (LR)}{(LL) + (LR)} = P_1P_2 \qquad ...(3)$$

If the second scattering is identical to the first, then $P_1 = P_2 = P$ and $\epsilon = P^2$. Hence the magnitude, not the sign, of P can be easily obtained from actual experiment. The sign may be determined by studying the interference of nuclear scattering with Coulomb scattering.

Low energy nucleon scattering cannot show any polarization, because only the S-wave contributes. In the energy range above 100 MeV, the scattering of protons and neutrons from heavier nuclei has been observed to polarize the nucleon beam. These scattering experiments provide valuable information concerning the spin dependence of nuclear forces. With a purely central interaction potential and an unpolarized target nucleus, the scattered particles would not show any left-right asymmetry, even with a polarized incident beam. This asymmetry can be explained, if the interaction has a tensor character with the spin-orbit coupling. The asymmetry is larger if orbital angular momentum *l* is large Thus, the polarization experiments become important at high energy, where high values of *l* contribute to the phase shift.

MESON THEORY OF NUCLEAR FORCES

Any attractive force between two particles is regarded as the exchange of an attractive property. The exchanged attractive property between two protons is an electron for an $(H_2)^+$ molecule and is the surrounding electric field for the Coulomb force between two charges. We may introduce similarly a new nuclear field surrounding each nucleon. The application of quantum mechanics to the electromagnetic field surrounding a charged particle leads to the conclusion that the electrical force is exerted by the transfer of a photon, which is referred to as *the field particle of the electromagnetic field,* from one charged body to another. Yukawa (1935) thought that t*he strong interaction between nucleons might be accounted for in a similar manner by postulating an appropriate field panicle of rest mass different from zero. This virtual particle has been given the name meson.* To show that the range of force is related to the mass of exchanged particle, assume that the π^o-meson is contained virtually in a proton. A temporary dissociation would be allowable if it does not take a time longer than Δt, given by uncertainty principle

$$\Delta t \approx \frac{\hbar}{\Delta E} \quad ...(4)$$

Here $\Delta E = (M_2 + m_\pi)\, c^2 - M_p c^2 = m_\pi c^2$

If this virtual particle travels with the velocity of light, as might be expected for a field particle, then the greatest distance the meson could travel in this time, also known as *range of the pion exchange force,*

$$R \simeq c\Delta t \approx \frac{\hbar}{m_\pi c} \quad ...(5)$$

Based on known nuclear dimensions, Yukawa assumed that the range of the nucleon force and of the field particle would be about 2×10^{-15} m and this led to a value of m_π roughly 200 times the mass of an electron. About two years after Yukawa's theory, particles of mass about 200 m, were discovered in cosmic radiation. These virtual particles are now known as mesons which were thought to be Yukawa field particles for more than ten years. Several different mesons with different charges and rest masses have been discovered. It was found that the Yukawa particle was not the μ-meson but its parent, the shorter lived μ-meson. The pion, which occurs in positive, negative and neutral forms has a mass about 270 times that of the electron and interacts very rapidly with matter. It has other properties spin and parity, which qualify it to be the field particle for nucleon forces.

The attraction between any two nucleons could arise from the transfer of a π^o-meson from one nucleon to the other. The force between a proton and a neutron could result from the transfer of a π^+-meson from the former to the latter or of a π^--meson in the opposite direction.

To interpret the nucleon-nucleon scattering data in terms of a potential function, let us compare the meson theory with the quantum theory of electromagnetic interactions. To obtain a pion wave-equation we express the total energy E of a pion in terms of the pion rest mass energy $m_\pi c_2$ and momentum p as

$$E^2 = c^2p^2 + m\pi^2c^2 \qquad ...(6)$$

The energy E and momentum component p are represented by the operators

$$E = \frac{i\hbar\partial}{\partial t},\ p_z = -\frac{i\hbar\partial}{\partial x},$$

$$p_v = -\frac{i\hbar\partial}{\partial y} \text{ and } p_z = -\frac{i\hbar\partial}{\partial z}$$

Introducing a pion wavefunction ϕ (a scalar), we obtain the Klein Gordon equation for a free particle of spin 0.

$$-\frac{\hbar^2\partial^2\phi}{\partial t^2} = -c^2\hbar^2\nabla^2\phi + m\pi^2c^4\phi$$

$$\nabla^2\phi - \frac{1}{c^2}\frac{\partial^2\phi}{\partial t^2} - \frac{m\pi^2c^2}{\pi\hbar^2}\phi = 0 \qquad ...(7)$$

From $m\pi = 0$, it reduces to the well known wave-equation

$$\left(\nabla^2\phi - \frac{1}{c^2}\frac{\partial^2}{\partial t^2}\right)\phi = 0 \qquad ...(8)$$

for the quanta of the electromagnetic field. The simplest type of electromagnetic field is the electrostatic field, where $\partial\phi/\partial t = 0$. The corresponding static pion field is the analogue of Laplace's equation in the absence of electric charges. The presence of charge requires the analogue of Poisson's equation $\nabla^2\partial\phi = ep/\epsilon_0$. where e is the magnitude of the electronic charge and p is the electron particle density.

In our present case we have nucleon charges interacting with the pion field. For a static pion potential due to a single point nucleon charge g at the origin, we have an equation

$$\nabla^2\phi - \left(\frac{m\pi^2 c^2}{\hbar^2}\right)\phi = 4\pi g\delta(r) \quad ...(9)$$

The solution of this eqn. which vanishes at infinity is:

$$\phi(r) = -\int -\frac{e^{-r}|r - r'|}{|r - r'|} g\delta(r')\delta\tau \quad ...(10)$$

Here $\mu \approx m\pi c/\hbar$ and $\delta\tau'$ is the volume element. If we have a point source of strength g at r_1, then we have,

$$\phi(r) = -g\frac{e^{-\mu}|r - r_1|}{|r - r_1|}$$

∴ Potcntial energy of a source g at is in this field is given by $V = g\phi(r_2)$. If the potential ϕ is due to another source g at r_1, then the interaction energy between two sources

$$V = -g^2\frac{e^{-\mu}|r_1 - r_2|}{|r_1 - r_2|} = -\frac{g^2 e^{-\mu r}}{r} \quad ...(11)$$

Comparing with the relation for Coulomb potential, we see that this nuclear potential decreases more rapidly with distance from the source than the Coulomb potential by an exponential factor. At separations small compared to μ^{-1}, this potential energy varies with r just as the Coulomb energy loss. Yukawa noted that this interaction energy may account for file fact that nuclear forces act only over a short range and that the range

$$R = \frac{1}{\mu} = \frac{\hbar}{m\pi c} \quad ...(12)$$

Substitution of the value of $m\pi$ as 270 m_e, gives R = 1.4 fermi.

Since in this theory the nuclear particle does not change its charge, we find that according to the theory n-n, n-p and p-p forces are equal. However, the theory does not explain the exchange nature of nuclear forces, which is established from high energy scattering experiments and is able to explain saturation of nuclear forces. The theory in its simple form cannot explain the spin dependence or the presence of non-central forces. The theory is modified by several theoretical physicists taking into account:

1. the tensoral character of the meson field wave functions,
2. the isobaric-spin character of the meson field wave functions,
3. the intrinsic nature of the source field coupling, and
4. the strength of the source-coupling constant.

The interaction potential between two nucleons so calculated gives approximately $g^2/\hbar c = 17$. It may be compared with the fine structure constant $e^2/4\pi\epsilon_0\hbar c = 1/137$. The large dimensionless coupling constant $g^2/\hbar c$ suggests that more than one meson is transferred simultaneously between two nucleons. For small distances, we have strong repulsive core arising from the δ function.

This core prevents the nucleons from coming close together. When two nucleons collide at very high energy (~BeV), the nucleons can penetrate or disturb each other's core. Thus we see that only the exchange of a relatively small number of mesons can have an appreciable effect on the nuclear force at low and medium energies. On the one hand, we see that the meson theory is qualitatively correct, but on the other hand, not a single quantity has been. calculated and measured which confirms quantitative correctness.

NUCLEAR FORCES

These attractive forces exceed the forces of repulsion. Even aside from the sign of the force, electrostatic forces are much too weak to account for the main effects, gravitational forces are weaker still and offer no assistance in the problem. It is therefore necessary to postulate an entirely new type of interaction, known as nuclear force. Let us collect following information on nuclear forces from study of complex nuclei.

Shape of the Potential

As the nuclear interactions do not extend to very large distances beyond the nuclear radius, and $1/r^n$ $(r > 1)$ character is not useful to solve the problem. To avoid this difficulty, the idea of cut-off in the potential is introduced. The parameters are taken to be the strength V_0 (the value of potential at the origin and the range α (the distance beyond which potential goes to zero rapidly). It is less than nuclear dimensions.

Saturation Property

We know that the nuclei have the same density $(R = R_0 A^{1/3})$ the binding energy per nucleon for nuclei with $A > 40$ is constant. These facts imply that the nuclear *force saturates*. Nucleons attract each other strongly only if they are in the same orbital state. According to the Pauli principle only two neutrons and two protons will be found in the same orbital state. Therefore, it is possible to find four nucleons strongly bound or α-particle structure, also confirmed by binding energy curve.

Charge Independence

We know that the atomic mass number A is approximately equal to twice the atomic number Z for the light and intermediate nuclei. It shows that light nuclei prefer to add nucleons in n-p pairs, *i.e.,* there is a strong interaction between neutrons and protons. The neutron excess in heavy nuclei confirms that n-n force is attractive, but it is sufficiently attractive to lead to a stable di-neutron. For a good approximation one can write

n-n $\simeq$ p-p $\simeq$ n-p or the forces *as charge independent.*

We use the principles of wave mechanics on nuclear problems and then by comparison with experimental data, find a consistent description of the nuclear forces acting between two nucleons (*two body problem*). There are two general methods of investigation, the study of n-p and p-p scattering events over a wide range of energy and the study of deuteron (only bound state of two nucleons).

DEUTERON

The deuteron does possess measurable properties which might serve as a guide in the search for the correct nuclear interaction. These properties are:

1. The parity of deuteron as measured, indirectly, by studies of nuclear disintegrations and reactions for which certain rules of parity changes exist, is even.
2. The angular momentum quantum number, often called the nuclear spin, of the ground state of the deuteron determined by a number of optical, radio frequency and micro-wave methods is one. It suggests that the spins are parallel (triplet state) and the orbital angular momentum of the deuteron about their common center of mass is zero. Thus, the ground state is ^{35} state.
3. The binding energy of the deuteron is very small. Its experimental value is 2.225 ± 0.002 MeV. Since the energy needed to pull a nucleon out of a medium mass nucleus is about 8 MeV, we must regard the deuteron as loosely bound.
4. The extraordinary stability of the alpha particle shows that the most stable nuclei are those in which number of neutrons and photons are equal. The deuteron consists of two particles of roughly equal masses M, so that the reduced mass of the system is 1/2 M.

5. A radio frequency molecular beam method has been employed to determine the quadrupole moment of the deuteron as $Q = 0.00282 \times 10^{-28}$ m^2. This shows the departure from spherical symmetry of a charge distribution. The +ve sign indicates that this distribution is prolate rather than oblate.

   The electric quadrupole moment and the magnetic moment discrepancy can be explained if the ground state is a mixture of the triplet states $^3S_1$ and $^3D_1$ having even parity. The percentage probability of finding the deuteron in D-state is $4 \pm 2\%$. As deuteron spends most of the time in the spherically symmetrical state (S-state), we will for the moment ignore the D-state contribution to the deuteron wave function.

6. The sum of the magnetic dipole moments of the proton (2.79275 μN) and neutron (–1.91315 μN), do not exactly equal to magnetic moment of the deuteron (0.85735 μN) measured by magnetic resonance absorption method.

7. Since the neutron has no charge, the force between the neutron and proton can not be electrical. This force can not be magnetic as magnetic moments are very small. It can not be gravitational force, as the masses are very small. So we must accept the nuclear force as a new type of force. This force is short range, attractive and along the line joining the two particles (central force). Since a central force can not account for the quadrupole moment of the deuteron. As a quadrupole moment is small, the assumption will be approximately correct.

8. The force depends only on the separation of the nucleons not on the relative velocity or orientation of the nucleon spins with respect to the line. This force can be derived from a potential. Since the force is attractive, V(r) is negative and decreases with decreasing r. Since it is short range, V(r) vanishes for $r > b$, where $b \sim 3$ fermi.

The Schrodinger wave equation for the two body problem is

$$\Delta^2\phi + (2m/h^2)(E - V)\psi = 0, \qquad ...(1)$$

where m is the reduced mass, E the total energy of the system equal to the binding energy of deuteron and V the potential energy describing the forces acting between the two bodies.

In this terms of spherical polar coordinates, eqn. (1) becomes

$$\left[\frac{1}{r^2}\frac{\partial}{\partial r}\left(r^2\frac{\partial\phi}{\partial r}\right)+\frac{1}{r^2\sin\theta}\frac{\partial\theta}{\partial}\left(\sin\theta\frac{\partial\phi}{\partial\theta}\right)+\frac{1}{r^2\sin^2\theta}\frac{\partial^2\psi}{\partial^2\phi}\right]$$

$$+ (2m/\hbar^2)\,[E - V(r, \theta, \phi)]\,\psi = 0 \qquad ...(2)$$

Let V (r, θ, ϕ) actually depends on r only and not on θ and ϕ. The solution of the above equation can be written as a product of a function of r only and one of θ and ϕ only as ψ(r, θ, ϕ) = ϕ (r) ϕ (θ, ϕ). Substituting it into equation (2) we get

$$\frac{1}{\psi(r)}\frac{d}{dr}\left(r^2\frac{d\phi(r)}{dr}\right)+\frac{2mr^2}{\hbar^2}[E - V(r)] = -\frac{1}{\psi(\theta,\phi)}$$

$$\left[\frac{1}{\sin\theta}\frac{\partial}{\partial\theta}\left(\sin\theta\frac{\partial\phi(\theta,\phi)}{\partial\theta}\right)+\frac{1}{\sin^2\theta}\frac{\partial^2\psi(\theta,\phi)}{\partial\phi^2}\right] \qquad ...(3)$$

The L.H.S. of this equation depends only on r, the R.H.S. depends only on θ and ϕ. For all values of the variables, each side of this must separately equal to the some constant, which comes out to be $l(l + 1)$. Thus we get

$$\frac{1}{r^2}\frac{d}{dr}\left(r^2\frac{d\psi(r)}{dr}\right)+\frac{2m}{\hbar^2}\left[E - V(r) - \frac{l(l+1)\,\hbar^2}{2mr^2}\right]\psi(r) = 0 \qquad ...(4)$$

where l is the angular momentum quantum number of the system.

The last term in the bracket appears as a straight addition to the actual potential V(r) and is known as centrifugal potential.

The Schrodinger equation for ^3S state (l = 0) of the deuteron is

$$\frac{1}{r^2}\frac{d}{dr}\left(r^2\frac{d\psi(r)}{dr}\right)+\frac{2m}{\hbar^2}[E - V(r)]\,\psi(r) = 0 \qquad ...(5)$$

In this case reduced mass m = 1/2 M. We expect the ground state to be spherically symmetric (S-state), so that ϕ(r) depends only or r. Substituting ψ(r) = u(r)/r in equation (5), where u(r) is called the radial wavefunction, we get

$$\left(\frac{d^2u}{dr^2}\right)+\left(\frac{M}{\hbar^2}\right)[E - V(r)]\,u = 0 \qquad ...(6)$$

The wavefunction of the bound state of the deuteron is not markedly dependent on the exact shape of the potential V(r) between a proton and neutron provided that a potential of short range is chosen. For simplicity, we represent V(r) by a *square well* of depth V_0 and radius b, where b is the range of the nuclear force. In it V(r) has a constant negative value $-V_0$ for separations less than a certain value b and the value zero for all greater separations. The other central force potentials may be of

Gaussian well type $$V(r) = -V_0 e^{-(r/\alpha)^2}$$

Exponential well type $$V(r) = -V_0 e^{-r/\alpha}$$

Yukawa well type $$V(r) = -\frac{V_0 e^{-r/\alpha}}{(r/\alpha)}$$

For the ground state of the deuteron, the total energy E is negative and equal to – B, where B is the binding energy of deuteron. Thus, equation (6) can be written as

$$\frac{d^2u}{dr^2} + \frac{M}{\hbar^2}[V_0 - B]u = 0 \text{ for } r < b \qquad ...(7)$$

and $$\frac{d^2u}{dr^2} + \frac{M}{\hbar^2}(-B)u = 0 \text{ for } r > b \qquad ...(8)$$

These equations can be written as

$$\frac{d^2u}{dr^2} + K^2u = 0, \quad r < b \qquad ...(9)$$

and $$\frac{d^2u}{dr^2} - \alpha^2u = 0, \quad r > b \qquad ...(10)$$

where $$K^2 = \frac{M(V_0 - B)}{\hbar^2}$$

and $$\alpha^2 = \frac{MB}{\hbar^2}$$

General solutions of eqns. (9) and (10) are given as

$$u = A_1 \sin Kr + B_1 \cos Kr \qquad ...(11)$$

$$u = A_2 e^{\alpha r} + B_2 e^{-\alpha r} \qquad ...(12)$$

The following boundary conditions must be imposed:

1. u $(r \rightarrow 0) = 0$, to keep wave function y finite.
2. u $(r \rightarrow \infty) = 0$, u must not diverge faster than r as $r \rightarrow \infty$.

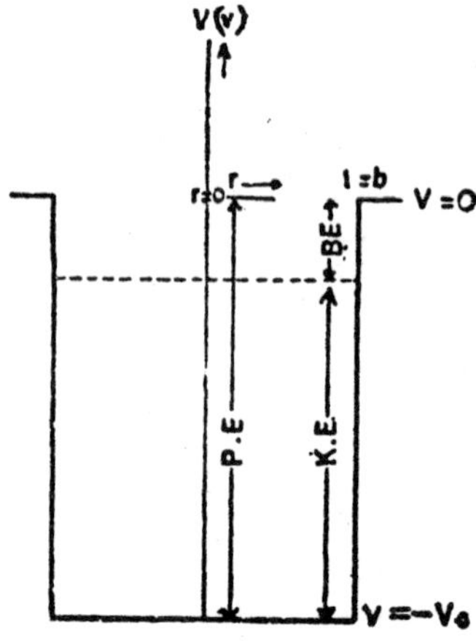

Fig. 22 : Square well potential of deuteron.

To satisfy the conditions at zero and infinity, the solutions reduce to

$u = A_1 \sin Kr$ for $r < b$ ...(13)

$u = B_2 e^{-\alpha r}$ for $r > b$ ...(14)

Since these two solutions join smoothly at r = b. Hence equating the values and first derivatives of u at r = b, we have

$$A_1 \sin Kb = B_2 e^{-\alpha b} \quad ...(15)$$

and $A_1 K \cos Kb = -B_2 e^{-\alpha b}$ ...(16)

$$\therefore \quad K \cot Kb = -\alpha \quad ...(17)$$

The constants A_1 and B_2 are obtained from the requirement that the integral of $|\psi|^2$ over all space must be equal to unity.

$$4\pi \int_0^{\infty} |\psi|^2 r^2 dr = 4\pi \int_0^{\infty} u^2 dr = 1$$

or $$4\pi \int_0^b A_1^2 \sin^2 Kr dr + 4\pi \int_0^{\infty} B_2^2 e^{-2\alpha r} dr = 1$$

$$A_1^2 \left[b - (1/2K) \sin 2Kb\right] + \left(\frac{B_2^2}{\alpha}\right) e^{-2\alpha b} = \frac{1}{2}\pi \quad ...(18)$$

Thus, A_1 and B_2 can be obtained from this equation with the help of eqns. (13) and (14). With the values of b and α as given above, the second term is about twice as large as the first. Hence the *nucleons in the deuteron spend only one third of the time within the range of nuclear force and thus the deuteron is loosely bound.* This can be seen from Fig. 4.2, where u(r) is plotted against r.

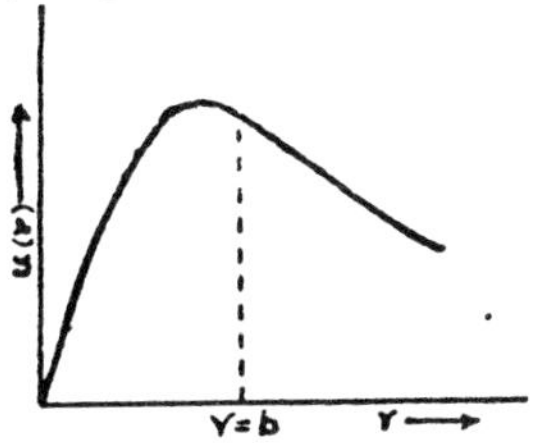

Fig. 23 : Ground state deuteron wave function.

No excited S-states—Equation (17) can be written as

$$x \cot x = -\alpha b \quad ...(19)$$

where x = Kb. Using B = 2.225 ± 0.002 MeV and b ≤ 3f we find that α = 0.232 and $\alpha b \leq$ 07. If we draw now curves y = cot x and y = $-\alpha b/x$, the intersections give the roots of eqn (19). From Fig. 24 it is clear that roots are slightly greater than $\pi/2$, $3\pi/2$, $5\pi/2$.... The correct solution is x = Kb $\approx$ 1/2 π, since if Kb were greater than π, the wave function would have a node at Kb = π and thus, u(r) and hence $\psi(r)$ would not be the wave function of the ground state—a *contradiction of our hypothesis*. We may thus put Kb= 1/2 π + $\in$ in eqn (19) and get

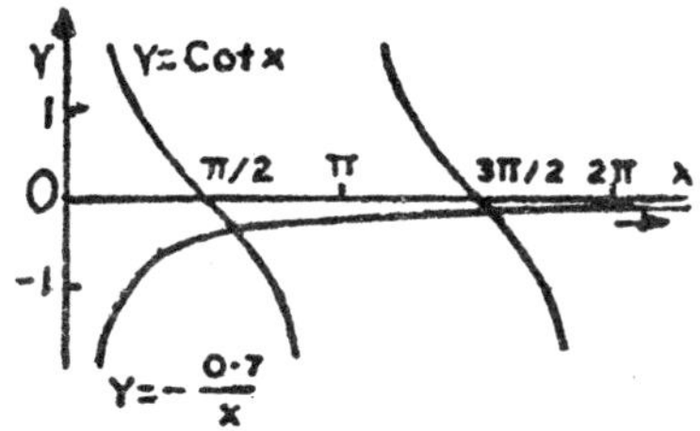

Fig. 24 : Solution of equation x cot x = –0.7

$$\left(\frac{1}{2}\pi + \in\right)\cot\left(\frac{1}{2}\pi + \varepsilon\right) = -\alpha b \quad ...(20)$$

As $\in$ is small, hence $\cot\left(\frac{1}{2}\pi + \in\right) \simeq -\in$, and $\in \simeq 2\alpha b/\pi$

Thus, $Kb = \frac{\pi}{2} + \frac{2\alpha b}{\pi}$. ...(21)

This shows that *there can not be any excited S-states.*

Range and Depth of Potential

Using $Kb = \frac{\pi}{2}$ and again neglecting B in the expression for K [as the depth of potential is very much greater than the binding energy *i.e.*, $K^2 >> \alpha^2$ which can be obtained from eqn (21)], we get

$$\frac{MV_0b^2}{h^2} = \frac{\pi^2}{4} \quad \text{or} \quad V_0b^2 = \frac{\pi^2\hbar^2}{4M} \qquad ...(22)$$

It is the relation between range b and potential depth V_0. Actually V_0b^2 is slightly greater than $\frac{\pi^2\hbar^2}{4M}$, as Kb is slightly greater than $\pi/2$. By accepting the approximate value of range b = 2 fermi, the value of potential depth is V_0 = 36 MeV.

Another result which does not depend on the form of potential is the wavefunction outside the range of nuclear forces. In this region function u(r) decreases exponentially with r and reduces to zero at infinity. The radial distance where the amplitude decreases to 1/e of its maximum amplitude is often called the *radius* of the deuteron.

$$\therefore \quad \text{Radius } R = l/\alpha = \hbar/\sqrt{(MB)} = 4.31 \times 10^{-5} \text{ m} \qquad ...(23)$$

It is about twice that of the range b. This explains that the *deuteron is a loosely bound system*. As b < R, hence the nuclear forces can be said to be short range.

$$\text{or} \quad \frac{d^2u_l(r)}{dr^2} + \left[K^2 - \frac{l(l+l)}{r}\right]u_l(r) = 0 \quad r \le b \qquad ...(24)$$

$$\text{and} \quad \frac{d^2u_l(r)}{dr^2} - \left[\alpha^2 + \frac{l(l+l)}{r}\right]u_l(r) = 0 \quad r > b \qquad ...(25)$$

where K^2 and α^2 are having their usual values.

The general solution of these equations involve spherical Bessel functions j_l and spherical Neumann functions n_l. As the latter approaches $-\infty$ as $r \to 0$, thus the solution of eqn (24) is

$$u_l = A\, j_l(Kr), \qquad r \le b \qquad ...(26)$$

where $$j_l(Kr)\left(\frac{\pi}{2Kr}\right)^{1/2} j_{l+1/2}(Kr) \quad ...(27)$$

The solution of eqn. (25) is

$$u_l(r) = B\ h_l(i\alpha r) = B[j_l(i\alpha r) + in_l(i\alpha r)] \quad ...(28)$$

where $$m(i\alpha r) = (1)^{l+1}\left(\frac{\pi}{2i\alpha r}\right)^{1/2} J_{-l-1/2}(i\alpha r) \quad ...(29)$$

Using boundary conditions that the function and its first derivative are continuous at the edge of the well, we get

$$\left[\frac{1}{u_l}\frac{du_l(r)}{dr}\right]_{inside} = \left[\frac{1}{u_l}\frac{du_l(r)}{dr}\right]_{outside,} \quad (r = b) \quad ...(30)$$

Using the relation $\frac{dj_l(\rho)}{d\rho} = j_{l-1}(\rho) = \frac{l+1}{\rho} j_l(\rho)$, we get

$$K\left[\frac{j_{l-1}(Kb)}{j_l(Kb)} - \frac{l+1}{Kb}\right] = i\alpha\left[\frac{h_{l-1}(i\alpha b)}{h_l(i\alpha b)} - \frac{l+1}{i\alpha b}\right]$$

or $$\frac{j_{l-1}(Kb)}{j_l(kb)} = \left(\frac{\alpha}{K}\right)\left[\frac{ih_{l-1}(i\alpha b)}{h_l(i\alpha b)}\right] \quad ...(31)$$

For $b < 1.43 \times 10^{-15}$m,$\alpha b < 1$ and since $\alpha << K$ the expression in the bracket on R.H.S is less than one, and is approximately zero. Thus

$$j_{l-1}(Kb) \approx 0 \quad ...(32)$$

This condition holds for all angular momenta except $l = 0$. We have already discussed the case $l = 0$. For $l = 1$, we get j_0 (Kb) $\approx 0 = \sin$ (Kr). Hence Kb $\simeq \pm \pi, \pm 2\pi, \pm 3\pi,$.... Thus, the minimum well depth is

$$V_0 \simeq \frac{\pi^2\hbar^2}{Mb^2} \quad ...(33)$$

If we choose $b = 2 \times 10^{-15}$, we get $V_0 = 144$ MeV, which is almost four times as large as the actual well depth in the ground state. Repeating this procedure for larger and larger values of l we find that a deeper and deeper well depth is required to produce a bound state. Thus, we conclude that no bound state exists for $l > 0$.

SOLVED EXAMPLES

Example 1:

In a certain betatron the maximum magnetic field was 4000 gauss, operating at 50 cycles/sec with a stable orbit diameter of 60 inches. Calculate the average energy gained per revolution and the final energy of the electrons.

Solution:

If B is the magnetic field at the end of the accelerating cycle at the orbit of radius R, the final energy

$$E = BeRc = 0.4 \times 1.602 \times 10^{-19} \times 30 \times 2.54 \times 10^{-2} \times 3 \times 10^{8} \text{ joule}$$

$= 91$ MeV. (Assuming $v = c$ for simplicity).

Total no. of revolution,

$$N = \frac{c}{4\omega R} = \frac{3 \times 10^8}{4 \times 2 \times 3.14 \times 50 \times 30 \times 2.54 \times 10^{-2}}$$
$$= 3.1 \times 10^5.$$

$\therefore$ Average energy gained per revolution

$$= 91 \times 10^6/3.1 \times 10^5 = 294 \text{ eV}.$$

Example 2:

Recall that b~λ/4'~π/2K for the ground level of the deuteron. From this, show that the radius of the deuteron, in the rectangular well model is approximately given by

$$R = \frac{2bV_0^{1/2}}{\pi B^{1/2}}.$$

Solution:

From eqns. (1) and (2), we have

$$K^2 = \frac{M}{\hbar^2}(V_0 - B) \text{ and } \alpha^2 = \frac{MB}{\hbar^2}$$

$$\therefore \quad \frac{K}{\alpha} = \frac{\sqrt{[(V_0 - B)}}{B}$$

Since $b = \dfrac{\pi}{2K}$ (given) and $R = \dfrac{1}{\alpha}$, hence

$$\frac{\pi/2b}{1/R}\sqrt{\left(\frac{V_0-B}{B}\right)} \text{ or } R\,\frac{2b}{\pi}\left(\frac{V_0-B}{B}\right).$$

As $B < V_0$, we can write for an approximation $V_0 - B \simeq V_0$.

$$\therefore \qquad \mathbf{R} = \frac{\mathbf{2bV_0^{1/2}}}{\pi \mathbf{B^{1/2}}}.$$

Example 3:

Show that for a square well of depth V_0 and range b, the scattering length a for a spinless neutron is given by the relation

Solution:

$K \cot Kb = (b - a)^{-1}$, where $K = (MV_0)^{1/2}/\hbar$.

For a square well of depth V_0 and radius r_0, we have

$$k_1 \cot k_1 b = k \cot (kb + \delta_0)$$

where
$$k_1 = \frac{\sqrt{M[(E+V_0)]}}{\hbar} \quad \text{and} \quad k = \frac{\sqrt{(ME)}}{\hbar}$$

Since $V_0 > E$, hence we can write $k_1 = \dfrac{\sqrt{[MV_0]}}{\hbar} = K$. As the phase shift δ_0 = m–ak (eqn. 48).

$$\therefore \quad K \cot Kb = k \cot (kb - ak) = k \cot k(b - a)$$

$$= \frac{\mathbf{k}}{\mathbf{\tan k\,(b-a)}} = \frac{\mathbf{k}}{\mathbf{k(b-a)}} = \frac{\mathbf{1}}{\mathbf{b-a}}.$$

Example 4:

Calculate scattering lengths a_t a, d, a_s. Given that σ_{para} = 4.19 and σ_{ortho} = 128 barns.

Solution:

The theoretical expressions for the total cross-sections are

$$\sigma_{para} = 7.69\,(3a_t + a_s)^2$$

$$\sigma_{ortho} = 6.69\,[(3a_t + a_s)^2 + 2(a_t - a_s)^2] + 1.74\,(a_t - a_s).$$

Substituting values of σ_{para} and σ_{ortho} in above relations and solving these equations, we get following four pairs:

$$\begin{matrix} a_t = \\ a_s = \end{matrix} \begin{vmatrix} 0.91 \\ -1.95 \end{vmatrix} \begin{vmatrix} 0.52 \\ -2.34 \end{vmatrix} \begin{vmatrix} -9.52 \\ -3.38 \end{vmatrix} \begin{vmatrix} -0.91 \\ -3.77 \end{vmatrix} \times 10^{-14}\text{ m}$$

Last two pairs can be discarded immediately because they have negative a_t values. The first pair can also be discarded on the basis that it is inconsistent with the n-p scattering data. Thus, the remaining solution is

$$a_t = 0.52 \times 10^{-14}\text{ m},\ a_s = 2.34 \times 10^{-14}\text{ m}.$$

Example 5(a):

Calculate the total cross-section for n-p scattering at neutron energy 2MeV (lab). Given at = 5.38F, a_s = –23.7F, r_{0l} = 1.70F and r_{0s} = 2.40F.

Solution:

Total cross-section $\sigma = \frac{3}{4}\sigma_l + + \frac{1}{4}\sigma_s$.

Using equation (70), this equation can be written as

$$\sigma = \frac{3\pi a_{t^2}}{\left[1 - \frac{1}{2}a_t\, r_{at}\, k^2\right]^2 + a_{t^2}k^2} + \frac{\pi a_{s^2}}{\left[1 - \frac{1}{2}a_s\, r_{as}\, k^2\right]^2 + a_{s^2}k^2}$$

In C.M. system $E_{C.M.} = \frac{1}{2}E_{Lab} = 1\text{MeV}$

$$\therefore \qquad k^2 = \frac{ME}{\hbar^2} = \frac{1.6748\times10^{-27}\times1.6\times10^{-13}}{(1.0549\times10^{-34})^2}$$

Substituting numerical values, we have

$$\sigma = 1.83 \times 80^{-28} + 1.074 \times 10^{-28} = 2.904\text{ barns.}$$

Example 5(b):

A linear accelerator for the acceleration of protons to 45.3 MeV is designed so that, between any pair of accelerating gaps, the protons spend one complete radio-frequency cycle inside a drift tube. The rf-frequency used is 200 Mc/sec.,

(a) What is the length of the final drift tube?

(b) If the first drift tube is 5.35 cm long, at what K.E. are the protons injected into the linac?

(c) If the peak accelerating potential is 1.49×10^6 volts, calculate the total length of the accelerator.

Solution:

Length of the final drift tube $L_F = \frac{v_F}{f} = \frac{1}{f}\sqrt{\left(\frac{2E}{M}\right)}$

$$= \frac{(2 \times 45.3 \times 1.6 \times 10^{-13})^{1/2}}{200 \times 10^6 \times (1.6724 \times 10^{-27})^{1/2}}$$

$= 0.4654$ m. $= 47$ cm

K.E. of the injected protons $= \frac{1}{2} Mv_1^2 = \frac{1}{2} ML_1^2 f^2$

$= 1/2 \times 1.6724 \times 10^{-27} \times (5.35 \times 10^{-2})^2 \times (200 \times 10^6)^2$

$= 0.60$ MeV.

$\therefore$ Increase in energy E = 45.3 – 0.6 = 44.7 MeV.

Energy increase in N gaps NqV = NeV.

$\therefore$ N = E/eV = $44.7 \times 10^6/1.49 \times 10^6 = 30$

Total length $L = \Sigma L_n = \frac{1}{f}\left(\frac{2qV}{M}\right)^{1/2} \int_0^N n^{1/2}\, dn$

$$= \frac{2}{3f}\left(\frac{2qV}{M}\right)^{1/2} N^{3/2}$$

$$= \frac{2(2 \times 1.6 \times 10^{-19} \times 1.49 \times 10^6)^{1/2} \times 30^{3/2}}{9 \times 200 \times 10^6 \times (1.6724 \times 10^{-27})^{1/2}} = 9.4\text{m}.$$

Example 5(c):

Show that the radius of curvature R of the path of a particle inside the dees of a cyclotron is proportional to the $N^{1/2}$, where N is the number of times the particle has been accelerated across the space between the dees.

Solution:

If there are N accelerations and the average potential difference between D's each time the ions cross the gap is V, the final energy will be

$$E = NqV = 1/2\ mv_m^2.$$

As the kinetic energy $E = q^2B^2R^2/2M$, hence we have

$$q^2B^2R^2/2M = NqV \quad \text{or} \quad R = (2MNqV)^{1/2}/qB$$

$$\therefore \qquad R \propto N^{1/2}.$$

Example 6:

Deuterons are accelerated in a fixed frequency cyclotron to a maximum dee orbit radius of 88 cm. The magnetic field, is 14000 gauss. Calculate the energy of the emerging deuteron beam and the frequency of the dee voltage. What change in magnetic flux density is necessary if doubly changed helium ions are accelerated. Given atomic masses :

$$H^2 = 2.014102 \text{ mu.},\ He^4 = 4.002603 \text{ mu.}$$

Solution:

Energy of the emerging deuteron $= = \frac{1}{2} B^2R^2q^2/M$

$$= \frac{1}{2} \times \frac{(1.4)^2 \times (.88)^2 \times (1.602 \times 10^{-19})^2}{2.014102 \times 1.66 \times 10^{-27}} \text{ joules}$$

$$= 36.3 \text{ MeV}$$

Frequency of the dee voltage = Frequency of oscillations

$$= \frac{Bq}{2\pi M} = \frac{1.4 \times 1.602 \times 10^{-19}}{2 \times 3.14 \times 2.014102 \times 1.66 \times 10^{27}}$$

$$= 10.67 \text{ Mc/sec.}$$

Magnetic field required for the acceleration of $(He^4)^{++}$

$$= \frac{2\pi Mf}{q} = \frac{2 \times 3.14 \times 4.002603 \times 1.66 \times 10^{-27} \times 10.67 \times 10^6}{2 \times 1.602 \times 10^{-19}}$$

$$= 1.39 \text{ web/m}^2.$$

$\therefore$ Change in magnetic field = 1.4 – 1.39 = 0.01 tesla.

Example 7:

A 100 MeV betatron has on orbit radius of 35 cm. in which the electrons acquire 480 volts per revolution. How far will the electrons travel in attaining full energy? The magnet i energized at 180 Hz and produces pulses of 2μ sec duration, each consisting of 3×10^9 electrons. What is the peak and the average beam current? What is the duty cycle?

Solution:

Total no. of revolutions $N = 100 \times 10^6/480 = 2.083 \times 10^5$.

$\therefore$ Total distance traversed by the electron $L = 2\pi RN$

$$= 2 \times 3.14 \times 0.35 \times 2.083 \times 10^{5} = 4.57 \times 10^5 \text{ m.}$$

As 3×10^9 electrons or 4.8×10^{-10} coulomb of charge passes in 2μ sec, hence peak value of current

$$I_p = 4.8 \times 10^{-10}/2 \times 10^{-6} \times 10^{-4} \text{ amp.}$$

In one second magnetic field reaches to its peak value 180 times, 4.8×10^{-10} coulomb of charge is flowing in every time, hence the flow of charge per sec or the average current

$$I_{av} = 4.8 \times 10^{-10} \times 180 = 8.64 \times 10^{-8} \text{ amp.}$$

As pulses are produced only for 2μ sec, and the time period of the field is 1/180 sec, hence the duty cycle

$$= \frac{dt}{T} = \frac{2 \times 10^{-6}}{1/180} = 3.6 \times 10^{-4}.$$

Example 8:

Deuterons are accelerated in the synchrocyclotron which has magnetic field of 15000 gauss at the centre and 14310 gauss at the periphery of the dee. Calculate the maximum frequency of the dee voltage. If the dee voltage frequency is modulated between this maximum and a minimum of 10 Mc/see., calculate the gain in energy of a deuteron.

Solution:

Maximum frequency of the dee voltage $f_0 = \omega_0/2\pi = B_0 q/2\pi M_0$

$$= \frac{1.5 \times 1.602 \times 10^{-19}}{2 \times 3.14 \times 2.014102 \times 1.66 \times 10^{-27}} = 11.44 \text{ Mc/sec.}$$

The relativistic mass of the particle at the periphery

$$M = \frac{B'q}{2\pi f'} = \frac{1.431 \times 1.602 \times 10^{-19}}{2 \times 3.14 \times 10 \times 10^6} \text{ kg}$$

$$= 2.199 \text{ mu.}$$

$\therefore$ K.E. of the particle $= (M - M_0)\, c^2 = (2.199 - 2.0141)$

$= 0.1849$ mu. $= 171.6$ MeV.

Example 9:

Assume that in the 70 MeV betatron synchrotron the radius of the stable electron orbit is 28 cm. Calculate:

(a) the frequency of the applied electric field;

(b) the value of the magnetic field intensity at the orbit for this energy, and

(c) the energy loss by radiation during a single revolution of an electron.

Solution:

Relativistic equation for energy of electrons is E = BeRc.

$$\therefore \quad B = \frac{E}{eRc} = \frac{70 \times 1.6 \times 10^{-13}}{1.602 \times 10^{-19} \times 0.28 \times 10^{8}} = 0.83 \text{ tesla}$$

Frequency of the applied electric field $= eBc^2/2\pi E$

$$= \frac{1.602 \times 10^{-19} \times 0.83 \times (3 \times 10^{8})^2}{2 \times 3.14 \times 70 \times 1.6 \times 10^{-13}}$$

$= 1.706 \times 10^8$ cycles/sec.

Energy radiated by an electron of energy E in one revolution

$$\Delta E = \frac{88.5E^4}{R} = \frac{88.5\,(0.07)^4}{0.28} = 7.5\text{eV}.$$

Example 10:

In the cosmotron proton synchrotron there are four quadrants of radius 9.144 m. and four connecting straight sections each of length 3.048 m. Protons are injected at 3.6 MeV.

Calculate the magnetic flux density at the moment of injection, the speed of the protons, and the frequency of the accelerating voltage. The protons are accelerated to a final kinetic energy of 3 GeV, calculate the maximum magnetic flux density and the maximum frequency of the accelerating voltage.

Solution:

Energy of the injected proton, when it moves in a circular path of radius R is given by E = $B^2R^2q^2/2M$.

$$\therefore\ B = \frac{(2ME)^{1/2}}{Rq} = \frac{(2 \times 1.67 \times 10^{-27} \times 3.6 \times 1.6 \times 10^{-13})^{1/2}}{9.144 \times 1.6 \times 10^{-19}}$$

$$= 0.02999 \text{ web/m}^2\ 300 \text{ gauss}$$

and $$v = \frac{BqR}{M} = \frac{0.03 \times 1.6 \times 10^{-19} \times 9.144}{1.67 \times 1027}$$

$$= 2.6 \times 10^7 \text{ m/sec.}$$

Frequency in a circular orbit

$$f_c = v/2\pi R = 2.6 \times 10^7/2 \times 3.14 \times 9.144$$

$$= 0.453 \text{ mega-cycles/sec.}$$

As proton moves in four quadrants, hence frequency reduces to

$$f_0 = \frac{2\pi Rf_c}{2\pi R + 4L} = \frac{2 \times 3.14 \times 0.453 \times 10^6}{2 \times 3.14 \times 9.144 + 4 \times 3.048}$$

$$= 370 \text{ kc/sec.}$$

For the protons of energy 3 GeV, we have to use relativistic.

$$B_m = \frac{3.33\,[T(T + 2M_0c^2)}{R} = \frac{3.33\,[3(3 + 2 \times 0.938)]^{1/2}}{9.141}$$

$$= 1.392 \text{ web/m}^2.$$

Maximum frequency can be obtained by combining.

$$f_0 = \frac{c^2eBR}{(2\pi R + 4L)\,(T + M_0c^2)} = 4.193 \text{ mega-cycles/sec.}$$

EXERCISES

1. The radius of the stable electron-orbit of a small electron synchrotron is 12 cm. and the maximum magnetic field is 9000 gauss. Calculate the frequency of the applied electric field and the maximum electron energy produced.

2. A synchrotron is to be designed to accelerate protons to 12 GeV kinetic energy:

   (a) Assuming a maximum field strength of 1.43 web/m^2 estimate the radius of curvature of the proton orbit.

   (b) If about 25 per cent of the protons path is spent in field free straight sections, what is the final revolution frequency?

   (c) Assuming one revolution per rf-cycle, at what kinetic energy must the protons be injected if the frequency over the entire acceleration cycle is to vary by a factor of 5? What type of device would you suggest for the injector?

3. Using the data given in Example 8, and the field index n = 0.6: Calculate the expected radial displacement of the protons from their correct orbit if the frequency error is +0.1 per cent

   (a) soon after injection and

   (b) when B reaches 1.4 weber/m^2.

4. Use the well parameters ^3a = ^1a = 2.8 F. $^3l_0$ = 21MeV and $^1V_0$ = 12MeV to compute the scattering cross-section for 2MeV (lab) neutrons.

5. Use the parameters a_t = 5.38F, r_{0l} = 1.70 F and as = –23.7 F to determine r_0s from the experimental observation that σ = 1.69 barns at energy 4.75 MeV (lab).

6. Worthington and others measured a p-p deferential scattering cross section of 0.1114 barns/steradian (C.M. system) for a proton energy of 4.203MeV (lab) and a scattering angle of 60° (C.M. system). Calculate the phase shift.

7. The H^+ and H_2^+ ions accelerated in a Van de Graaff generator to 5 MeV are to be magnetically separated from one another. Approximately over what distance must a 10,000 gauss field be applied if the two beams are to diverge by 20^0?

8. Estimate the total electromagnetic energy in the accelerating tube of the SLAC two miles linac during the acceleration pulse when it is operated to give 40 BeV electrons. The diameter of the wave guide is 8.2 cm. What is the total energy in joules of the electrons in one pulse after acceleration?

9. A standard cyclotron of 120 cm. pole diameter is operated with a 10 Me oscillator:

(a) What magnetic field is required for the acceleration of deuterons?

(b) What will be the final deuteron energy?

(c) With the same rf frequency, what is the maximum K. E. to which He^3 ions could be accelerated in this cyclotron?

(d) Under the assumption that the magnetic field calculated in (a) is the maximum available, what oscillator frequency would be required to obtain the highest possible H^3 energy and what is that energy ?

10. By what fraction would the time required to pass through a dee be increased as compared with the time required at non-relativistic speed, if the cyclotron is designed to accelerate particles to half the speed of light?

11. A 98 MHz oscillator supplies an effective voltage of 50 kV to the dees of a cyclotron accelerating α-particles to a maximum energy of 28 MeV. Calculate the closest approach of adjacent turns of the spiral ion path, assuming the beam to lie between radii of 10 and 75cm.

12. In a certain betatron the maximum magnetic field was 0.4 tesla, operating at 60 Hz with a stable orbit diameter of 66 inches. Calculate the maximum K.E. of an electron, injected with energy 50 kV and the approximate total time of flight.

13. A synchrocyclotron is being designed to accelerate protons to 700 MeV. What percentage frequency change will be required in the rf-oscillator?

14. Protons are accelerated to 740 MeV in the Berkeley synchro-cyclotron which has a B = 2.3 tesla. What is the oscillator frequency at the time a pulse of protons is injected at the centre? What is the oscillator frequency at the time the protons have reached the rim ?

15. Prove that $(\vec{\sigma}_1 \cdot \vec{\sigma}_2)^2 = 3 - 2\vec{\sigma}_1 \cdot \vec{\sigma}_2$ and that $(\vec{\sigma}_1 \cdot \vec{\sigma}_2)^2 = 0$.

16. Show that p-p scattering can take place only in ^1S, ^2P. ^1D....states, that can you say about the total isospin of the n-p system in these states ?

3

Different Processes in the Scintillation Detector

ABSORPTION PROCESS

The electrons emitted will give up their kinetic energy in ionization or excitation in the scintillator material. This absorbed energy appears either as heat energy or as luminescence photons. If the dimensions of the scintillator are large compared to the range of the incident particle, the incident charged particle can dissipate all its energy to the scintillator. If the γ-ray is incident on a scintillator it may interact with the scintillator material in three ways: *photo-electric absorption, Comption scattering and pair production* each cases, electrons are produced. The all or part of the energy of incident γ-ray is transformed into the kinetic energy of these electrons.

Scintillation Process

In the latter cases the scintillator material de-excites by light emission within 10^{-8} seconds. This emitted light is known as scintillation. If τ is the mean life of the scintillator, the number of light photons in time t after the ionization radiation has arrived is given by

$$n = \text{const}\ (1 - e^{-t/\pi}) \qquad ...(1)$$

Formation of Electrical Pulses

The light produced in scintillator falls on the photocathode of the photomultiplier tube, producing photoelectrons. These electrons fall on the first dynode, producing a bunch of secondary electrons. This process is repeated and a gain upto $10^{-7} - 10^{8}$ is achieved. The electrons finally fall on the electrode, known as anode, and thus, produce an electrical pulse across the load resistance R. The pulse height V at the output related to the amount of charge Q collected at the output through the relation

$$V = \frac{Q}{C} \qquad ...(2)$$

where C is the capacity of the output point.

If E_i be the energy of the incident charged particle or γ-rays and A the probability of amount of energy absorbed by the phosphor.

$\therefore$ Energy dissipated in the phosphor = E_iA.

This energy is converted within efficiency C_{IP} into photons of average energy E_p.

$$nr = E_iA\ C_{IP}/E_p \qquad ...(3)$$

If T_p is the optical transparency and G is the fraction of the photons falling on the photocathode. Hence, the number of photons incident on the photocathode.

$$n_p/ = T_pGn_p = E_iAC_{IP}T_pG/E_p.$$

These photons are converted into photoelectrons with an efficiency C_{PE} f(v), where C_{PE} is the photoelectric conversion efficiency of the cathode and f(v) is the relative response at frequency v.

$\therefore$ No. of photo electrons n = $C_{PE}/(v)$ np/

If f_d be the fraction of photoelectrons reaching the first dynode and M the electron multiplication factor, then the total number of electrons reaching the anode

$$N = f_dMn = C_{PE}f(v)\ E_iAC_{IP}T_pGf_dM/E_p.$$

Hence total charge Q = eN and the pulse height

$$V = Q/C = eE_i\ AC_{IP}T_PGC_{PE}f(v)f_dM/CE_p \qquad ...(4)$$

Here, we see that the pulse height is proportional to the incident particle if the particle is absorbed in the scintillator completely.

Resolving Power

The height of the output pulse exhibits variations about the mean value for a given particle energy E. This variation is due to the statistical fluctuations in the various factors of eqn. (4). The resolution R of a scintillation counter can be defined as follows

$$\frac{1}{R} = \frac{\overline{Q}^2 - (\overline{Q}^2)}{(\overline{Q}^2)} \qquad ...(5)$$

where $\overline{Q}$ is the mean charge collected at the anode and $\overline{Q}^2$ is the mean squared charge collected at the anode. The smaller the half width of the photo peak, the higher will be energy resolution R.

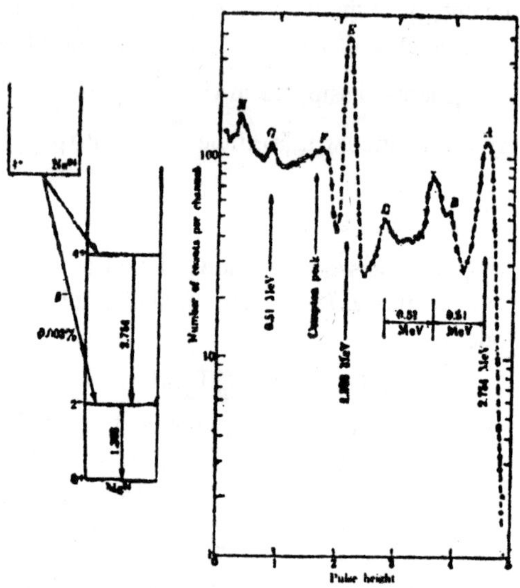

Fig. 1 : Decay scheme for Na^{24} (Left). Pulse height spectrum for gamma rays (Right).

The scintillation counter can be used as a spectrometer in conjunction with an electronic pulse height analyzer, which has found many applications in low energy, as well as in high-energy nuclear physics. Suitable circuits have been designed to sort out particles with in a given range of energy so that distribution in energy of the incident radiation may be measured. Unfortunately, even mono-energetic γ-rays produce a complex pulse height spectrum. The pulse height spectrum produced by γ-rays from the lower excited states of Mg^{24} is shown in Fig. 1 (right). This isotope is produced in low lying states by β-emission of Na^{24} according to the decay scheme shown in Fig. 1 (left). Two γ-rays of energy, one at 2.754 MeV and other at 1.368 MeV are emitted. The peak A in this spectrum corresponds to the full energy of 2.754 γ-ray and is due to pair production process only.

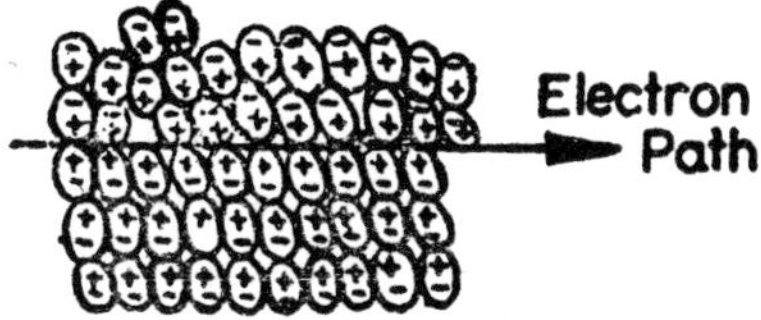

Fig. 2 : Polarization of molecules.

The next peak B corresponds to the electrons produced by compton process. The peak C is caused by pair-production processes in which one of the two annihiliation photons has escaped and peak D corresponds to the processes in which both the annihiliation photons have escaped. The central sharp peak E corresponds to the energy 1.368 MeV and is due to pair production or photo electric process. The compton peak corresponding to this peak is F. As 1.368 MeV pair production is relatively unimportant, hence first and second escape peaks corresponding to the energies 1.368 – 0.51 = .858 and 1.368 – 1.02 = .348 are not observed. Peak G corresponds to 0.51 MeV annihiliation photons produced outside the scintillator but absorbed by it. The peak H arises from back scattered photons from compton events.

The scintillation counter has a much higher counting efficiency for gamma rays due to the greater amount of energy dissipation by the gamma rays. The scintillation counter offers greater stability, greater accuracy, shorter resolving time and higher efficiency than G.M. counter. Provided all the particle energy is dissipated in the scintillator, the voltage pulse height produced can be used to measure the particle energy, once the counter is calibrated by means of a standard source. Because of their extremely rapid response scintillation counters are having many applications in atomic energy field. These are used for the accurate timing of nuclear particles moving with very high speeds.

CERENKOV COUNTERS

When a charged particle moves through a transparent dielectric medium with a velocity greater than the velocity of light in that medium, weak electro-magnetic radiation is emitted. These radiations are named Cerenkov radiations. The radiation is emitted mainly in the direction of the charged particle and has a continuous spectrum from red to ultra-violet at least. A charged particle traversing a dielectric medium causes polarization of the molecules adjacent to its path, as the charged particle attracts the oppositely charged component of a molecule and repels the similarly charged component of the molecule. For slowly moving particles, polarized regions are formed in step with the motion in the vicinity of the particle. The resultant dipole field is zero as the polarization field around the electron is symmetrical in the axial and azimuthal directions, hence no radiation is emitted. For fast moving particles the polarization is *axially asymmetric* but azimuthally symmetric and the resultant dipole field acts even at large distances from the electron's track. The radiation pulses are thus emitted in succession by the molecules along the path of the charged particle. These radiations are like the radiations emitted from a *damped oscillating dipole moment.*

In a dielectric medium of refractive index μ, photons move with a velocity c/μ. Fig. 3 (left) shows the three interactions between a charged particle moving with velocity v (< c/μ) and the electrons in the medium. The wavefronts of the electromagnetic waves resulting from interaction do not interrelate and these interactions are entirely independent. Fig. 3 (right) shows the interactions when v > c/μ. The wavefronts from 1 and 2 lie inside 3 at the time of the third interaction.

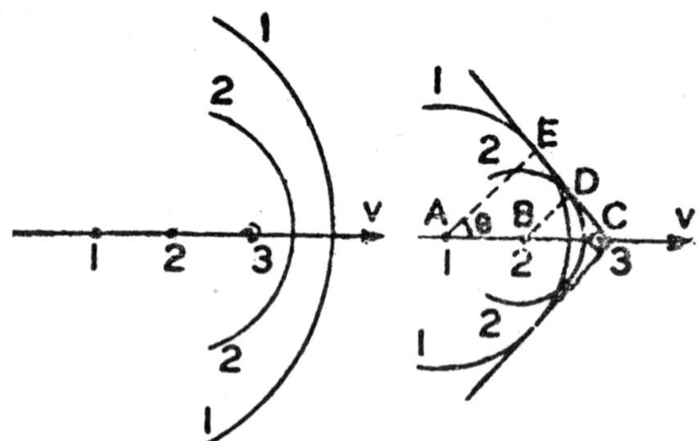

Fig. 3 : Wavefronts resulting from radiation losses: with v < c/μ (left) and with v > c/μ (right).

The Huygens construction of elementary wave optics shows that a conical wavefront can be drawn tangent to the spherical surfaces and the Cerenkov radiation can be observed at a particular angle with respect to the direction of motion of the particle. The particle moves along ABC with velocity v It radiates energy in all directions from the points along its path ABC. The time in which particle travels a distance AC, the radiations will move a distance AE, CE is thus the wavefront. The angle which the Cerenkov Radiation makes with the direction of charged particle is given by

$$\cos\theta = \frac{AB}{AC} = \frac{BD}{BC} = \frac{ct/\mu}{vt} = \frac{c}{\mu v} = \frac{1}{\beta\mu} \qquad ...(1)$$

The Cerenkov radiation is not confined to one plane but is propagated along the surface of a cone whose axis coincides with the direction of motion of the charged particle and whose semivertical angle is θ, given by eqn. (1), which shows that:

1. θ depends on the frequency (*i.e.,* wavelength) of the Cerenkov radiation. The radiation arc mainly of the high frequency region of the visible spectrum.
2. Unlike the bremsstrahlung, θ is independent of the mass of moving charged particle.
3. The emission angle increases with increasing the particle velocity β.
4. As P cannot exceed unity the maximum angle of emission.

   $$\theta_{max} = \cos^{-1}(1/\mu)$$

5. For a given constant refractive index μ, there is a threshold velocity $\beta_{th} = l/\mu$ below which no Cerenkov radiation is emitted and at which the radiations are along the particles direction of motion.

Radiation (33) for the constructive interference should satisfy the following conditions:

(a) The particle track length must be large compared with the wavelength λ of the Cerenkov radiation.

(b) The particle must move with the constant velocity in the medium.

Frank and Tamm have shown from classical theory that the total energy radiated in a short element of the particle's path ds is given as

$$\frac{dT}{ds} = \frac{\pi z^2 e^2}{\epsilon_0 c^2}\int\left(1 - \frac{1}{\beta^2\mu^2}\right)v\,dv$$

where ze is the charge on the moving charged particles, v is the frequency of the emitted radiation, and the integral extends overall frequencies for which $\beta\mu > 1$. This loss is quite small (*e.g.*, about 10^3eV/cm for singly charged particles moving through glass or lucite) and is negligible compared to the ionization losses. Consider the Cerenkov radiation emitted between two frequencies v_1 and v_2 as composed of quanta whose average energy is hv = 1/2h ($v_1 + v_2$), the average number of quanta emitted per unit length is

$$\frac{1}{h_v} = \frac{dT}{ds} = \frac{\pi z^2 e^2}{\in_o hc^2}(v_2 - v_1)\left(1 - \frac{1}{\beta^2 \mu^2}\right) \qquad ...(2)$$

where μ if the average refractive index of the medium over the frequency interval v_1 to v_2.

Relation (35) shows that for extreme relativistic electrons passing through glass ($Z = -1$, $\beta \simeq 1$, $\mu \simeq 1.5$) the (Cerenkov angle is 48° and the Cerenkov photons emitted per cm within the visible spectrum ($\Delta v = v_2 - v_1 \simeq 3 \times 10^{14}$ sec^{-1}) are 200. As the number of quanta per unit path length varies as I/λ^2, the short wavelength are thus predominantly present in Cerenkov radiation.

The properties discussed above have been extensively utilized in Cerenkov detectors which can be used not only to detect swiftly moving charged particles but also to determine their velocity. These detectors possess the following advantages:

1. Dependence of intensity on particle velocity.
2. Directional emission of light, with an angle dependent on velocity.
3. High efficiency and high counting rate.
4. The duration of the light pulse at any point with in the Cerenkov cone is vanishingly short (~ ≤ 50^{-10} sec).

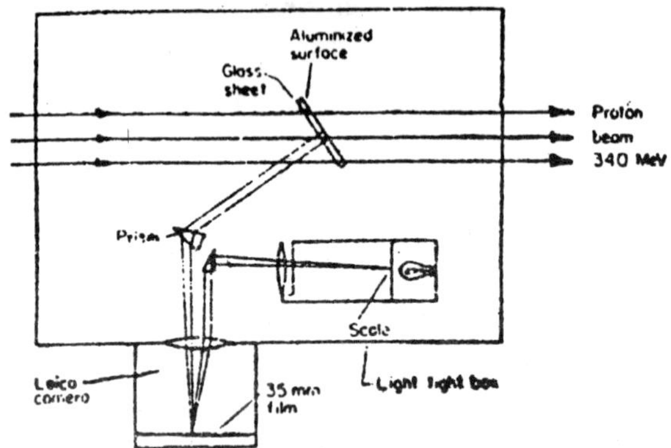

Fig. 4 : Measurement of velocity of high energy protons.

R.L. Mather has observed Cerenkov radiations radiated by a collimated beam of 340 MeV protons passing through an optically flat sheet of dense flint glass of 0.67 mm thickness with a 35 mm Leica camera. The glass sheet, whose rear surface was alumunized, was set with respect to the proton beam so that part of the Cerenkov cone radiated in the glass would incident normally on the aluminium surface and reflect in such a way that there would be no refraction at the front surface of the glass.

The angle of emission of radiations varies with wavelength λ because the refractive index depends on λ. This dispersion effect was decreased to nearly zero by Mather after introducing a prism. The recorded Cerenkov radiation on the camera film corresponded to a θ value of 38.5°. Using the relation (33) β comes out to be 0.680. The kinetic energy E_{kin} of the protons can be obtained as

$$E_{kin} \; m_p c^2 (1 - \beta^2)^{-1} - m_p c^2 + 341 \text{ MeV} \qquad ...(3)$$

This value was found to have 1% accuracy.

J. Marshall designed a good detector, in which the Cerenkov light was focused on to the cathode of two photomuitiplier tubes for recording high energy π-mesons (pions). A collimated beam of pions was allowed to pass along the axis of a large hemispherical perspex lens. The radiation radiated in this lens was reflected by a surrounding cylindrical mirror M_1 on to one of the plane mirrors M_2, M_3 and then on to the cathodes of the photomuitiplier tubes (lP 28), which were operated in coincidence. The position of the radiator lens was adjusted so that the Cerenkov light cone was correctly focused on to the photo tubes to give maximum coincidence counting rate. For 145 MeV pions the energy resolution is 13%.

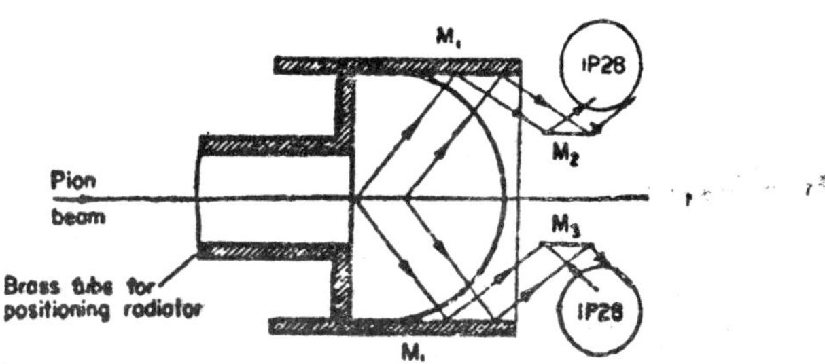

Fig. 5 : Marshall's focusing Cerenkov detector.

Cerenkov counters have been designed for charged particles in a narrow velocity range. Cerenkov detectors in which the light is radiated by a gas have been used for highly relativistic charged particles. Because the radiation in a given medium is produced only when the speed of the charged particles is greater than the velocity of light in that medium, this counter can be used as a *threshold detector, i.e.*, to detect only particles whose speed exceeds a certain value.

These counters are essentially *directional counters* became the Cerenkov radiations are emitted in a forward cone of an angle given by eqn. (2). The direction of the incoming particle can, therefore, be measured. The angle θ depends on the velocity v of the incident particle. This counter can thus be used to measure the velocity of incoming particle. Particles of *different masses* can be distinguished in two ways: first, for the same velocity, *i.e.,* the same Cerenkov angle, the brightness of the flash is proportional to the square of the atomic number and, second, for the same K.E. or range, the velocity of the particle is inversely related to the square root of the mass number. The particles of different charge can also be distinguished by these counters.

The blue glow seen around nuclear reactors of the swimming pool type reactor is almost due to *Cerenkov radiation*, produced by high speed electrons (Compton electrons produced in γ-ray scattering events). Since the core or fuel elements are in extended source and the emitted electrons show a distribution over a wide range of energies and angles. The radiation thus appears diffuse. The energy and momentum of charged particles (electrons) can thus be evaluated.

CLOUD CHAMBER

During the stay in the Scottish hills, in 1984, he started experimenting under laboratory conditions and intended to study the optical interesting phenomena appearing from illuminated fogs. After two years he discovered the fundamental principle of the cloud chamber and designed the first cloud chamber based on the principle in 1911.

Air mixed with saturated vapour of water or any other liquid, *e.g.*, alcohol or ether contained in a cylinder fitted with a piston can be expanded rapidly by a fast but relatively short motion of a piston. This adiabatic expansion will result in fall of temperature and super-saturation of vapour. The vapour will condense in cloud of water droplets. The production of such a cloud is however, impossible unless nuclei, on which the water vapour may condense are provided.

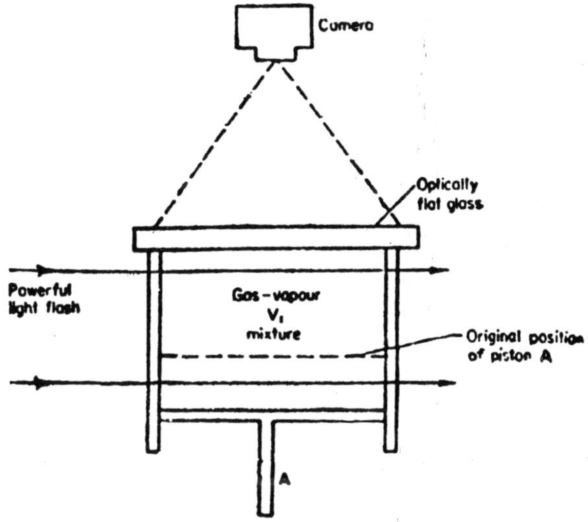

Fig. 6 : Schematic diagram of Wilson expansion cloud chamber.

If particles of dust are present in the air (as an ordinary cases) they will act as condensation nuclei. But in dust free air this phenomenon of condensation and formation of cloud will, therefore, be extremely difficult. C.T.R. Wilson discovered that the condensation cloud could be produced even in dust free air provided the air was ionized by an ionizing agent such as the X-rays, cathode rays or any nuclear radiations. If ionizing particle entered the chamber either immediately before or immediately after the expansion, the ions left in its path would act as condensation nuclei. This experiment shows that as super saturation increases, the negative ions first serve as centres of condensation, then as the volume increases both positive and negative ions serve as the nuclei of droplets. A close array of fine droplets, *i.e.*, a kind of linear cloud, called a *cloud track*, will thus be formed. By using suitable strong illumination from the side, the track appears as a white line on a dark background. This can be photographed by means of a camera so that a record can be obtained.

Let V_1 be the volume, P_1 the saturated vapour pressure and T_1 the temperature of the vapour. Its mass m_0 (in moles) is determined by the relation

$$P_1V_1 = m_0RT_1 \qquad ...(1)$$

If the vapour is expended adiabatically from V_1 to V_2. The new temperature T_2 is given by

$$T_1 V_1 \gamma^{-1} = T_2 V_2 \gamma^{-1} \qquad ...(2)$$

If no vapour condenses, then the new pressure is given by

$$P_2 V_2 = m_0 R T_2 \qquad ...(3)$$

The vapour is in supersaturated condition. If ions are present in the space, the droplets will condense on them till the pressure falls to P_2. If m is the mass of the vapour remaining uncondensed, then

$$P_2 V_2 = m R T_2 \qquad ...(4)$$

The degree of supersaturation, is the main determinant governing the growth of droplets about ions and may be defined as the ratio of the actual density of vapour to the saturation density, Thus, we have

$$S = \frac{m_0}{m} = \frac{P_2}{P_1} = \frac{P_1}{P_2}\left(\frac{V_1}{V_2}\right)^{\gamma} = \frac{P_1}{P_2} \times \frac{1}{(\text{Expansion ratio})^{\gamma}} \qquad ...(5)$$

The pressure of vapour in equilibrium with small drops of liquid is not the same as that for a plane liquid surface. For a drop of radius r and charge ze the saturated vapour pressure P_r is given by

$$\log \frac{P_r}{P_\infty} = \frac{M}{\rho T R}\left[\frac{2\theta}{r} - \frac{z^2 e^2 (k-1)}{32 \in_0 \pi^2 k^{\gamma 4}}\right] \qquad ...(6)$$

The formula for ze = 0 was derived by Lord Kelvin and the electrostatic term was later introduced by J. J. Thomson. In this relation P_∞ a and M are respectively the normal saturated vapour pressure for plane surface and the molecular weight of the vapour and ρ, ∞ and R are respectively the density, surface tension and dielectric constant of the liquid.

The Wilson cloud chamber makes it possible to study the behaviour of individual atoms, to photograph the actual path of ionizing radiations and to analyse at a spare time the complicated interactions which may take place between charged particles and individual atoms. Since the year 1911 the apparatus has been improved in many ways, although the fundamental principle remains unchanged. P.M.S. Blackett and G.P.S. Occhialini invented the Geiger counter controlled expansion cloud chamber for the study of cosmic rays. Such chambers have been used extensively in T.I.F.R. Bombay and Bose Institute Calcutta, in India, for cosmic ray work. One such design is shown in Fig. 7.

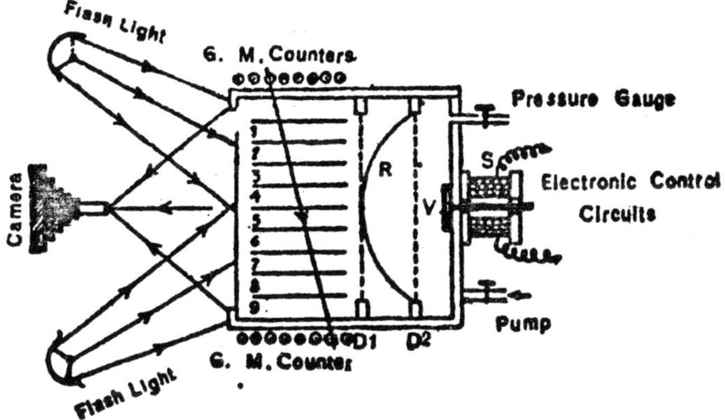

Fig. 7 : Schematic diagram of automatic.

In this arrangement two sets of G.M. counter trays are placed one at the top and other below the bottom of the active volume of the chamber. These trays are connected in coincidence so that any event due to the passage of a particle is triggered by these coincidence counters. The chamber is kept at a higher pressure with the help of compression pump which presses the membrane of rubber R. R is kept under tension and rests on the perforated metal disc D_1. R magnetic valve V, electrically controlled with the help of a solenoid S, is kept in position by energising the solenoid. When the counters (in coincidence) register an event, the electric pulse produces an impulse of current. The impulse is given to the solenoid in opposite direction. The valve V is now released and the gas escapes out bringing the membrane of rubber R to the position of the perforated disc D_2. After expansion, light is automatically turned to illuminate the tracks, the chamber volume is photographed. The ions are swept from the chamber by application of a clearing field and both chamber and camera are reset to await the next event.

In Wilson Cloud Chamber water is used to saturate the air but it becomes more common to employ ethyl or propyl alcohol or a mixture of alcohol and water. The use of alcohol gives better condensation on positive ions and the extent of expansion necessary for droplet formation is diminished from 1.25 to about 1.10 at ordinary pressure. Although, air is the usual gas, cloud chambers containing argon are sometimes employed. The higher pressures are desirable for the study of high energy particles. In cosmic-ray work dense metal plates have often been placed across the cloud chamber to ensure interactions of the high energy particle.

The heavy, slow particles like α-particles produce broad, densely packed, straight line tracks. Near the end of their path the particles may suffer sharp deflection as the result of impacts with the nuclei of oxygen or nitrogen present in the air. Slow electrons produce narrow, beaded, tortuous tracks, while fast particles both light and heavy produce narrow, beaded, straight tracks. By counting the drops in the cloud track the specific ionization can be determined and the nature of the particle identified. The sign of the electric charge and the momentum p of the particle can be determined if the chamber is placed in a strong magnetic field. If a stream of a particles, each carrying a charge q, moves initially in a straight line with velocity v and a magnetic field B is applied in a direction at right angles to the direction of motion, the panicles will be forced by the field to follow a circular path of radius R. The magnetic force Bqv is exactly balanced by the centrifugal force mv_2/R. Thus

$$Bqv = mv^2/R \text{ or } p(= mv) = BRq \qquad ...(7)$$

The kinetic energy E of the particle can be easily calculated, if the rest mass energy m_0c_2 of the particle is known, by the relation

$$E = \sqrt{[p^2c^2 + (m_0c^2)^2]} - m_0c^2 \qquad ...(8)$$

The *disadvantage of the cloud chamber* lies in the fact that it needs a definite time to recover after an expansion and hence it is not possible to have a continuous record of events taking place in the chamber. A continuously sensitive, known as *diffusion cloud chamber* has been developed by A. Langsdorf in the United States in 1939. It consists of a vessel containing air or other gas, kept warm at the top and cold at the bottom. The liquid, usually methyl or ethyl alcohol, vaporises in the warm region, where the vapour pressure is high.

The vapour diffuses downwards continuously through a region in which a vertical temperature gradient is maintained by cooling the bottom of the chamber. As the vapour diffuses into the colder regions of the gas, the saturation vapour pressure decreases rapidly and there is a volume near the cold base of the chamber where the supersaturated condition is maintained and thus eliminating the insensitive cycling time of the conventional cloud chamber. If some ions are present in this region, vapour forms droplets around them. In order to increase the frequency of occurrence of nuclear events, the pressure of the gas in the chamber is increased. *The main drawback of the diffusion chamber is that the sensitive region is no more than 7.5 cm deep.*

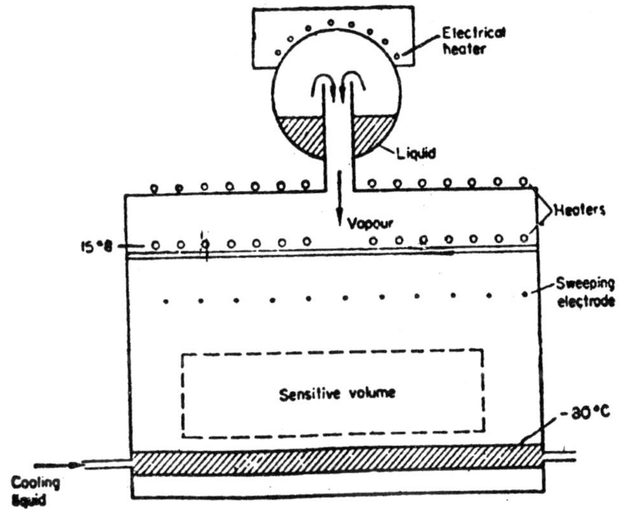

Fig. 8 : Schematic diagram of diffusion cloud chamber.

The diffusion cloud chamber is a continuously sensitive chamber, has many uses in the study of high energy particles obtained from accelerators. Its interesting application is the determination of the half-life of the free neutron. The discovery of the positron by electron positron pair production method, K° meson by observing the decay in flight into pions, hyperon Λ° by the observation of the decay in flight and that of muons in secondary cosmic radiation were made with the cloud chamber. The first reasonably accurate determination of the rest mass of the muon was made by W, B. Fretter with the help of cloud chamber.

BUBBLE CHAMBER

To detect very energetic particles the Cloud Chamber is not very suitable since the interaction of energetic particles cannot be completely observed in the chamber. To overcome this disadvantage it is necessary to use a chamber filled with a substance having a much greater stopping power. This has led to the development of the Bubble Chamber. In 1952, D.A. Glaser, at the University of Michigan, designed and made the first bubble chamber and this has now become one of the largest and at the same time one of the costiliest type of particle detector in use.

We know that normally the liquid boils with the evolution of bubbles of vapour at the boiling point. If the liquid is heated under a high pressure to a temperature well above its normal boiling point, a sudden release of pressure will leave the liquid in a super heated state. If an ionizing particle traverses the liquid with in a few milliseconds after the pressure is released, the ions left in the track of a particle act as condensation centres for the formation of vapour bubbles. The vapour bubbles grow at a rapid rate and attain a visible size in a time of the order of 10 to 100μ sec. Upon this principle chamber operates. The bubbles are like the droplets in a cloud chamber, visible under strong illumination. If nuclear reactions take place in the liquid of the bubble chamber, the sets of tracks diverge from the collision centre as in the cloud chamber. The tracks are photographed against a dark background.

The time sequence of events for a typical chamber is shown in Fig. 9. At the instant A, the high pressure P_1 (400 units) applied to the bubble chamber liquid is suddenly reduced to the lower pressure P_1 (100 units). The pressure starts to rise due to the bubble formation. The liquid is recompressed at C and the chamber returns to its initial position at A'. The time interval between A and A' gives the cycling rate of the chamber. The expansion is timed so that the particle enters the chamber at X and the bubbles are illuminated and photographed at the moment Y.

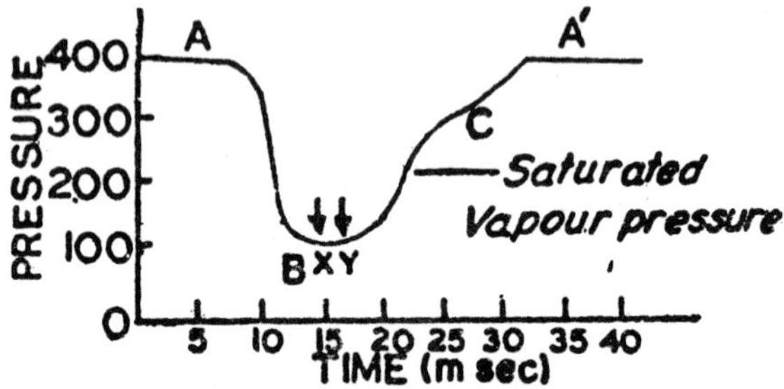

Fig. 9 : Time sequence of events in bubble chamber.

The liquids suitable as the working substance of a chamber are transparent and having a low surface tension and a high vapour pressure. The medium in first bubble chamber was ordinary ether, *i.e.,* diethyl ether, but liquid hydrogen, deuterium, helium, propane, xenon and other liquids have been utilized successfully for different purposes. In some experiments it is desired *to study the interaction of particles with protons*, the liquid in the chamber is then hydrogen. Propane (C_3H_8) is commonly used in the bubble chamber to avoid the need for low temperature (–246°C) operation with liquid hydrogen. It has almost 1.3 times as many hydrogen nuclei (*i.e.*, protons) per unit volume, but the carbon nuclei present are sometimes a drawback. For *observations on neutron interactions* liquid deuterium with a neutron fairly loosely bound to a proton if employed.

Liquid helium chambers operating at 4°K and 1 atm. have been used for studying high energy interactions with the nucleons in the He^4 nucleus. A liquid containing heavy atoms is favoured where stopping power is the main objective, *e.g.*, *in studying the interactions of neutral particles and gamma rays.* Examples are the freon type compound, trifluoro-bromomethane (CF_3Br) and liquid xenon. The former can be used just above room temperature (30°C, 18 atm.) however there are three distinct complex nuclei present in this liquid. Liquid xenon operating at 14°C is ideal for studying interactions with high energy γ-quanta and neutral pions. For satisfactory bubble tracks a small amount of ethylene is to be added. The operating conditions for bubble chambers with different fluids are given in the following Table 1.

Table 1 : Operating conditions for Bubble Chamber fluids

Fluid	Temp. °C	Pressure (atmospheres)	Mean free path for 100 MeV γ-rays (m)	Density (Kg m^{-3})
Hydrogen	–246	5	27	60
Deuterium	–241	7	20	130
Helium	–269	1	18	130
Propane	58	21	2.2	430
Pentane	157	23	—	500
Xenon	–20	26	0.07	2300

Liquid hydrogen is an ideal substance for studying high energy interactions with protons, as it is a pure proton target having very small Coulomb scattering. L W. Alvarez at the Lawrence Radiation Laboratory, California University, thought that if the chamber was sufficiently large and the pressure of the compressed liquid was decreased rapidly, good tracks could be obtained in the interior in spite of the fact that bubbles were formed at the walls. This conjecture proved to be correct and a chamber six feet long and of cross-section 1½ feet, containing 520 litres of liquid hydrogen, was first operated in 1959 successfully. A schematic diagram of a hydrogen bubble chamber, operating at a temperature of 27°K and developed by Alvarez and his co-workers is illustrated in Fig. 2.27. The vessel was made of stainless steel with glass ports at the top for the viewing cameras. A box of thick glass walls was filled with liquid hydrogen and connected to the expansion pressure system. To maintain the chamber at constant temperature it was surrounded by liquid nitrogen and liquid hydrogen shields. From the side window W the high energy particles were allowed to enter the chamber. Liquid hydrogen was kept under pressure first but when a particle has passed through it the pressure was released so that it was super heated at a lower pressure. A sudden release of pressure from the expansion valve we followed by light flash and camera took the steroscopic view of the chamber.

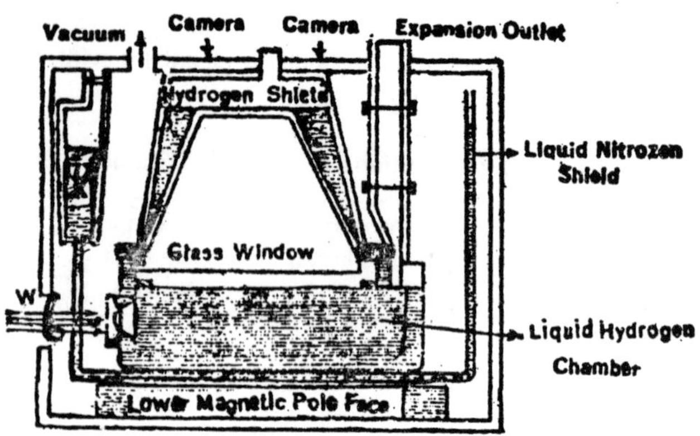

Fig. 10 : Schematic diagram of Alvarez bubble chamber.

This was the first of the so-called dirty bubble chambers. Subsequently, 80 inch hydrogen bubble chamber was built at the Brookhaven National Laboratory and similar one is in use at CERN (*Conseil Europeen Pourala Recherche Nucleaire*). Even larger bubble chambers, upto 14 feet in length, are being designed. Similar to the original cloud chamber bubble chamber is not continuously sensitive.

(i) A dirty bubble chamber is sensitive for only a few milliseconds after the pressure reaches its minimum.

(ii) The relatively slow expansion of the liquid does not permit the process to be initiated by an ionizing particle continuously.

(iii) The bubbles formed spontaneously at discontinuities and they interfere with the photography of the particle tracks.

(iv) Once the liquid has been expanded, it takes some time to recover sufficiently for another set of tracks to be recorded.

These drawbacks are overcome by using bubble chambers in connection with high energy particle accelerators, which function in a pulsed manner. The chamber can be made sensitive each time a pulse of particles emerges from the accelerator, and the large density and large volume of the chamber permit appreciable numbers of interesting collision processes even when the incident particles have energies of thousands of MeV. The timing is such that the particle beam from the accelerator enters the bubble chamber just prior to the pressure minimum. A few milliseconds later, powerful lights flash on and the tracks are recorded on a photographic film which moves forward automatically after each exposure. For accelerators, which emit particle pulses at more frequent intervals, smaller bubble chambers, with shorter recovery times, are employed.

A magnetic field is usually provided at the chamber in order to distinguish the sign of the charged particles and to measure their momenta from the radius of curvature of the bubble track. For particles of high momentum, the curvature is small and hence is difficult to measure unless a strong magnetic field is used. Thus, to analyse the collisions it is necessary to maintain the whole chamber in a strong uniform magnetic field (~20,000 gauss). To achieve this a magnet weighing several hundred tons is required with drastic cooling arrangement. Consequently, there has been much interest in the possibility of constructing superconducting magnets for use of bubble chambers.

The problem in the construction of such superconducting magnets of appreciable size has been the difficulty of producing cable with satisfactory physical and mechanical properties at a reasonable cost. In recent years, certain metallic compounds (alloys) of niobium, *e.g.*, with tin, titanium, or zirconium, have been developed for fabrication into super conducting cables. The first successful operation of a bubble chamber, 10 inches in diameter, with a super conducting magnet was achieved at the Argonne National Laboratory in 1966. The analysis is laborious and is greatly assisted by computing machines. A simplified account of the methods used is given below. The first step in the analysis, involving the selection of events of significance to the experimenter, is known as *scanning*. The magnified image of the film is inspected and the locations of interesting events are recorded. The next stage in track analysis, known as *tracking*, involves various degrees of automation. The tracking operation is greatly expedited by utilizing a flying spot digitizer. A small spot of light is moved rapidly in this system either mechanically or electronically, across and down the image of the bubble chamber event.

Each time the moving spot of light crosses a track, co-ordinates of the point are expressed in digital form. The complete co-ordinates of the event of interest are next treated by a computer to yield what is called the *geometry* of the event. For the next stage, known as *kinematic analysis*, it is required to know the masses of the particles that produce the observed tracks. The kinematic analysis, made by the computer at each vertex of an event is based on the conservation laws of momentum and of kinetic energy.

Nevertheless, cloud and bubble chambers both are important detectors of radiation from a historical point of view since they have played such an important part in the study of fundamental particles and their interaction. They are used mainly in Nuclear Physics research. It has been estimated that a comparatively small bubble chamber is equivalent in stopping power to a cloud chamber some 140 ft. long.

SPARK CHAMBER

The spark chamber utilizes an incipient electrical discharge in a gas, like that used in a Geiger counter, to give track geometry information, like that provided by a bubble chamber. This has found increasing use in the field of high energy physics. The basic principle of operation is as follows:

A high voltage is maintained between two plates spaced in a vessel containing a gas but the electric field is not quite enough to permit the passage of a spark. Now if an ionizing particle enters the gas space, a spark will pass and it will tend to follow the path of the ion pairs produced by the ionizing particle. The passage of a spark through a gas that had been ionized by nuclear radiation was used by H. Greinacher, in 1935, for the detection of α-particles. F. Bella and C. Franzinetti in 1953, in Italy, confirmed the localization of the spark discharge and the first photographs were given in 1955 by P.G. Henning in Germany. The modern spark chamber stems from the work described by the British physicists T. E. Cranshaw and J. F. de Beer in 1957. In 1959, S. Fukui and S. Miyamoto in Japan found that with neon several particle tracks could be observed simultaneously in the spark chamber and with air only one particle track could be formed at a time. Spark chambers were used by Sree Kantan and coworkers at T. I. F. R. Bombay in 1960, to trigger high energy cosmic-ray events.

In its simplest form spark chamber consists of a series of large thin parallel metal plates (aluminium), several square feet in area, spaced about 1 cm. apart in a chamber filled with a neon gas at atmospheric pressure. All the plates are isolated from each other. The first, third, fifth, etc., plates are grounded and the second, fourth, sixth, etc., plates are connected to a high voltage direct current pulse generator, which gives them a high potential in short bursts of the order of a microsecond each. This potential is just enough to cause sparks to occur between the plates in such regions as are ionized by a particle entering the chamber. The chamber is sensitive for about half microsecond preceding the application of the voltage. Any charged particle passing through the chamber during this period produces along its path a visible and audible spark discharge, which can be located by photographing or by other methods. The photograph of all the individual spark reveals the sections of the trajectory of the particle between the plates. Ordinarily two cameras at right angles to one another are used to photograph.

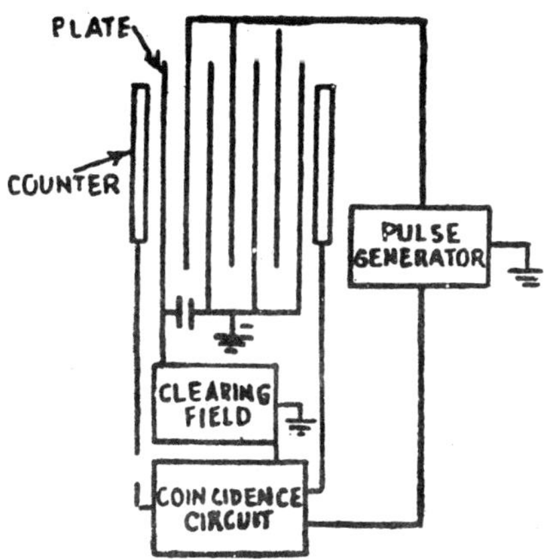

Fig. 11 : Schematic representation of spark chamber circuit.

In order to prepare the chamber for recording the path of another particle, any ion-pairs present between the plates must be removed, by applying a clearing field of about 200 volts per cm. continuously in the direction opposite to that of the high voltage spark field. Due to this field a spark can form only if the high voltage is attained within about a microsecond after the entry of an ionizing particle. In practice this delay time is about 0.2 microsecond. The recovery time depends on the gas, the design of the chamber and the energy of the spark. It is usually between one to 20 milliseconds, which is still very short in comparison with the recovery times of cloud and bubble chambers.

The chamber can be operated in a magnetic field and the momentum and charge sign of the ionizing particle can then be determined from the curvature of the track. The number, thickness, spacing and material of the plates in a spark chamber are varied to suit the experimental requirements. The plates are commonly made of very thin aluminium to minimize energy absorption from the particle and are closely spaced to define the track more accurately. The sparks tend to be perpendicular to the plates in their passage between the adjacent plates. The photograph of spark chamber tracks and their interpretation is shown in Fig. 2.29. Sometimes it is desired to study the interactions of specific particles with matter. The plates are then thicker usually and are made of either carbon or a metal, such as iron, lead, tantalum or brass. The interactions take place almost within the solid plates as the density of nuclei in the gas between the plates is small.

One of the main advantages of the spark chamber over the other chambers is that the triggering and removal of ions by the clearing field are comparatively simple. With the help of this chamber it is possible to study an individual event produced by one particle among the many passing through the chamber at a rate that may be as high as a million per second. The spark chamber event of interest is selected by an arrangement of counters, called a *Logic circuit.*

Analysis of spark chamber photographs can be carried out by ordinary calculations or by computer techniques. The preliminary scanning stage may not be required because spark chamber events of interest have been selected by the logic system. The tracking, geometry and kinematic analysis arc performed in the manner as with bubble chamber pictures.

There are *two main drawbacks* to a spark chamber as compared with a bubble chamber. First, in the bubble chamber the liquid may be chosen so as to restrict the interactions of the incoming particle to those of a special type. Second, the co-ordinates of tracks and of the points of interaction in a bubble chamber can be defined with much greater accuracy than is possible in a spark chamber. As a general rule, bubble chambers are preferred where good space resolution is desirable. On the other hand, spark chambers permit the selection of events of interest with a high repetition rate, *i.e., they have better time resolution.*

Since 1963, various "filmless" spark chambers have been developed in which photography of the spark discharge is avoided. In such devices, the co-ordinates of the spark tracks are obtained directly in a form suitable for a computer to determine the geometry of the event. The grounded plate of every pair is replaced by an array of closely spaced parallel wires. For the high voltage plates a similar arrangement of wires could be used but solid plates are generally employed since they are more convenient and easier to fabricate. The wires arc made from a magnetostrictive material (50% Fe and 50% Co alloy) in *one type of wire spark chamber*. A passage of a spark from a particular wire produces a local magnetic field and a deformation occurs at the affected point. The deformation travels then along the wire to suitable detectors at each end

The ratio of the times between the firing of the spark and arrival of the deformation at the ends gives one co-ordinate, the other co-ordinate of the spark track is obtained by identifying the particular wire at which signals were received. The most promising wire spark chamber, referred to as a *spark-chamber hodoscope*, differs from the acoustic wire chamber in respect that the wires constituting the alternate grounded plates are alright angles to each other. At one end each wire is threaded through a ferrite core, such as is used in computer memory devices. A ferrite material has two stable states of magnetization and by a magnetic field (from an electric current) either one can be converted to the other. The magnetism of ferrite cores associated with the wires from which the spark passes is changed from one state to the other by the spark current or we can say that ferrite cores are nipped.

By interrogating the cores in a manner similar to that employed in a computer the locations of the affected wires are determined and flipped cores are reset at the same time. Since successive planes of wires are at right angles, the information from the cores can give the co-ordinates of the spark track in three dimensions in digital form. It has a advantage of a very short recovery time. The amount of energy required to flip the core is very small and a feeble spark, which would not give off enough light to be photographed, is adequate for this purpose.

Sometimes it is desirable to make observations of interactions at very-very short intervals with high energy particles. *The scintillation counter hodoscope* can be used for this purpose. Although the accuracy of determining the co-ordinates of particle tracks decreases. It consists of an array of long thin rectangular strips (25 × 0.5 × 0.5 cm) of a plastic scintillator material. These strips are set side by side to produce a sensitive planer area. A photoelectric tube is associated with each strip and the passage of a charged particle through a given strip is indicated by the output signal from the corresponding photo-tube. This provides one co-ordinate of the particle track.

For the determination of co-ordinates in the other two directions, it is necessary to build a hodoscope from several layers of scintillators arranged with the strips in successive layers at right angles. The output signals can be fed directly to an on-line computer for analysis. This type of hodoscope can record events at a much faster rate than wire spark chamber hodoscope can. It is because of the very short recovery time of a scintillation counter. The accuracy with which the track co-ordinates can be determined in a scintillation counter hodoscope is limited as it is dependent on the width of the scintillator strip.

NUCLEAR EMULSIONS

The discovery of radioactivity by Becquerel (1896) can be explained as:

As charged particles pass through a photographic emulsion of a photographic plate they interact with the tiny silver halide crystals producing a line of latent images along their path. In 1911, M Reinganum in Germany recognized that individual α-particles traversing a photographic emulsion produced separate tracks in a photographic emulsion as a series or closely spaced black specks of silver. Because of the relatively high density and stopping power of the photographic emulsion, the tracks produced by charged particles were very short compared with those obtained in a cloud chamber but do not differ greatly from bubble tracks.

Miss Blau and Wambacher, in 1937, discovered 'Star' due to the interaction of high energy cosmic rays with the atomic nucleus in the emulsion.

The optical photographic emulsions is not suitable for quantitative work with nuclear radiations. The sensitivity is low and the tracks due to charged particles have non-clear range because the developed crystal grains are large and widely spaced. The composition of the emulsion was changed so as to make it more suitable for study of various ionizing particles, such as alpha particles, protons, mesons and even electrons.

The nuclear emulsions differ from the optical emulsions in that they have considerably higher silver halide content and smaller grain size In nuclear emulsion the thickness is greater than that of optical emulsion. The smaller the grain size, the more sensitive the emulsion to ionizing particles. Thus different commercially available emulsions, differing chiefly in grain size, can be used to discriminate between different particles. The least sensitive emulsions will show only fission fragment tracks, whereas the most sensitive ones reveal the tracks of singly charged minimum-ionizing particles. A series of nuclear emulsions has been developed for detecting particles of different specific ionization. D-l emulsion is the suitable type for detecting fission fragments and alpha particles. The C-2 Ilford emulsion is a useful general purpose emulsion, suitable for recoiling mesons, protons and heavy nuclei. The Ilford G-5 is used to record high energy electrons. The tracks produced in an emulsion are very short.

Table 2 : Composition in kg/m^3 of Dry Nuclear Emulsions

↓ Element→ Type	H	C	N	O	S	Br	Ag	I
Ilford	49	300	73	200	11	1465	2025	57
Kodak Ltd.,	38	270	80	160	–	1440	1970	36
Eastman Kodak	43	340	110	170	–	1220	1700	54

They can be magnified and photographed. High quality microscopes are required for sinning the developed plates. The nuclear emulsion with their high silver halide concentration shrink during the fixation process unfortunately. A knowledge of this shrinkage factor is necessary when we want to determine the angle of dip of a track in the emulsion.

Fast neutrons can be detected by the knock on protons in unloaded emulsions. Slow neutrons can be detected by using emulsions impregnated with lithium, boron or uranium.

Table 3 : Nuclear and Optical Emulsions

Property	Optical Emulsion	Nuclear Emulsion
AgBr : Gelatin (mass)	4 : 53	80 : 20
AgBr : Gelatin (volume)	15 : 85	45 : 55
Grain size (micron)	1—3.5	0.1—0.6
Thickness (")	2—3	25—2000
Sensitivity to light	Very high	Poor
Response to α-particles	Dense blackening	Individual tracks
" " β "	Moderate "	Faint fog
" " γ "	Faint "	Almost none

Table 4 : Characteristics of Nuclear Emulsions

Ilford Emulsion type	D1	E-1(K1)	C_2(K2)	B-2	G-5
Mean grain diameter (mm)	0.12	0.14	0.10	0.21	0.18
Highest detectable velocity β	0.2	0.31	0.46	all	
Highest detectable energy of protons in MeV		20	50	120	all

Gamma rays of energy above about 3 MeV can be detected through the track of the protons, released in photo-disintegration reaction, in emulsions loaded with deuterium compounds. Under favourable conditions it is possible to determine the charge, mass and incident K.E. of a charged particle brought to rest in a nuclear emulsion. The range of the particle, the grain density and its variation along the track and the small scattering angle due to collisions of the particle with nuclei in the emulsion are all relevant to these determinations. We shall confine ourselves here to discussion of the main relationships which have been established by experiments.

The number of grains deposited per micron dn/dx of track is closely related to the energy loss per unit of track dT/dx keV/micron. These two quantities are related by the empirical relation

$$\frac{dn}{dx} = a\left\{1 - e^{bz(dT/dx^{1/2} - c^{1/2}}\right\}, \quad ...(1)$$

where a, b and c are empirical constants and z is the integer characterizing the charge of the ionizing particle. If m (the mass in units of proton mass) and z of a particle are known, its kinetic energy T (MeV) may be obtained from its total range R (microns) by an equation

$$T = Kz^{2n} m^{1-n} R^n, \qquad ...(2)$$

where K and n are empirical constants.

The total number of silver grains deposited and the total particle range are related by an equation

$$N = K' z^{2n'} m^{1-n'} R^{n'}, \qquad ...(3)$$

where K' and n' are another constants.

Fig. 12 gives the range of protons, deuterons, α-particles, pions and muons in Eastman Kodak NTA emulsions as a function of particle energy.

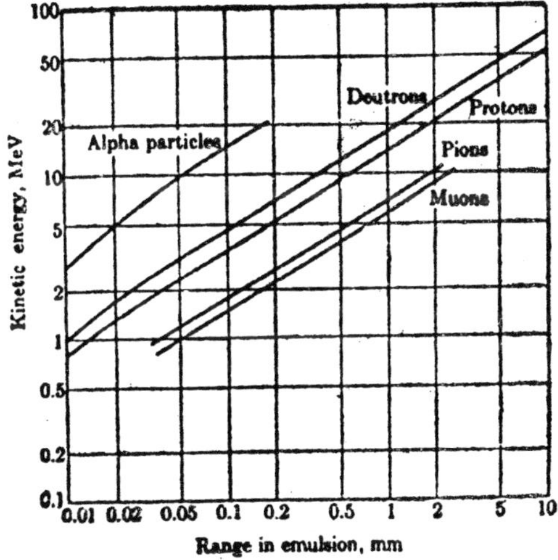

Fig. 12 : Range V/s energy for various particles in nuclear emulsions.

As compared with track chambers, like bubble and cloud chambers, photographic emulsions have following *drawback*. It is very difficult to determine the sign of the charge and the momentum of a particle from observations of the curvature of the path in a strong magnetic field. The actual path length in the emulsion is small compared to that in a cloud chamber and the multiple Coulomb scattering in the emulsion stimulates the curvature. By the use of very strong magnetic fields, the deflection of long (but not short) tracks in emulsions has been observed. This gives the sign of the charge, but the large amount of scattering makes the deflected tracks useless for the momentum calculations.

Some of the *advantages* of the nuclear emulsion detector are:

1. It produces a visible track following the complete trajectory of a charged particle. Particles of different ionizing powers produce tracks which appear quite different.
2. The emulsion is continuously sensitive and is consequently always available to record an event.
3. The energy of product particles can be determined from their range in the emulsion and the differential scattering cross section from counts of the number of tracks formed in each emulsion.
4. The emulsion is relatively light and cheap. This makes it better for high altitude cosmic ray experiments.
5. Due to high stopping power of emulsions, many short lived particles can be brought to rest in the emulsions, before they decay.
6. The measurement of the ranges of muons emitted from stopped pions, before they decay into electrons, enables the mean lifetime of the muon to be found.
7. They were widely employed in cosmic ray studies and led to the discovery of the π- and K-mesons. They are also employed in observations on hyperfragments.

The properties and applications of various nuclear detectors are listed

DIFFERENTIATING AND INTEGRATING CIRCUITS

Differentiating and integrating circuits are used for pulse forming and sharpening or for sorting out pulses. The need for a sharp pulse of voltage that repeats periodically at a high repetition rate is common in practical applications, *e.g.*, in electronic circuits including television, transmitters and receivers. Sharp pulses are used to initiate action of trigger circuits.

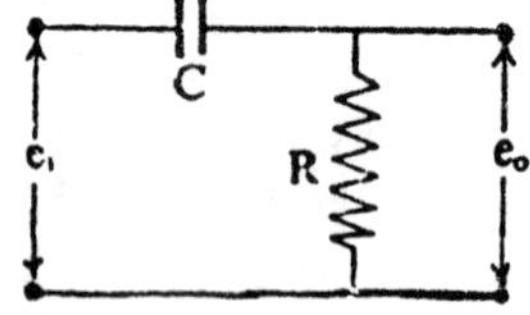

Fig. 13 : Differentiating Circuit.

Consider the CR circuit shown in Fig. 2.30. For an initial potential across C zero, the input potential

$$e_i = \frac{1}{C}\int i\,dt + iR$$

or $$\frac{e_i}{R} = \frac{1}{RC}\int i\,dt + i \qquad ...(1)$$

If RC is small with respect to the time of integration, then the second term is small enough and can be neglected w.r. to the first. We thus have current

$$i = C\,de_i/dt.$$

Thus, the output potential $e_0 = CR\,de_i/dt$. ...(2)

Hence for a small value of RC, the output voltage is proportional to the derivative of the input signal. In this way we obtain the rate at which the voltage is changing with respect to time across the output. The differentiation of square wave by RC circuit is shown in Fig. 14. When e_i goes from zero to its positive value V_0, its rate of change is large and positive. It appears at once across the resistor, since the voltage across the capacitor cannot change instantaneously. As capacitor then charges, the potential across R decreases exponentially, $V = V_0\, e^{-t/CR}$ and becomes zero. When input potential decreases from V_0 to zero, its rate of change is large and negative but equal in amount to the former positive rate of change so the negative value of e_0 at that instant represents this rate of change

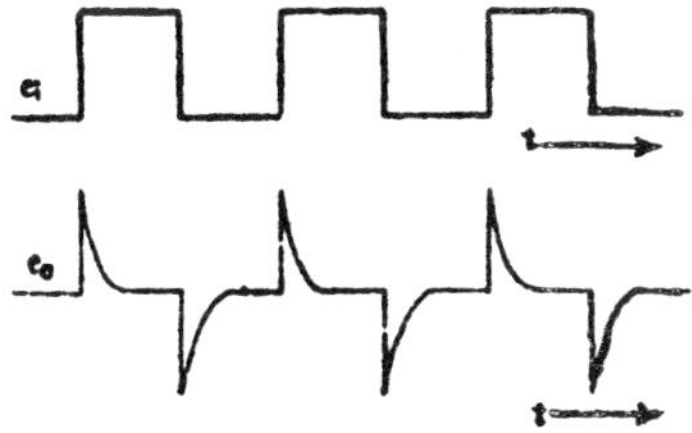

Fig. 14 : Differentiating Input (above) and out-put (lower).

A sine-wave input becomes a cosine-wave at the output and a cosine-wave input becomes an inverted sine wave at the output. The rate of change with respect to time is zero at the maximum and minimum points of the sine and cosine waves.

The integration of a voltage wave may be done by the circuit shown in Fig. 15. In order to calculate the voltage change at the output over any time interval, the input volt-time area is simply divided by the time constant. The time constant *RC* has to be very large in comparison with the period of the input signal Mathematically it can be shown that for this circuit, output potential is same as obtained after integrating input potential.

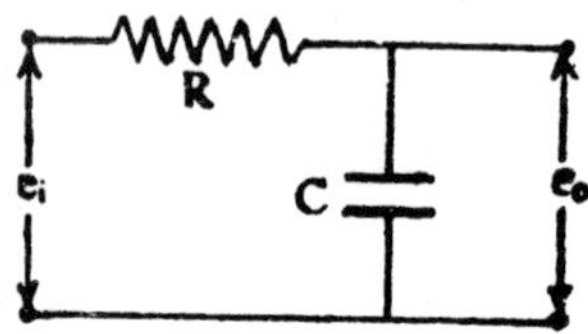

Fig. 15 : Integrating circuit.

From eqn. (48), for large value of RC, we have

$$i = e_j/R$$

∴ Output voltage

$$e_0 = \frac{1}{C}\int idt = \frac{1}{RC}\int ei\,dt$$

The output of an integra-ting circuit is illustrated in Fig. 16 along with the input. From above results we sec that the output of a differen-tiating circuit is independent of the d.c. level of the input, but the output of an integrating circuit must attain a d.c. level equal to that of the input. Thus, a true integration is attained for input signal having an average value of zero.

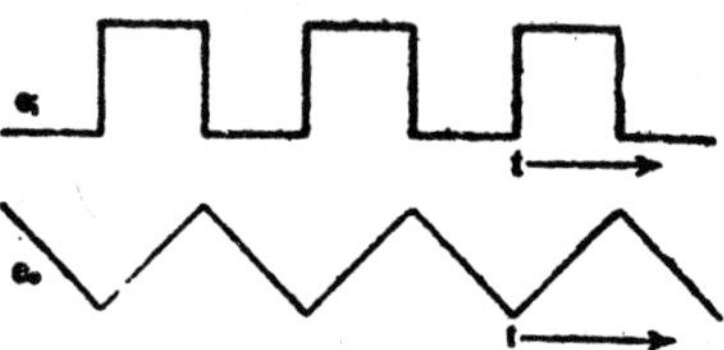

Fig. 16 : Integrating input (above) and output (lower).

PULSE AMPLIFIER

In nuclear radiation detectors, *e.g.*, in an ionisation chamber, the pulse height is very small and is of the order of micro volts. Since output voltages of 10 to 100 volts are often required for electronic pulse height discrimination, the gain requirement might be 10^5 to 10^6 for the amplifier. If we arc only interested in counting the ionization pulses then it is not necessary for the height of the output of the amplifier to be accurately proportional to the height of the counter pulse. In order to determine the energy of the ionizing particle, we require linear amplifier. To maintain proportionality between input pulse height and output, the stability of the amplifier is important. Linearity is obtained by proper choice of operating points for the amplification elements and by the use of negative feedback.

The pulse amplifier usually consists of three units in cascade. The first unit is the pre-amplifier which is attached to the nuclear detector by a very short cable This reduces distributed capacitance, which decreases rise time. Its gain is usually about 1 to 100. The output stage of the pre-amplifier is a cathode follower with its characteristic low output impedance. It is, therefore, practical to connect the output terminals of the pre-amplifier by a coaxial screened cable several feet long to the input terminals of the main amplifier, comprising cathode follower and negative feedback.

The coupling network CR between pentode valves may be used as the short time constant differentiating network. The third stage of the unit is a triode connected pentode or a triode with a cathode load resistance shunted by a small trimming capacitor. The output signal

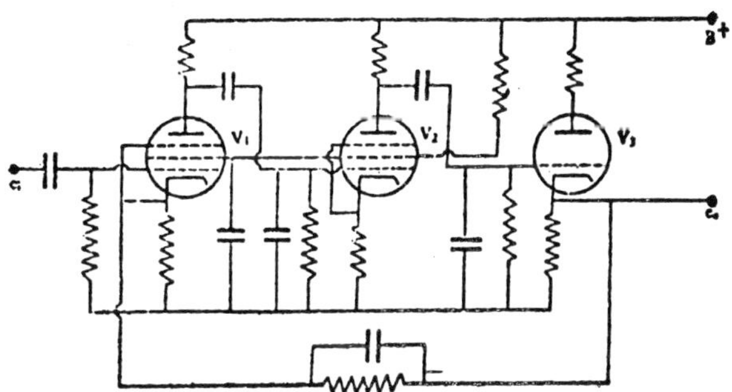

Fig. 17 : Linear Pulse Amplifier.

is feedback into the input stage through the direct connection between cathode resistances of pentode V_1 and triode V_3. The trimmer capacitor is adjusted so that no transient overshoot occurs with a fast rising input pulse to the amplifier.

THE TRIGGER (FLIP-FLOP) CIRCUIT

A closed loop regenerative system having two or more clearly defined stable or quasi-stable slates, each of which is maintained without recourse to external forcing, is known as *multivibrator*. Multivibrators are utilized in one of three basic configurations: Free running, monostable, or bistable. *Free running or a stable multivibrator* oscillates continuously to provide outputs used as clock pulses, or as sweep waveforms when a continuous time base is required. The *monostable or one-shot multivibrator* oscillates through one cycle to produce variable-width pulses for use as gate or timing pulses.

The *bistable or flip-flop multivibrator* does not oscillate but is driven through each switchover action and is used in counting or toggle circuits when the stability characteristic in either of two states is important. It is a two stage device with one stable slate described as having one stage on or conducting and the other stage off or non-conducting. The second stable state is the reverse with the first state conducting stage off and the first stale non-conducting stage on.

The-most famous circuit of this class is the *Eccles-Jordan Trigger circuit.* We assume that V_1 is conducting, it then follows that V_2 is not conducting because its grid bias is kept below cut off. For this purpose first assume that the bias battery E_{bb} is not present but that the bottom ends of the grid leak resistors are connected to the grounded cathodes. The voltage-divider action of R_1 and R_2 would cause the lowered plate potential at A to put a small +ve bias on the grid of V_2 As E_{cc} is larger than this +ve bias voltage, when it is put back in place it will make the grid negative with respect to the cathode.

If a positive trigger pulse is applied to the input, it will not alter the current in V_1 which is already conducting, but it will derive V_2 into conduction. This reduces the potential at B and at the grid of V_1, reducing the plate current of V_1 and raises the potential at A and the grid potential of valve V_2. This action further turns on valve V_2 and the process continues until V_1 is cut off and K_2 is conducting fully. The circuit will remain in this second stable condition until another +ve pulse conies along and reverses the action. Capacitors C_1 and C_2 are not needed to hold the grids of triodes negative between switching but are used to transfer instantaneously to opposite grids the changes in plate voltage of the triodes.

Negative pulses may also be used to trigger the circuit. They will not affect the cut off tube and will reduce the current in the conducting tube. The only requirement on the triggering pulses is that the signal must be strong enough and last long enough that the plate current in the on-coming tube exceeds that of the off going tube so that the regenerative action will continue the switching to completion.

VOLTAGE DISCRIMINATOR

It is often necessary to discriminate between pulses of different amplitudes. A voltage discriminator rejects all pulses below a predetermined height. It should have high stability and sensitivity to discriminate between input pulses differing in height by a fraction of one volt, should have short rise and recovery times, should generate an output pulse of standard height if the pre-determined level is exceeded. A circuit, which meets all these requirements, is *Schmitt trigger circuit.* It is a modified bistable multivibrator, *i.e., cathode coupled bistable multivibrator*, illustrated. This circuit is also useful as a squaring circuit.

In this circuit the direct coupling from the plate of V_1 to grid of V_2 is the same as for conventional bistable but reverse is eliminated. The common cathode resistor R_6 provides the other necessary coupling for regeneration between stages. When the input signal is below a preset voltage value, obtained by adjusting discriminator level potentiometer R_2, one tube conducts and the other is cut off.

The moment the voltage exceeds the preset value, there is a rapid transition of states. The circuit remains in the new states as long as the input remains above a voltage level that is slightly lower than the initial triggering voltage. As soon as the voltage drops below this second level, there is another rapid transition of states back to the original states. Input signals below the preset voltage level do not provide an output signal.

The triggering level can be set by a potentiometer R_2 which gives an average value of E_{g1} of tube V_1. With E_{g1} sufficiently low, V_1 is cut off and R_8 is adjusted so that V_2 is heavily conducting. The plate current of V_2 causes a potential drop E_K across the common cathode resistor R_6. The difference $E_{g1} - E_K$ is now the grid bias of V_1. Since the transition of states occurs the moment the cut off voltage E_{co} is exceeded, hence the triggering level can be given by $E_{co} - (E_{g1} - E_K)$. At this point V_1 starts conducting. Its plate voltage and grid voltage of V_2 decrease, causing a decrease in plate current of V_2. The resulting drop of cathode voltage increases the plate current of V_1 and the regeneration process continues until V_2 is *off* and V_1 is on. The output from the plate of V_2 jumps to the plate supply value because of the transition.

SCALARS

The simplest type of operating or indicating device in a counting system is an electromechanical register with a suitable driving circuit. It consists basically of an electromagnet energised by the amplifier output. The energised electromagnet causes an armature to move which rotates a numbered drum through a pawl and ratched system. Each voltage pulse could momentarily close the switch contacts and advance a register one count. The drum carries ten numerals and on completion of one revolution it causes the adjacent drum to move forward. This type of registers are used when the pulse repetition frequency is not too high of the order of 5 per sec.

For most nuclear counting work, the register is preceded by a scaling circuit. The simplest one is the Eccles-Jordan scale of two. It has been seen in the flip-flop that at point B, a +ve pulse is obtained as output when V_2, cuts off and a –ve pulse appears when V_2 turns on. The second flip-flop circuit, which responds only to negative pulses, therefore, responds only to one pulse for every two pulses sent to the flip-flop circuit. The circuit is given the name of a scale of two circuit. Thus the count-down feature has a factor of 2^n, where n is the number of flip-flops.

A small cold cathode neon indicating lamp is across the plate-load register of each of the odd numbered tubes V_1, V_3, V_5. Each will light up whenever its tube is conducting. When the tube is cut off, the potential difference across the neon lamp is well below the striking potential. The plates of V_2 and V_4 are capacitively coupled to the common plate lead of the tube that follows so that the signal will be transferred to the next flip-flop. A reset switch R is provided to put positive potential on the grids of V_2, V_4 and V_6, so that they will be on and odd numbered tubes will be off at the start of the count and ail indicator lamps are off.

The first input pulse (remembering that all input pulse are negative) reduces the potential at points A and B momentarily. Since V_1 is off there is no current between these points. The coupling capacitor sends this negative pulse to the grid of V_2, reducing its current. Thus the conduction is transferred to V_1, the light of lamp 1 goes on and potential of point

C rises abruptly. This positive pulse at C is transferred to the second flip-flop, where nothing happens because V_4 is conducting already and V_2 has too much negative bias from the common cathode resistance.

The second pulse switches conduction back to V_2 with a consequent drop in potential at C and triggering of the second flip- flop. V_3 now conducts and the light of lamp 2 goes on. The potential of point D rises hut causes no change in V_5 and V_6. The third input pulse turns off V_2, turns on V_1, and lamp 1, but the positive pulse sent forward from C does not do anything to V_3 or V_4. The fourth pulse at the input causes C to drop in potential and this negative pulse turns on V_5. Point D drops in potential and this turns on V_5 and the light of lamp 3 while V_6 goes off. The fifth pulse at the input causes C and D to rise in potential but no change occurs in V_3 or V_4. The sixth pulse at the input causes C to fall in potential and this turns off V_4 and rises potential at D. The state in the fast flip-flop, where V_5 on and V_6 off, does not change. The seventh pulse at the input causes C to rise in potential with no change beyond this point. V_4 is off. The eighth pulse at the input causes C to fall, turning V_4 on and causing D to drop in potential. This negative pulse turns off V_5 and turns on V_6. The drop in potential at E is the negative pulse, the output needs to actuate the, mechanical counter. The increase in the register reading from start to finish must be multiplied by 8 to get the total number of input pulses.

Thus, we see that lamp 1 is on for one pulse, lamp 2 is on for two pulses and lamps 3 is on for four input pulses. With 7 pulses all three lamps arc lit and their indicated total value is

$$(1 \times 1) + (1 \times 2) + (1 \times 4) = 7$$

Scale of 10 counting is possible when feed back is used in the circuits. This is more convenient because our system of numbers is based on the factor 10.

DEKATRON

An alternate method of counting is provided by the multi-electrode cold cathode scaling tube known as dekatron. It is designed so a glowing spot moves successively around its circumferences with successive input pulses. It is a neon filled (13 cm. Hg pressure) tube containing an anode in the form of a flat disc of diameter ~1 cm. Surrounding the anode are thirty identical electrodes. They are symmetrically spaced and are about 0.05 cm. from the anode. These are grouped into three categories based on the operation principle of the device. Ten of the electrodes act as cathodes, A, to A, connected to a common point and K_0 the output cathode. Adjacent to each cathode is a pair of identical electrodes known as transfer electrodes or guides. The electrode nearest the cathode in a given direction is labelled guide 1 or 1G and the next electrode in the same direction is labelled guide 2 or 2G. The ten 1G electrodes are connected to a common terminal as are the ten 2G electrodes. The guide 1G and 2G electrodes are biased 20 V above the potential of the cathodes so that normally the discharge rests on one of the cathodes. Switch is normally closed but is opened for resetting to zero.

When the switch is open, K_0 is the most negative electrode and the gas discharge occurs between K_0 and the anode. With switch closed and a large negative input pulse applied to 1G, the slow discharge transfers from K_0 to $1G_0$. The duration of pulse should be greater than 60 μ sec. to ensure transfer. A delayed negative pulse is applied to the 2G electrode so that $2G_0$ remains negative after $1G_0$ returns to its normal bias potential of +20 V and the glow discharge moves smoothly to $2G_0$. The subsequent removal of the negative pulse from $2G_0$ moves the discharge to the next closest and most negative electrode K_1 and thus lights up the number 1 on the face plate. A second pair of negative pulses moves the discharge to $1G_1$, $2G_1$ and then to K_2 where by the number 2 is illuminated. Thus we see that the net result of a single drive pulse is to transfer the discharge from one cathode to the adjacent cathode. Two guide electrodes are required to ensure that the discharge always transfers to the next digit cathode and does not jump back. The current for the discharge is supplied by the anode voltage source. The input pulses are used only to stimulate the movement of the discharge from one cathode to the next. Every tenth input pulse transfers the discharge from K_9 to K_0. A load resistor R_{K0} produces a positive output or carry pulse. The carry pulses are amplified to produce large negative drive pulses for the next Dekatron stage.

Direct readout of the count can be accomplished within each tube by placing numerals around the periphery, adjacent to the glow discharge positions. Other techniques include stacking discs with number shaped perforations to makeup the common anode and by forming each cathode into the shape of a number.

DETECTORS OF THE NUCLEAR

The detection of radiation or particles in nuclear physics is bases on the fact that charged particle. While passing through matter leaves along its path and counted. Most of the detectors measure the ionization produced by the passage of a charged particle through a suitable material. When an electric field is maintained across the material, the ion will be set in motion resulting an ionization current.

The various instruments which are used in the detection of nuclear radiations are as classified:

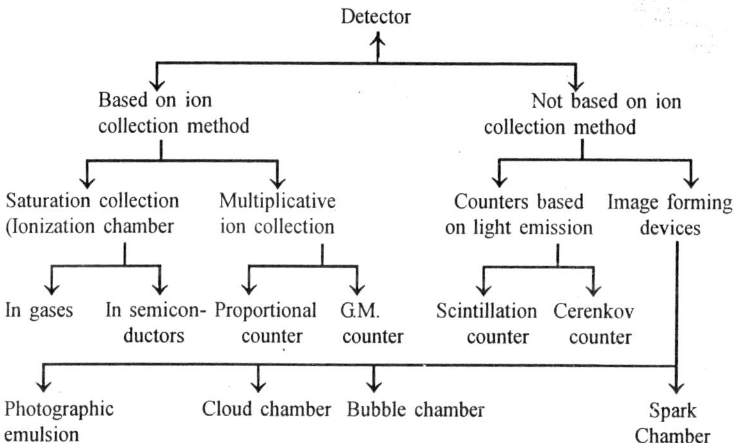

Fig. 18

Detector are also as *electric* (Ionisation chamber, Proportional counter, G.M. counter, semiconductor counter), and *optical* (scintillation counter, Cerenkov counter, Photographic emulsion, Cloud chamber, Bubble chamber, Spark chamber). The same detector can be used in studies of various radiations. Interactions of the radiations in the detector may depend on different phenomena.

IONIZATION CHAMBER

An ionization chamber, in its simplest form consist of vessel usually field with gas and into which it inserted an insulated electrodes. The gas becomes ionized when charged particles pass through it. If no voltage is applied to the container there will be no directing force acting the ions and no current will flow through the gas. If we applied a small potential difference to the electrodes, the –ve ions (or electrons) will be attracted to the anode and the +ve ions will go to the cathode. At the surface of the anode there will be a high concentration of negative ions. Recombination in this region will be a small because the rate of recombination is proportional to the product n^+n^- and here n^+ is very small. All positive ions produced in this region move out promptly towards the cathode. Therefore, the recombination will be maximum at the center of the space. A small current will flow through a resistor connecting the electrodes externally As the voltage between the electrodes is increased, the ions will move faster, the number of recombination will be minimum and the pulse size will thus be increased. The pulse size increases with the voltage, until it reaches to a value corresponding to the no recombination of ions. This value of current is called the *saturation current* and the chamber is said to reach in *saturation.*

The ionisation chamber works in the region of constant pulse size. In this region the applied voltage is high enough to prevent recombination of ions and low enough to prevent multiplication.

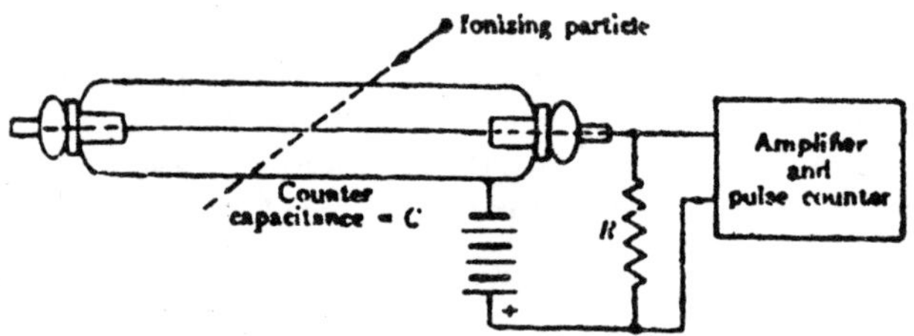

Fig. 19 : Ionization chamber.

In the initial stages, the gold-leaf electroscopes of great sensitivities or the quadrant electrometers were used. The next step in the development was a modification of the electroscope, designed by Lauritsen, where the gold leaf was replaced by a tiny metal coated quartz fibre. It could not give quick response to particles that follow one another in rapid succession. This difficulty was overcome to a great extent by a device known as linear or dc amplifier, which increases, without distorting, the feeble ionisation current from a single particle to a value which can easily register on an oscillograph or an electrical counter, dc amplifiers are more susceptible to disturbance and drift and more difficult to arrange with several successive stages of amplification than those designed for amplifying alternating currents.

Ionization chamber instruments may be grouped into two categories, depending upon the value of the time constant RC of the system relative to the frequency of arrival of the ionizing events. If RC is long compared to time between ionizing events a steady state is reached and a *direct current* may be measured. On the other hand, if RC is small each pulse produced by an ionizing event may be detected separately. These two types are also known as *integrating and non-integrating* respectively and the instrument as current and pulse ionization chambers.

Let a track of n ion pairs is formed parallel to and at a distance x_0 from the axial wire The electrodes are parallel and separated at a distance d. A voltage V_0 is applied to the electrodes, so that electrons and positive ions drift in opposite directions in a uniform electric field V_0/d. Let us first assume that the time constant RC (where C = capacitance of the chamber + input capacitance of the amplifier) is much greater than the collection time of ions. So that the current through R may begin as soon as the electrons and +ve ions move apart in the chamber.

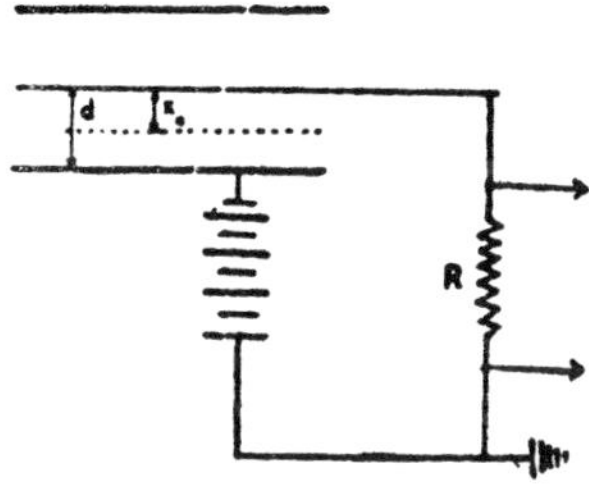

Fig. 20 : Ionization in an ionization chamber.

The change of potential at the positive electrode at the time t, due to a track of n ion pairs formed at t = 0 is given as

$$V(t) = n\,[q_e(t) + q_p(t)]/C, \qquad ...(1)$$

where $q_e(t)$, the charge induced on the positive electrode at time t by a single electron, and is given as

$$q_e(t) = -e.\frac{\text{P.D. between the electron at time t and the -ve electrode}}{\text{P.D. between the electrodes}}$$

$$= -e.\frac{d - x_e(t)}{d} = -e\frac{[d - (x_0 - v_e t)]}{d} \qquad ...(2)$$

and $q_p(t)$, the charge induced on the positive electrode at time t by a single positive ion, is given as

$$q_p(t) = +e.\frac{\text{P.D. between the +ve ion at time t and the -ve electode}}{\text{P.D. between the electrodes}}$$

$$= e.\frac{d - x_p(t)}{d} = e\frac{[d - (x_0 - v_p t)]}{d} \qquad ...(3)$$

The relations (2) and (3) are only true for parallel plate geometry in which potential gradient is uniform.

From eq. (2), (3) and (1) we get.

$$V(t) = -ne(v_e t) + v_p t)/Cd. \quad ...(4)$$

As the electron drift velocities v_e are about a thousand times greater than the heavy ion velocities v_p, hence, for the time between 0 and t_e is the electron collection time ($x_0 = v_e t_e$), the above relation transform to

$$V(t) = -nev_e t/Cd. \quad ....(5)$$

For $t > t_e$, there is an abrupt decrease in the rate of rise of potential, the new rate being v_p/v_e times the original rate. The change of potential

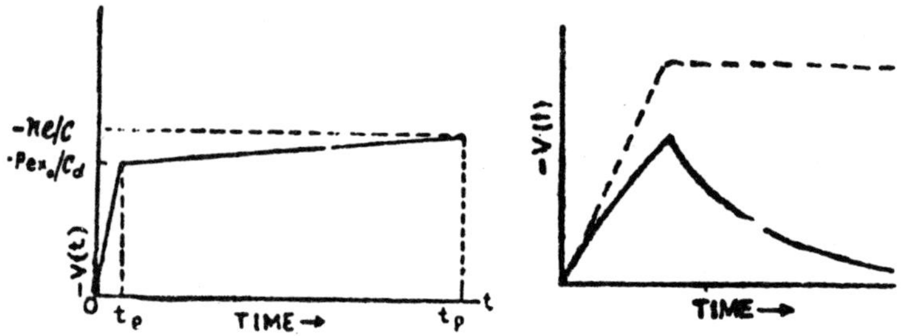

Fig. 21 : (a) Shape of potential pulse. (b) Pulse cliping.

becomes—ne/C when $t = t_p$, the time in which positive ions reach the negative electrode. This limit is proportional to n and is independent of where the ions were formed in the chamber. The shape of the collector potential curve does depend upon where the ions are formed.

With sufficient amplification, a large pulse will appear at the amplifier output terminal. To have the height of the output pulse proportional to the amount of ionization produced by the particle in the chamber, two kinds of proportional or linear amplifiers can be used. One, known as slow *amplifiers*, in which the shortest time constant is chosen long compared with the drift time of the +ve ions, have several disadvantages. The others, known as fast amplifiers, in which time constant RC is made short enough so that the induced potentials by the +ve ions are not of interest and clip the pulses Thus the slow linear rise, after the collection of electrons, is removed and the height of the recorded pulses depends on $V(t_e)$ and not on $V(t_p)$. If the complete range of an *ionizing particle is confined to the region between the negative electrode and grid*, the grid screens the +ve electrode from the electrons and +ve ions. Electrons, while drifting through the grid, induce a change of potential at the +ve electrode. As positive ions are screened from the +ve electrode by the grid, hence have no effect. Thus, the pulse height increases due to the induction effect of the moving electrons. When all the electrons are collected it attains a maximum value given by

$$V_{max} = -\frac{ne}{C} \quad ...(6)$$

It is also independent of the position and orientation of the ionization track.

The height of the amplified pulses from the ionization chamber

$$V = -\frac{AEe}{WC} \quad ...(7)$$

where A is the voltage gain of the amplifier, E the energy of the ionizing particle, e the electronic charge and W the mean energy required to produce an ion pair in the gas filled in the chamber.

SOLID-STATE DETECTORS

Crystal counters are essentially the ionization chambers using solid dielectrics between parallel plate electrodes. The first solid state nuclear detector was given by P.J. Van Heerden in 1945. A crystal of silver chloride (*intrinsic semiconductor*) was placed between two electrodes to which a voltage was applied. When the crystal was exposed to the radiations, the some of the valence band electrons got energy and entered the conduction band, leaving the (*positive*) holes. Under an electric field the electrons moved toward the negative electrode and holes moved toward the positive electrode and a pulse of current equivalent to the number of electrons transferred to the conduction band was produced. The first junction type detector was demonstrated by K.G. Mckay in the United States in 1949. A collimated beam of α-particles was allowed to fall on a n-p-germanium junction. He showed that nearly whole of energy of *α-particles absorbed in it was utilized in producing electron hole pairs and the average energy required to produce one electron hole pair was about 3eV which was very small in comparison to 34eV, the energy required to form an ion pair in a gas.*

J.D. Mayer and B. Gossick showed that *the size of the output pulse was proportional to the alpha-particle energy. In this respect, a solid state detector behaves like an ideal pulse type ionization chamber.* Most semi-conductor junction type detectors fall into three categories, according to the formation of junction.

Diffused Junction Detector

In this detector, a thin layer of the n-type is formed on the surface of a p-type silicon (or *germanium*) crystal, by allowing a donor impurity (*phosphorous*) to diffuse into one face of the crystal at a fairly high temperature. There is a tendency for electrons to diffuse from the n-region (*where they are in excess*) into p-region (where the electrons are in deficit).

An equilibrium is reached when the fermi-level is constant throughout the material. Due to this an intermediate volume around the interface will be cleared from carriers of both signs and this region is called the *depletion layer*. This layer is of low conductivity and high field, hence offers ideal conditions for the detection of ionizing radiation.

When the negatively charged electrons diffuse away from the n-type region, they leave the material positively charged. Similarly, the filling of the hole in the p-region generates a corresponding negative charge. The small potential difference of about a volt, established in this manner, is enhanced by an additional potential of several hundred volts, referred to as a reverse-bias voltage. In the presence of this electric field the positive holes in the p-type side are pulled towards the –ve electrode, the electron towards the p-n junction and thus, a depletion layer is created in the p-type material. The result is an extremely low leakage current. The thickness of the depletion layer increases with the reverse-bias voltage and the limit is set by electrical break-down in the semi-conductor.

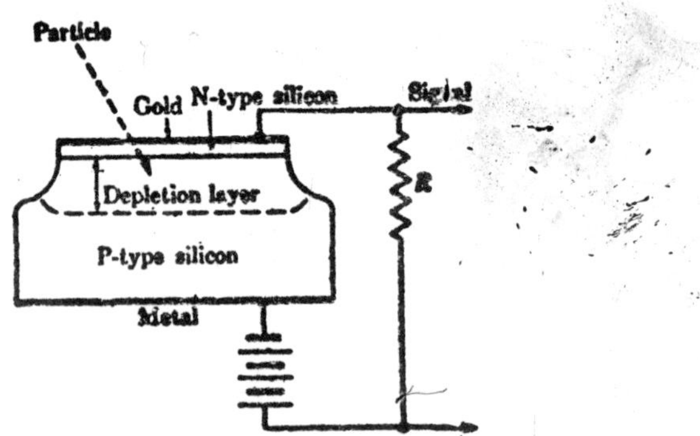

Fig. 22 : A p-n junction detector.

Thickness of Depleted Region

The thickness of the depleted region and the capacitance can be calculated by Poisson's equation. Assuming the variation of V in the direction ⊥ to the surface separating the n- and p-regions, we have

$$\frac{d^2V}{dx^2} + \frac{\rho}{\in} = 0 \qquad ...(1)$$

ρ the volume charge density
given by

$$\rho = q(p + N_{\&} - n - N_a) \qquad ...(2)$$

where p, n, $N_{\&}$ and N_a denote, respectively, the concentrations of holes, electrons, donors and acceptors. In the completely depleted region n and p are negligible, hence we have

$$\rho = q\, N_{\&} \quad 0 < x < x_1$$

$$\rho = -q\, N_a \quad -x_2 < x < 0$$

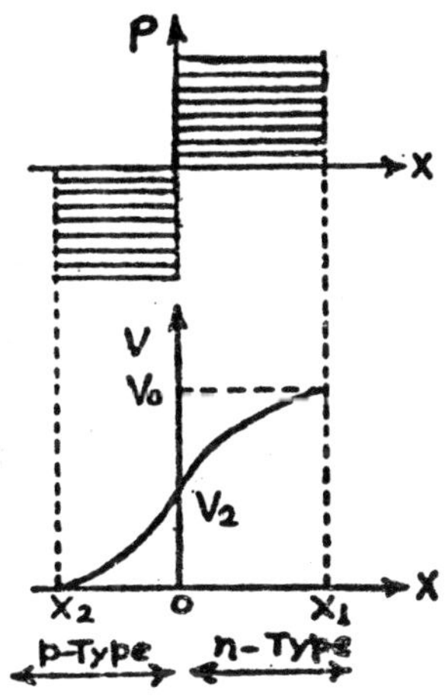

Fig. 23 : Charge and potential disturbance.

After successive integrations of eqn. (2), we get

$$V = -\frac{qN_d}{\in}\frac{x^2}{2} + \frac{qN_d}{\in} xx_1 + V_2$$

$$\therefore \qquad V(x_1) = V_0 = qN_d x_1^2/2\in + V_2 \qquad ...(3)$$

or $V_0 - V_2 = qN_d x_1^2/2\in$ ...(4)

Similarly $V(-x_2) = 0$ and $V_2 = qN_a \frac{x_2^2}{2\in}$ ...(5)

Hence the potential barrier height becomes

$$V_0 = q(N_d x_1^2 + N_a x_2^2)/2\in \quad ...(6)$$

As charge is conserved, hence $N_d x_1 = N_a x_2$ and

$$V_e = \frac{q}{2\in}\left(1 + \frac{N_a}{N_d}\right) N_a x_2^2 = \frac{q}{2\in}\left(1 + \frac{N_d}{N_a}\right) N_d x_1^2$$

The thickness of the depletion layer

$$X_0 = x_1 + x_2 = \left[\frac{2\in V_0}{q}\frac{N_a + N_d}{N_a N_d}\right]^{1/2} \quad ...(7)$$

If p-type region is more heavily doped, than the other region, we have $N_a > N_d$ and $x_2 < x_1$. Hence, we have

$$X_0 \simeq x_1 = \left[\frac{2\in V_0}{qN_d}\right]^{1/2} = [2\in \rho_n \mu_n V_0]^{1/2} \quad ...(8)$$

where ρ_n is the resistivity of n-type region and μ_n the electron mobility.

The same treatment will give the value of X_0 in the presence of reverse bias, as

$$X = \left[\frac{2\in(V_0 + V)}{qN_d}\right]^{1/2} = [2\in\rho_n \mu_n (V_0 + V)]^{1/2}$$

If the applied voltage $V > V_0$, we get

$$X \simeq [2\in \rho_n \mu_n V]^{1/2}, \quad ...(9)$$

where a is a constant of a medium.

The depletion layer having area A has a capacitance C given as

$$C = \frac{\in A}{X} = A\left[\frac{q\in N_a N_d}{2(N_a + N_d)(V_a + V)}\right]^{1/2}$$

For $N_a > N_d$ and $V > V_0$, we have

$$C \simeq A\,(\in/2\ \rho_n\mu_n\ V)^{1/2}. \qquad ...(10)$$

If ρ is in ohm cm, V in volts, X in microns (μ), and C in pf/cm² then we have following values:

G_e	n-type p-type	$X = (\rho V)^{1/2}$ $= 0.65\ (\rho V)^{1/2}$	$C = 1.37 \times 10^4\ (\rho V)^{-1/2}$ $= 2.12 \times 10^4\ (\rho V)^{-1/2}$
Si	n-type p-type	$X = 0.5\ (\rho V)^{1/2}$ $= 0.3\ (\rho V)^{1/2}$	$C = 2.2 \times 10^4\ (\rho V)^{-1/2}$ $= 3.7 \times 10^4\ (\rho V)^{-1/2}$

These detectors should always be kept in complete darkness, otherwise the photosensitivity will bring out a *photoelectric current.*

Surface Barrier Detector

It has been very useful for charged particles detection. It is made by exposing a chemically etched surface of an n-type silicon crystal to air. The etched surface is thus oxidised and this oxidation laye-acts like a very thin p-type layer.

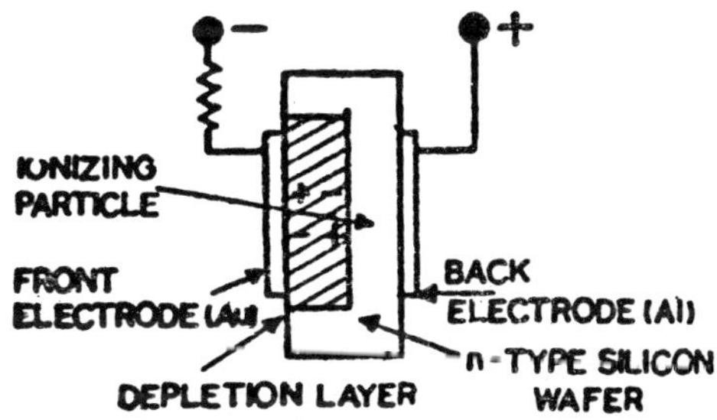

Fig. 24 : Surface barrier detector.

In analogy to an N-P junction, a depletion layer is formed, but it is clear that this layer only exists with in the semiconductor. The thickness of the depletion layer X_0 (or X with the reverse bias V) and the capacitance can be written as given in the case of N-P junction.

Using 4000Ω silicon with the reverse bias of 1000 volts, this thickness reaches to 3 mm. In contrast to N-P junction, the temperature attained during the various stages of the construction of the surface barrier detectors does not exceed the boiling point of water, and the life time remains as that of the original material.

Detection Characteristics of Surface Barrier Detectors

(a) *Resolution:* The high energy resolution means a narrow distribution of pulse size around the mean, *i.e.,* FWHM (full width at half maxima). With monoergic radiations,, Siffert, *et al.*, found that the value of FWHM was smaller for protons and deuterons than for α-particles. The broadening of peak for α-particles was due to nuclear collisions. The FWHM is found to be ~10 keV for α-particles of several MeV. Mayer and others suggested FWHM of 2.1keV at 135°K. The cooling increases the resolution. It is thus usually necessary to cool the detector to –30°C.

(b) *Linearity:* Many experiments verify the linearity of the response to various particles with different energies. In addition to the defect related to a thin space charge region, other deviations arise for heavy particles and fission fragments.

(c) *High Conversion Efficiency:* The energy required to produce ion pair in silicon is 3.5eV. Thus, the number of pairs for a particular incident energy is about 10 times the number of pairs in air.

(d) *Differential Sensitivity:* The surface barrier detectors are relatively insensitive to neutrons or photons. They can detect charge particles in the presence of other radiations.

(e) *Small Size:* These detectors are of very small size and are important. In some cases the small sizes are a drawback.

(f) *High Speed of Response:* The carrier mobilities are high (~1500cm^{2}volt/sec) for electrons and 500 for holes. The ion pairs move only a short distance before the collection and the pulse widths are in the order of nanoseconds.

Lithium Ion Drifted Junction Detector

The above limitation was reduced by E.M. Pell in the United States in 1960 after introducing the lithium-ion drifted junction detector. The technique is to allow the lithium, which acts as an electron donor, to diffuse into the p-type silicon or germanium at an elevated temperature (120 to 150°C) and under reverse bias. Under these conditions the ions drift for a considerable distance into the crystal where the lithium donor exactly compensates the existing acceptor impurities in the p-region and an effective intrinsic semi-conductor layer is produced between the n- and p-regions.

Principle and Characteristics

The first step of Pells method is to form an N-P junction, by diffusing Lithium from a layer deposited on the surface of a p-type material. If a reverse bias V is applied to this diode, +vely charged Li-ions will move from n-region to p-region. The electric field E is assumed to be large enough to make drifting much more important than the diffusion. To obtain a sufficiently large mobility μ, the temperature of the material may be reached. In the case of a long drift time, the thickness

$$X \simeq (2\mu\ V^{t})^{1/2} \qquad \ldots(11)$$

In the presence of a reverse bias, the electrons move towards N-region and holes towards P-region. A space charge is thus produced which varies linearly with distance, if recombination is neglected (Fig. 25b). This space charge causes a distortion of the electric field. The field becomes minimum at the centre of the intrinsic region (Fig. 25c), when the mobilities are equal. The distribution of Li-ions must be such that the space charge is compensated in the intrinsic region. It varies with distance as shown in (Fig. 25d).

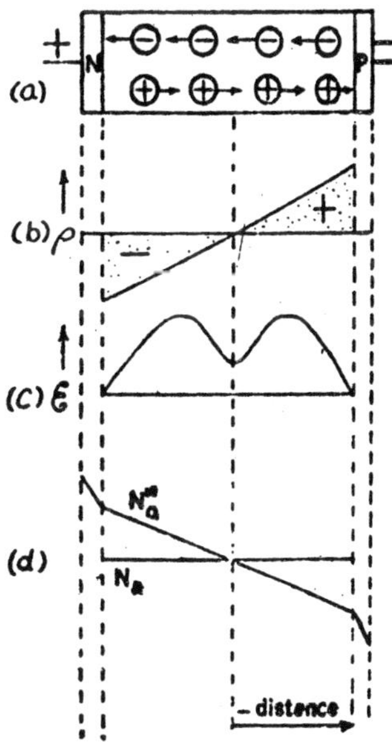

Fig. 25 : NIP-diode.

When the diode is cooled after the drift the thermally generated carriers become very small but the lithium distribution remains unchanged. Lithium-drifted germanium detectors are more suitable than silicon detectors for the detection of γ-rays. It may be recalled that the photo-electric absorption cross section for γ-rays is proportional to Z^5 and therefore germanium is more efficient than silicon for γ-ray

Advantage

1. The elimination of windows and the requirement of low applied voltage to operate the counter are advantages over the gas counters.
2. These are more suitable than the grid ionization chambers coincidence experiments.
3. Very good energy resolution is possible.
4. Counting rates high as 5×10^4 counts/sec. are possible.
5. If the particle range is less than the junction thickness, the pulse height is proportional to the K.E. of the incident protons, deuterons, α-particles or any other ionizing particles.
6. These detectors have low sensitivity to gamma background radiation.

REGIONS OF MULTIPLICATIVE OPERATION

Ion current or pulse height in an ionization chamber increases above the saturation value at sufficiently high applied voltages because electrons moving in these high fields acquire enough energy to cause secondary ionization. The size of the current pulse depends on the number of initial ion pairs produced by the ionizing particles and on the applied voltage. For a given ion pairs, the pulse size is shown in Fig. 26 on a logarithmic scale as a function of the applied voltage.

By these results shows that the curve can be divided into six more distinct regions.

In region I, the voltage applied is so low that recombination of ion pairs takes place. As the voltage is increased there is a competition between the loss of ion pairs by recombination and the removal of ions by collection on the electrodes. The region is known as ion recombination region.

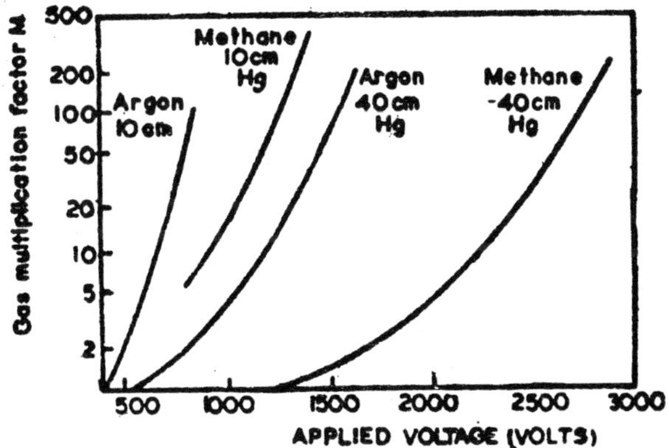

Fig. 26 : Variation of pulse size (log) with applied voltage.

In region II, the voltage is sufficiently high so that only a negligible amount of recombination takes place and the ion pairs move to the electrodes so rapidly that virtually every ion pair reaches the electrodes.

Thus, the pulse size remains independent of the potential as long as the voltage is increased upto the point at which secondary electrons arc produced by collision in the gas. This region is known as *ionization cham-ber region or saturation region.*

In region III, voltage is sufficiently high so that electron approaching the central wire attains suffi-cient energy to produce new ions, due to collision with gas molecules. The original electron and the new one, thus, produced, are again accelerated by the field and more ions are still produced by collision. This process is cumulative and an avalanche of electrons is thus produced. The phenomenon fit frequently called the *Townsend avalanche*, in honor of J.S. Townsend who did pioneer work in this subject. The number of the ion pairs formed by collision as each electron travels toward the central wire is known as *multiplication factor M*. The voltage region in which M remains constant is called the *proportional region*. The apparatus operates in this region as a proportional counter. In this pulse height remains proportional to the amount of energy lost in the chamber by the primary ionizing particle.

If the voltage is increased further (*region IV*) pulse heights continue to increase. The heights are still related but no longer proportional to initial ionizing intensity. The voltage region in which M depends some what on the pulse size is called the region of *limited proportionality* and is not much used for measuring. The limitation may be because:

(a) Ultraviolet photons may be formed.

(b) New electrons may be formed when positive ions reach the cathode,

(c) The space charge may distort the electric field.

For still higher voltages the *region V* is reached, in which pulse height is very large and also substantially independent of the initial number of ions. This is called *Geiger region.* The device operated in this region is called a Geiger-Müller counter. In the Geiger region the avalanche associated with the drift of a single electron to one point on the anode initiates avalanches over the entire anode length.

At the end of the Geiger region the counter goes into continuous discharge region,

PROPORTIONAL COUNTER

In the parallel plate ionization chamber if the electrode voltage is increased by beyond the saturation interval up the primary ions produced secondary by collisions with the gas: consequently, the primary ionization is multiplied by a factor that depends on the Geometry of the apparatus and on the applied voltage v the chamber under these condition is ion own as proportional counter.

For a cylindrical geometry, the electric field required to produce secondary ionization is given by

$$E = \frac{V}{r \log_e b/a}$$

Where,

a = radius of central wire

b = inner radius of cathode cylinder

V = electrode or central wire voltage

r = distance of electric field E from the central wire.

The electrons ejected from gas molecules by ionizing particle drift towards the positive electrode. If they move in a region of high electric field E, they will produce additional ionization by collisions with molecules of the gas. In drifting a short distance ax, n electrons will increase the number of elections to n + dn, where

$$dn = \alpha \, n \, dx \qquad ...(1)$$

Here a is the first *Townsend coefficient.* If n_0 is the number of electrons set free by ionizing particle and x is the distance through which these electrons drift to reach the positive electrode, the number of electrons reaching this electrode is given by

$$n = n_0 \exp.\left[\int_0^x \alpha \, dx\right] \qquad ...(2)$$

The ratio n/n_0 is called the *gas amplification or gas multiplication factor M.* The radius of the cylinder and a the radius of an axial wire, the electric field strength at a radius x is

$$E_x = \frac{V_0}{[x \log_e b/a]} \qquad ...(3)$$

Let us assume that every electron produced in the primary ionization gives rise to n-secondary electrons by collisions. The production of secondary electrons during the collision is associated with the photons which in turn produce photoelectrons. If p is the probability of production of photoelectrons, there will be np photoelectrons in all. If each photoelectron produces n-electrons by further collisions, we shall have second-generation avalanche of n^2p electrons, which in turn will give rise to more electrons.

$\therefore$ Total number of electrons $M = n + n^2p + n^3p^2 + ...$ ...(4)

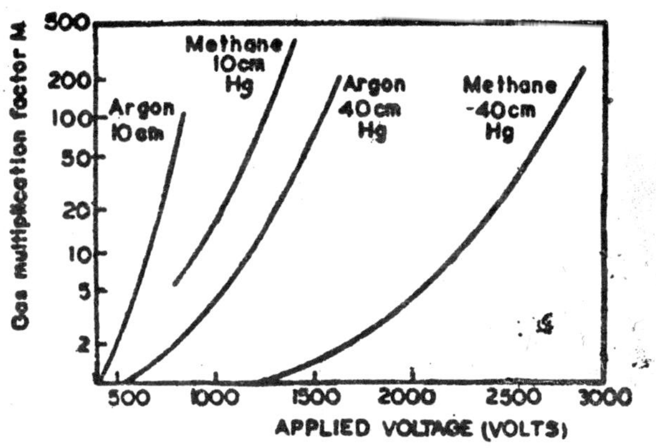

Fig. 27 : Variation of M with applied voltages.

For practical proportional counter np < 1, hence M = n/(1– np). It represents that in a proportional counter the total number of secondary electrons is proportional to the number of initial electrons. The variation of M with voltage in argon and methane at two different pressures is shown in the given Fig. 27.

The voltage-time dependence of the pulse is shown in Fig. 28. The upper curve shows the behaviour for large values of the time constant. The initial rise of the pulse is due to the high velocity of the positive ions as they recede from the anode. Positive ions then move in a weak field and the pulse approaches its final height extremely slowly. For the case of a finite time constant the pulse has the shape shown by lower curves. In practice, suitable clipping time constants (RC ~ 0.1μ sec.) are used in the output circuit so that better time resolutions are available.

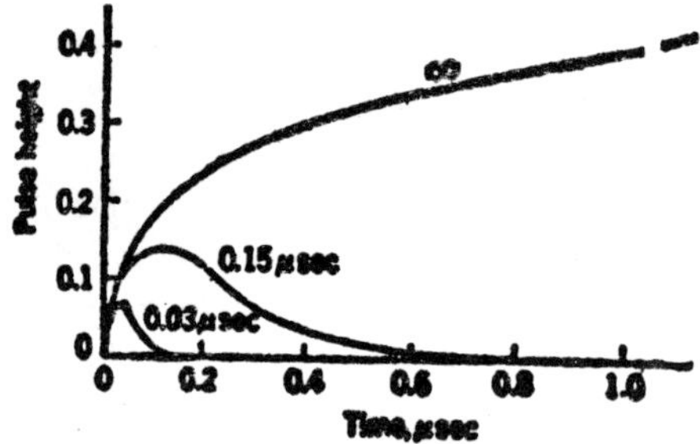

Fig. 28 : The pulse space.

If the primary ionizing particles arrive in rapid succession, mechanical counter can not record all the pulses. The output from the amplifier is therefore first passed through a scaling circuit before being fed into a mechanical counter. A sealer is an electronic circuit arrangement that feeds a predetermined fraction of the actual particle counts to the mechanical counter.

GEIGER-MÜLLER COUNTER

Geiger and Müller in Germany developed the counter in 1928. This counter is thus named as Geiger-Müller counter or G.M. counter. It works in Geiger region. The main difference between proportional and Geiger counter is that in the *former the avalanche is formed only at one point whereas in the latter the avalanche spreads along the whole length of the central wire. The amplification does not therefore depend on the initial ionization produced by the ionizing particle.*

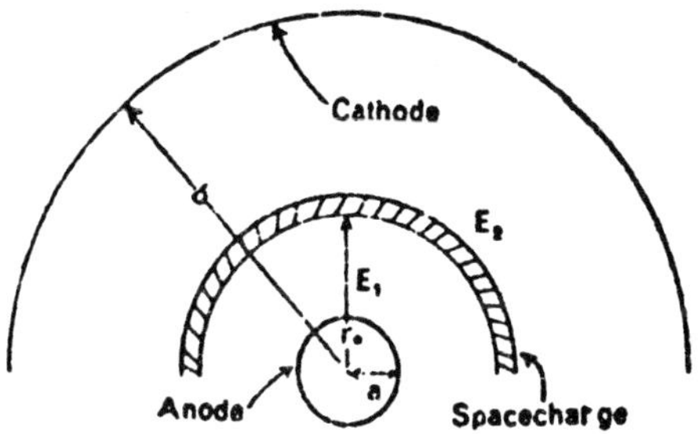

Fig. 29 : Positive ion sheath.

Pulse Formation and Decay

Let us Consider the situation when the space charge has moved outside to a radius r_0. Let there be q charges per unit length on the wire and Q charges per unit length in the space charge. The field strengths inside and outside the sheath are

$$E_1 = q/2\pi\epsilon r \text{ and } E_2 = (q + Q)/2\pi\epsilon r.$$

Voltage applied to the counter

$$V = \int_a^{r_0} \frac{q}{2\pi \epsilon r} dr + \int_{r_0}^{b} \frac{q+Q}{2\pi \epsilon r} dr$$

Integrating we have

$$-\frac{q}{2\pi \epsilon} \log_e \frac{b}{a} + \frac{Q}{2\pi \epsilon} \log_e \frac{b}{r_0}$$

$$\therefore \text{ Charge} \quad q = \frac{2\pi \epsilon}{\log_e b/a}\left[V - \frac{Q}{2\pi \epsilon} \log_e \frac{b}{r_0}\right] \quad ...(5)$$

The negative term in the bracket shows that q and hence E_1 is reduced during a pulse. As the positive ion sheath moves toward the cathode, r_0 increases, negative term decreases and the field strength attains its maximum initial value. The positive ion sheath will reach the cathode in about 10^{-4}sec. As the positive ions reach the wall and are about 10^{-7}cm from it they cause electron to be liberated from the metal of the cathode by *field emission*. These electrons then neutralize the positive ions, leaving the excited molecules. From where they promptly go to the ground state radiating characteristic spectral lines. Some of these lines are in the ultraviolet region of the spectrum and hence liberate photoelectrons from the cathode. These secondary electrons are accelerated toward the central wire.

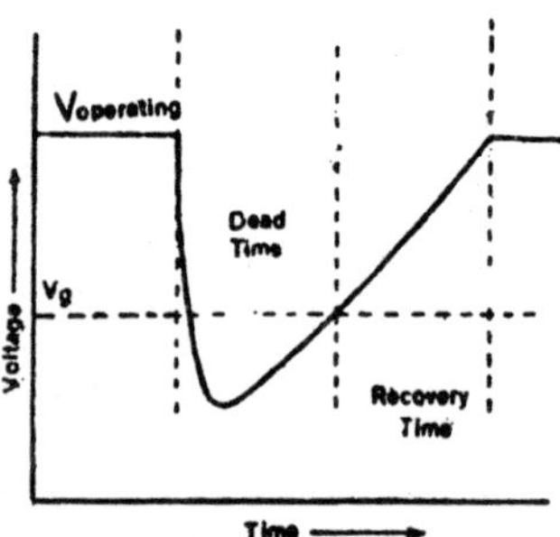

Fig. 30 : Dead time and recovery time.

When the electrons reach the wire, they start a new avalanche which produces a new sheath and the discharge will continue indefinitely. This continuous discharge is not really continuous in time but consists of a series of pulses at close and some what overlapping intervals. The growth of the potential pulse at the anode is determined by the motion of the +ve ion sheath as it is repelled away from the high field region. When the ion sheath has travelled a distance equal to about half the radius of cathode cylinder, the field close to the anode wire is below the Geiger threshold. Hence the second pulse can not be recorded. The time, during which the counter is incapable of responding to a second ionizing event, is known as *dead time* (~200μ sec.). As the space charge moves further towards the cathode, the counter sensitivity gradually returns to its original value and the second pule can now be recorded.

This time, during which pulses of reduced size are produced, is known as *recovery time*. The dead time plus recovery time may corresponds to the time which +ve ion sheath takes to reach the cathode, and is known as the *resolving time* of the counter. The high counting rate is generally reduced by reducing the high potential supply to the counter for a definite time Interval called *paralysis time*. It is achieved by an external electrical circuits generally a relaxation oscillator triggered by the step rise of Geiger pulse;

Quenching of the Discharge

The process used to prevent a continuous series of pulses from taking place in the G.M. counter are called quenching. There are three types of quenching (*suppressing*) involved in the operation of a counter:

(a) quenching of photons in the initial avalanche,

(b) electrostatic quenching of the avalanche by the +ve ion space charge, and

(c) quenching of the secondary emission when the +ve ions reach the cathode. The quenching of the discharge involves both the electrostatic quenching and the quenching of secondary emission. There are two procedures for quenching the discharge in order to improve the resolving power of G M. counters.

In the *nonself-quenched* (externally quenched) type of G. M. counter, the quenching of the discharge is achieved by means of an external resistance or by use of an auxiliary electronic circuit. In the first method counter can be quenched by using the load resistor ($10^{10}\Omega$) in series with the tube. High voltage is applied across the tube which reduces as the current pulse flows through the resistance, the gas amplification factor is thus decreased preventing a second discharge. With a chamber of 10pf capacitance this method results in a counter dead time of the order of 0.1 second. In the second method the load resistance is about $10^7\Omega$ and at the same time the voltage across the chamber is reduced, by the application of a negative voltage pulse to the anode from a monostable multivibrator, every time a count takes place. The dead time of the counting circuit can be adjusted electronically to be a chosen value.

In the *self-quenched* (internally quenched) type of G.M. counter, quenching is achieved by the inclusion of a suitable gas in the counter, such as:

(a) organic molecules, and

(b) the halogen gases. In the first case the filling is a mixture of argon and 10% of a polyatomic organic gas or vapour such as methane, ethane or ethyl alcohol as a quenching gas. As the quenching gag ionizes more readily than argon due to its lower (11.3 eV instead of 15.7 eV) ionization potential, hence the ion cloud reaching the cathode consists almost entirely of alcohol ions. These ions are neutralized by pulling electrons from the metal surface of cathode and again form largely excited molecules. *These excited molecules usually prefer decomposition rather than photon emission and hence no photons are available to start a second avalanche.* The photons if emitted by excited molecules are not likely to travel more than a millimeter through the gas before being absorbed by the photoelectric process. The process of charge exchange during collisions prevents any argon ion reaching the cathode, thereby preventing the release of any electrons from the cathode which would trigger off a second discharge.

The each transfer of ionization from argon to alcohol is accompanied by the emission of photons of energy 15.7 – 11.3 = 4.4eV. This is too low to eject an electron from a cathode.

A satisfactory quenching gas must nave three main properties:

1. When in an excited state it must prefer to dissociate rather than to de-excite by the emission of photons.
2. Its ionization potential should be lower than that of the main counting gas in the tube.
3. It must have broad and intense ultraviolet absorption bands.

The halogen quenching action seems to be similar to that of the organic agents in that some of the molecules dissociate at each discharge. However, the dissociated halogen molecules combine at the end of the discharge, so the tube has unlimited life. Halogen molecules are chemically so active that the materials of which the tube and its electrodes are made must be chosen carefully. The halogen quenched counters usually operate at 300 to 700 volts. They can be used over a temperature range of from –50°C to +60°C, whereas the organic quenched counters often cease to work at 0°C because the organic gas condenses. They are not damaged by a continuous discharge whereas poly-atomic organic gases filled counters damaged. Due to their lower voltages, long life and robustness, the halogen quenched counters are recommended for use in colleges.

In the self quenching type counters the discharge is quenched due to the internal mechanism and hence no electronic quenching circuit or high resistance is necessary. The disadvantage of high resistance is that it lengthens the recovery time of the counter. The equivalent electronic circuit permits a short time constant to be used but adds the inconvenience of an extra stage in the circuits. The advantage gained in the elimination of the quenching resistor or electronic circuit is secured at the price of stability.

G.M. Tube Construction

G. M. *tubes* have been made in a great variety of sizes and shapes, from 1 cm. to 100 cm. in length and from 0.3 cm. to 10 cm. in diameter. The walls can be of metal (copper), a metallic cylinder may be supported inside a glass tube, or the interior surface of the glass tube may be coated with a thin layer of an electrical conductor) *e.g.*, silver). The central wire is usually of tungsten with a thickness of 2×10^{-3} to 5×10^{-3}cm. Various observers have tried a large number of different compounds in counters and found that any polyatomic gas molecule would produce the quenching action. The molecule need not be organic. There does not appear to be any great difference in merit between various possible polyatomic molecules, each having certain advantages and disadvantages as compared to other.

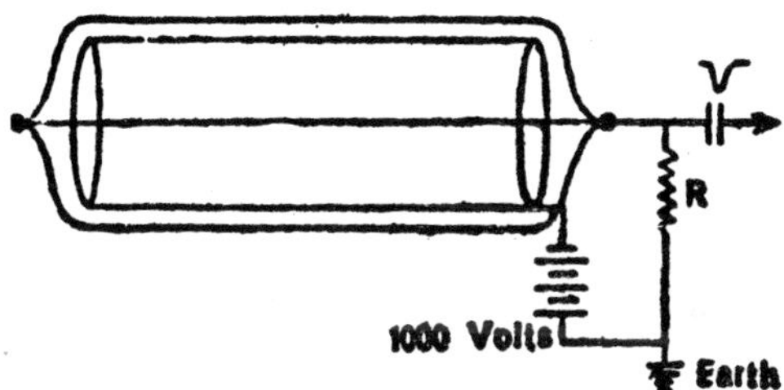

Fig. 31 : Basic circuit of G.M. counter.

For example, amyl acetate shows a some what longer useful life but a slower resolving time than does methane. To extend the life time of the tube, bromine vapour can be introduced instead of the alcohol. To detect radiation of very small penetrating power counters with very thin end windows are used. Thin end windows of mica, cellophane or glass, only 1 or 2 mg/cm^2. thick will allow a- and P-particles to pass into the counter. For detecting P-radiation thin walled glass counters (the cathode is often a thin layer of colloidal carbon deposited on the inner wall of the glass envelope), which allow the entry of β-rays from all directions, can be used. Detection of γ-rays is mainly through the ejection of photoelectron from the cathode cylinder. The probability of this process varies as Z^4. hence, counters using high Z (bismuth or lead) cathode are used for the detection of γ-rays. For detecting cosmic rays more robust counters sometimes two or three feet long are used. For our purposes:

(i) *G.M. tube should be capable of responding to the radiation to be detected,*

(ii) *It should operate at fairly low voltages,*

(iii) *It should have long life, and*

(iv) *Its cost should be minimum.*

In India G.M. tube of different shape and size are designed in Hyderabad. These are marked as I 1000, I 1002 (Helogen filled); I 1010 (Thin walled); I 1030, I 1031, I 1032 (end windows); I 1200 (liquid sample); I 1300 (gas flow detector windowless and I 1350 (gas flow detector end window).

Characteristic Curves

There are two important characteristic curves exhibited by Geiger counters. The first is the familiar *plateau curve.* This is a plot of the recorded counting rate when the voltage applied to the counter is varied. The G. M. counter, similar to proportional counter, is connected with a device capable of indicating only relatively large pulses but no small ones. Until the voltage reaches the value indicated as the *starting potential* the pulses are too small to be detected. But with rising potential, the gas amplification increases and number of pulses rise rapidly to a flat portion of the curve called *the plateau.* This is the Geiger tube region for which the *count rate is nearly independent of the potential difference across the tube.* Beyond the plateau the applied electric field is so high that a continuous discharge takes place in the tube and the count rate increases very rapidly. It should be emphasized that the shape of the knee of the plateau curve is partly a property of the associated recording circuit.

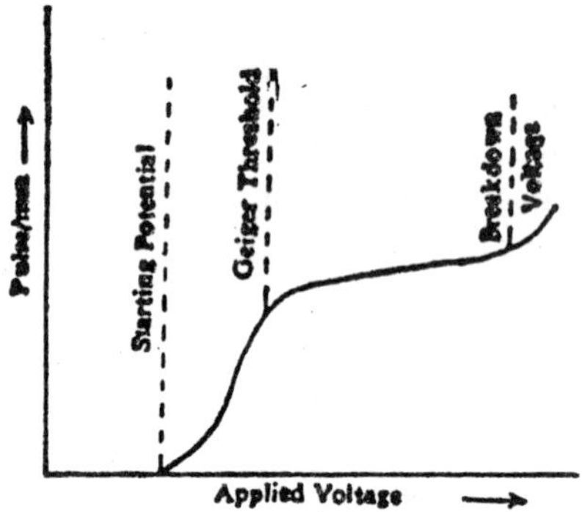

Fig. 32 : Plateau of G.M. tube.

$$\text{Slope of plateau } \frac{n_2 - n_1}{n_{av}} \times \frac{100}{V_2 - V_1} \times 100\% \text{ per 100 volts}.$$

The plateau has a width (or length) of a few 100 volts. A good tubes often have a plateau length of 100 to 200 volts or more with a slope of about 5% counts per 100 volts applied. *The slope for halogen filled tubes is however greater than that for organic gase filled tubes.* When the plateau becomes notably shorter and steeper, it is a sign that a counter tube is nearing the end of its useful life. The other type of curve is that of the starting potential as a function of the pressure of the gas in the counter. It can be shown from these curves that the *starting potential increases, practically linearly with the pressure of the filled gas.*

The *efficiency of the counter* is defined as the ratio of the observed counts/sec to the number of ionizing particles entering the counter per sec. Counting efficiency is defined as the ability of its counting if at least one ion pair is produced in it.

$$\therefore \qquad \text{Counting efficiency} = (1 - e^{-s/p}).$$

where s is the specific ionization at one atmosphere, p the pressure in atmospheres and *l* the path length of the ionization particle in the counter.

Precautions in Using G.M. Tube

1. The tube must be disconnected when measurements arc over.
2. Introduction of light must be prevented in order to avoid photo electric effects.
3. The background counting rate must be subtracted.
4. The tube should never operate at a voltage higher than its normal voltage.
5. In case a continuous discharge is produced, the voltage should be decreased.
6. The high voltage must be relatively stabilized.

Main Features of a G.M. Counter

(i) Constant output pulse size, independent of initial ionization.

(ii) Sensitivity to the production of even a single ion pair,

(iii) A long insensitive time to allow entry of each particle.

(iv) Infinite life (specially with halogen quenched),

(v) Can detect α, β, γ and cosmic rays.

Pulse Counting Circuits

The pulses obtained from the G.M. tube are too small to count direct. It requires amplification, which is done frequently in two steps. The counter is immediately followed by a *preamplifier*, which has little gain but produces an output signal across a relatively low impedance. This signal is then fed into the main amplifier, which has a gain strictly independent of the size of the input pulse. The output signal is then fed into the scaling circuit.

Resolving Time and Actual Counts

Let us assume that a counting system with a resolving time τ responds at a rate n counts per unit time when exposed to N initiating events per unit time. In unit time the total insensitive time will be $n\tau$ and the number of counts missed will be $Nn\tau$.

No of counts missed = error in counting or $Nn\tau = N - n$

$\therefore$ Actual count rate $N = n/(1 - n\tau)$

Thus, the actual count rate can be calculated by knowing τ. The most usual method for it involves the measurement of four counts using two radioactive sources. The background count (counts without any source) rate B is first found. With one source, count rate is determined. Let the observed count rate be $n_1 + B$ and expected $N_1 + B$. With the first, source the second source is introduced and adjusted until the count rate is approximately doubled. We now have $N_1 + N_2 + B$ expected and $n_{12} + B$ observed. The first source is removed and the second is left in position. If the counts rate is $N_2 + B$ expected and $n_2 + B$ observed. Solving equations

$$N_1 = \frac{n_1}{(1+n_1\tau)}, \; N_1 + N_2 = \frac{n_{12}}{(1-n_{12}\tau)} \text{ and } N_2 = \frac{n_2}{(1-n_2\tau)},$$

We get

$$\tau = \frac{n_1 + n_2 - n_{12}}{2n_1n_2} \qquad ...(1)$$

Standard Deviation

Because of the random nature of radioactive decay it is found that if a long lived source is counted for a number of equal intervals of time the result shows considerable variations. This deviation is the square root of the average of the squares of the deviations from the mean. If N is the total number of counts in time t, then the

$$\text{Standard deviation } \sigma = \sqrt{\frac{(N)}{t}} \qquad ...(2)$$

$$\text{The coefficient of Variation } V = \frac{\text{Standard deviation } \sigma}{\text{Rate of counting R}}$$

$$= \frac{\sqrt{(N)/t}}{N/t} = \frac{1}{\sqrt{(N)}} \qquad ...(3)$$

Account must be taken of the count rate due to background, unless the count rate of the sample is very large. Hence actual counting rate

$$RT = R_s + R_B,$$

where the subscripts refer to the total, sample and background respectively.

The standard deviation $\sigma s = \sqrt{(\sigma T^2 + \sigma n^2)}$

$$\text{or} \qquad \sigma s = \sqrt{\left(\frac{N_T}{tT^2} + \frac{N_B}{t_B{}^2}\right)} = \sqrt{\left(\frac{R_T}{tT} + \frac{R_B}{t_B}\right)} \qquad ...(4)$$

$$\text{and } V_S = \frac{\sigma s}{Rs} = \frac{\sigma s}{R_T - R_B} \qquad ...(5)$$

SCINTILLATION COUNTERS

One of the earliest methods of detecting nuclear radiations was by the luminescence, they produced in certain substances. In 1903, Crookes in England and Elster and Geitel in Germany independently reported that alpha particles impinging on zinc sulphide screens produced individual flashes which could be observed through a microscope. The first attempt to count α-particles, by observing flashes of light (scintillations) generated in diamond by them, was made by E. Regener in Germany in 1908. In the same year *Rutherford and Geiger compared the number of scintillations seen with the number of pulses recorded by an ionization chamber. The numbers were same in both cases. Hence, if each particle caused a single pulse in the counter then it also gave rise to one scintillation.* In this way this method was available for counting individual α-particle. The observations of scintillations through low power microscope were very tedious and limited to relatively low counting rates. For this reason early forms of scintillation method soon became obsolete and more attention was paid to the development of gas counters.

Since 1945, this old technique, is now playing a most important role. This has been due to the development of the photo-multiplier tube and the intensive study of the luminescent properties of many inorganic solids as well as organic compounds either in the solid state or in solution. A purely schematic diagram of a scintillation counter unit is shown in the following figure. A scintillation counter consists of the luminescent material (known is *scintillator*), reflecting layer such as aluminium foil enclosing the luminescent substance to facilitate the collection of light, light pipes, the photomultiplier tube, amplifier, voltage discriminator and an electrical circuit to record the output pulses. Before considering the performance of the scintillation counter let us first examine the behaviour of some of its components.

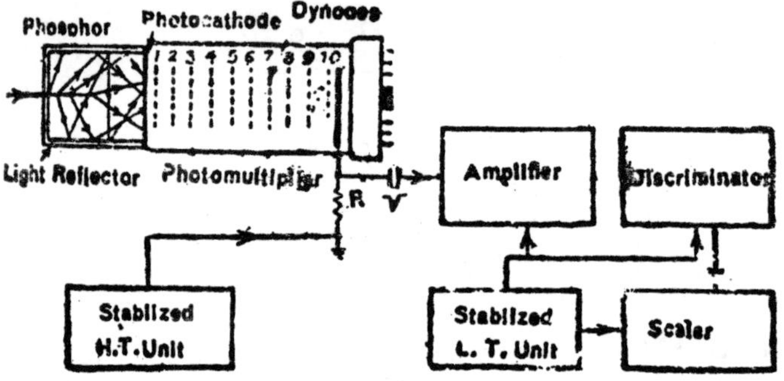

Fig. 33 : Schematic diagram of Scintillation Counter.

Photomultiplier Tube

In modern scintillation counters human eye is replaced by an electronic device called a *photomultiplier tube*. It is a highly sensitive photocell, converting light energy into electrical energy. In one of the commonest types of photomultipliers a semi transparent Cs – Sb cathode is deposited on the inside of the end of a high vacuum envelope. Cathode's of this type have a maximum response to light at the blue end of the spectrum. The electrons ejected from this cathode are accelerated towards the electrode D_1 and are collected by the electrode where they liberate more secondary electrons. These are focused on an electrode D_2. The process is repeated again and again These collecting electrodes D_1, D_2, etc., are called *dynodes*.

A photomultiplier tube may have as many as thirteen or sixteen dynodes, each one being maintained at a positive potential of 100 volts with respect to the previous one. The total voltage drop across the whole system is supplied from the rectifier power pack.

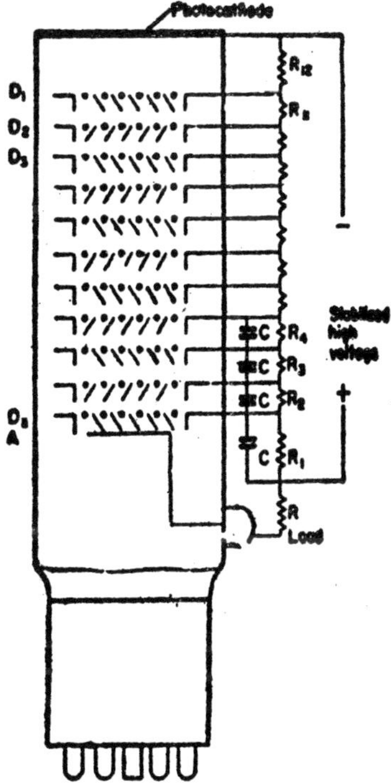

Fig. 34 : Photomultiplier tube.

The dynode has a *venetian blind structure*, and is covered with a layer of material with a high secondary emission coefficient, such as Cs – Sb or Ag – Mg. The number of electrons reaching the anode A is thus a million or more times the number of electrons reaching the first dynode. The anode A is connected to the positive voltage supply through a series resistance R. Because of the voltage drop across the load resistor, a negative output pulse is produced by a flash of light falling on the photo cathode. At room temperature there will be an appreciable emission of thermionic electrons from the cathode to produce undesirable current, known as *dark current*.

Scintillators

The principal methods of excitation are *incandescence and luminescence.* The luminescence caused by photons are named as photo luminescence and are further divided into phosphorescence and fluorescence. When the visible or ultraviolet light is emitted within 10^{-8}sec or less after the radiation absorption the emission is called fluorescence. Phosphorescence refers to delayed light emission, which may follow the radiation absorption by minutes, day or even years. For the moment we shall consider fluorescence response. The substances scintillate when bombarded by radiations are named as scintillators and are mainly classified as:

(a) Organic Liquids

Some solutions of organic compounds are also capable of acting as phosphors, although, they are not as efficient as the best pure crystals. The liquid scintillator has two main components: the solvent, (*e.g.*, toluene or xylene) and the primary solute (*e.g.*, a few percent of diphenyloxazole or terphenyl). Most of the nuclear radiation energy is absorbed by the solvent which is not a scintillator and is then transferred to the solute which emits the light actually. When the scintillations occur in the extreme ultraviolet region one uses a light shifter to increase the wavelength of the emitted light.

(b) Organic Crystal

Kallmann employed a large clear crystal of the organic substance naphthalene as phosphor. Following this, it has been found that other related organic compounds, consisting of several linked benzene rings are better scintillators. Of the solid organic scintillators anthracene and stilbene appear to be the best. The organic crystalline scintillators have a faster decay time ($\sim 10^{-8}$ sec), high transparency and poorer conversion efficiency especially for heavy particles. For β-particles response is linear. Because of low Z and ρ, these crystals have low cross-sections for photoelectric and pair production processes, hence can not be used for the detection γ-rays and X-rays.

(c) Inorganic Crystals

Unlike the organic compounds, inorganic substances do not scintillate when they are pure. Best one, especially for gamma rays, is crystalline *sodium iodide activated with thallium* (about 0.1 per cent). A less sensitive alternative is the nonhygroscopic cesium iodide, also activated with thallium. Calcium iodide with a trace of europium is more efficient than NaI, although, it is also hygroscopic. Silver activated zinc sulphide is an excellent scintillator but can be used only in thin layers, because it soon becomes opaque to the light emitted by it.

(d) Plastics

Organic scintillator system, which lies between a solid and liquid has a plastic as the base. A solution is made of the primary solute in a solvent like vinyl toluene or styrene which can be readily converted into a solid plastic. The resulting transparent material can be made in large pieces and cut to any desired shape. Plastic scintillators are widely used in high energy physics.

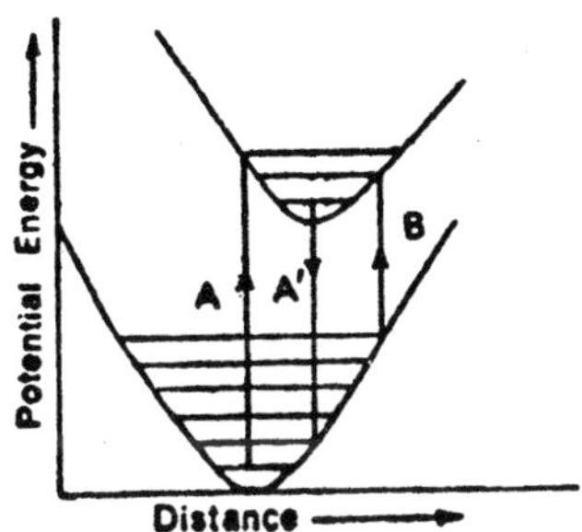

Fig. 35 : Luminescence proccess in NaI (Tl).

In a crystal such as NaI (T*l*) an ion Tl^+ replaces an ion Na^+ and forms an emission center. The T*l* ion thus occupies a position of equilibrium in the lattice such that a displacement from that position produces an increase in potential energy for the entire lattice system near the Tl^+ ion. The absorption of radiation changes the Tl^+ ion to an excited state for which the potential energy curve has a minimum at a position different from that in the ground state. The absorption (transition A) leaves the system in a state of high vibrational energy. The excited particle makes many collisions and with in 10^{-5} to 10^{-8} sec dissipates its surplus vibrational energy to the lattice. The most intense emission thus corresponds to transition A'. If the temperature is high enough, so that the higher vibrational levels of the ground state have appreciable population. Now, it is possible that the absorption takes place along B which is followed by emission of radiation.

ZnS (activated with Cu or Ag)

1. The output is proportional to the energy of the incident particle.
2. C_{IP} is independent of the incident energy.
3. It has highest light conversion efficiency C_{IP} of any known phosphor.

4. The efficiency and emission spectra depend on the preparation and activator used.

For ZnS (activated with Cu)—conversion efficiency C_{IP} = 28% and the emission spectra from 4000-6000Å has a peak at 4500Å.

For ZnS (activated with Cu)— C_{IP} = 25% and λ_{peak} = 5200Å.

It has following limitations

1. Decay time is $\sim 10^{-6}$ sec or longer.
2. A large crystal can not be prepared.
3. It has low transparency. Beyond thickness 25 mg/cm^2, the detector efficiency decreases.
4. It can not be used for the detection or γ-rays and electrons. Only the detection of slightly heavier particles such as a, p, d is possible.

NaI (activated with Tl)

1. Its absorption spectra lies in ultraviolet region, with the peaks at 2340 and 2930Å. Due to this reason, the crystals are highly transparent.
2. Its emission spectra has a average value of wavelength ~4100Å, and FWHM = 850 ± 100Å.
3. Its can be grown into large single crystal.
4. It has apart from ZnS, the highest.
5. Because of high values of p and Z, it has large cross sections for pair-production, photoelectric effect and compton effect. Due to this reason it is very suitable for the detection of gamma and X-rays.
6. The decay time at room temperature is lowest ($\sim 2.5 \times 10^{-7}$sec.).

C_{IP} decreases with increasing specific ionization causing a nonlinear response to α-particles. The response is linear to p, d and e. The crystal must be protected from moisture with surface film of liquid paraffin or be placed in dry atmosphere.

(c) Gases

Inert gases of the atmosphere, *e.g.*, argon, krypton and xenon when exposed to nuclear radiations produce scintillations in the ultraviolet region of the spectrum. To increase photomultiplier response the flashes are converted into visible light by means of a suitable wavelength shifter. An important advantage of the inert element scintillators iff that the light output is proportional very closely to the energy of the nuclear radiations.

Choice of Phosphors

NaI (Tl) is a highly effective scintillator as far as light output is concerned, but the light pulse has a relatively high decay time, compared with the organic phosphors. Hence *NaI (Tl) is less satisfactory when a very rapid response is required.* Iodine present in it gives a much greater stopping power for γ-rays. The size of the output pulse is proportional to the energy.

Organic scintillators give comparatively smaller light pulses, but have a shorter decay time. Hence, a higher counting rates are possible. The decay time in the liquid and plastic scintillators are even shorter than in the crystals Because of the small stopping power, the organic scintillators cannot be used for γ-rays unless the counter is large and are used for β-detection.

Screens with activated ZnS, in very thin layers, are useful for the detection of α-particles. Neutrons can be detected through the emission of a suitable charged particle resulting from a nuclear reaction initiated by the incident neutron.

SOLVED EXAMPLES

Example 1:

Calculate the capacitance of a silicon detector with the following characteristics : area 1.5 cm^2, dielectric constant 12, depletion layer 50 microns. What potential must be developed across this capacitance by the absorption of 4.5 MeV α-particle which produces one ion pair for each 3.5 eV expended?

Solution:

Silicon detector is equivalent to a parallel plate capacitor.

∴ Capacitance of the detector

$$= \frac{\varepsilon A}{d}$$

$$= \frac{12 \times 8.85 \times 10^{-12} \times 1.4 \times 10^4}{50 \times 10^{-6}}$$

$$= 318 \text{ pf.}$$

No of ion pairs produced by α-particle

$$= 4.5 \times 10^6/35 = 1.286 \times 10^6.$$

∴ Charge associated with these ion pairs

$$= 1.286 \times 10^6 \times 1.6 \times 10^{-19}$$

$$= 2.057 \times 10^{-13} \text{ coul.}$$

Hence the potential applied across the capacitor

$$V = q/C = 2.057 \times 10^{-13} \times 10^{-12} = 6.5 \times 10^{-4} \text{ volt.}$$

Example 2:

Compute the approximate number of quanta of visible light emitted as Cerenkov radiation in the frequency range corresponding to wavelengths in vacuum of λ = 4000 to 7000 A.U. given off by an electron of energy 20 MeV traversing 1 cm of Incite (μ 1.49).

Solution:

Relativistic kinetic energy is given by $E = m_0c^2\, [(1 - \beta^2)^{1/2} - 1]$

$$\frac{1}{\sqrt{(1-\beta^2)}} - 1 = \frac{E}{m_0 c^2} = -\frac{20 \times 1.6 \times 10^{-12}}{9.109 \times 10^{31} \times (3 \times 10^8)^2}$$

$$\therefore \quad (1 - \beta^2)^{-1/2} = 40 \text{ or } \beta^2 = 0.9995$$

Hence the number of quanta emitted as Cerenkov radiation per metre

$$N = \frac{2\pi z^2}{137}\left(\frac{1}{\lambda_2} - \frac{1}{\lambda_1}\right)\left(1 - \frac{1}{\beta^2 \mu^2}\right)$$

$$= \frac{2 \times 3.14}{137} \times \frac{3000 \times 10^{+10}}{4000 \times 7000} \times \left[1 - \frac{1}{0.9995 \times (1.49)^2}\right]$$

$$= 26980 \text{ quanta/metre} = 270 \text{ quanta/cm (Approx.)}$$

Example 3(a):

Estimate the number of ion pairs produced in a proportional counter by a 10 MeV proton if the counter size and pressure are large enough *for all the proton's energy to he absorbed. If the gas multiplication factor is 10^3, how many coulombs flow in the counter when this proton is absorbed? If the pulse of current flows for about 10^{-3} sec and if the resistance is 10^4 ohms, estimate the height of voltage pulse.*

Solution:

Energy required for the production of one ion pair = 35 eV, hence the no. of ion pairs produced by a proton of energy 10 MeV

$$= 10 \times 10^6/35 = 2.86 \times 10^5$$

After amplification the total number of ion pairs = nM

$$= 2.86 \times 10^5 \times 10^3 = 2.86 \times 10^8.$$

Charge carried by these ions $\Delta Q = 2.86 \times 10^8 \times 1.6 \times 10^{-19}$

$$= 4.57 \times 10^{-11} \text{ coulomb.}$$

Hence the current passes through the resistor ΔI

$$= 4.57 \times 10^{-11}/10^{-3} = 4.57 \times 10^{-8} \text{ amp.}$$

$\therefore$ The height of the pulse ΔV = I.R. $= 4.57 \times 10^{-8} \times 10^4$

$$= 4.57 \times 10^{-4} \text{ volt.}$$

Example 3(b):

A slow neutron entering a U^{235} fission chamber releases fission fragments of a total energy of 200 MeV. If the capacity of the collector system is 25 $\mu\mu f$, calculate the resultant pulse height.

Solution:

Since the energy required for the production of one ion pair is 35 eV, hence the total number of ion pairs produced

$$N = 200 \times 10^6/35 = 57.14 \times 10^5$$

$\therefore$ Total charge produced = Ne $= 57.14 \times 10^5 \times 1.6 \times 10^{-19}$ coul.

$$= 91.42 \times 10^{-14} \text{ coul.}$$

$$\text{Pulse height } \Delta V = \frac{\Delta q}{C} = \frac{91.42 \times 10^{-14}}{25 \times 10^{-12}} = 3.65 \times 10^{-2} \text{ volt.}$$

Example 3(c):

A.G.M. tube with a cathode 5.0 cm. in diameter and a wire diameter of 0.012 cm is filled with argon to a pressure such that the mean free *path is 7.8×10^{-4} cm. Calculate the value of the voltage that must be applied to just produce an Avalanche.*

Solution:

Electric field strength E at any radius r between two concentric cylinders, when a potential V is applied is given by

$$E = \frac{V}{r \log_e (b/a)}$$

Avalanche will just, start when the integral of this equation, taken over one mean free path is equal to the ionisation potential of the filling gas, *i.e.*,

$$E_{a\lambda} = \frac{V\lambda}{a \log_e (b/a)} = I \text{ or } \frac{Ia \log_e (b/a)}{\lambda}$$

$$V = \frac{15.7 \times 0.006 \times 10^{-2} \times 2.3026 \log_{10} (0.025/0.00006)}{7.8 \times 10^{-6}}$$

$$= 730 \text{ volts.}$$

Example 4(a):

A Geiger-Müller counter has a plateau slope of 3% per 100 volts. If the operative point is at 1100 volts, what is the maximum permissible voltage fluctuation if the counting is not to be affected by more than 0.1?

Solution:

As the plateau slope $= \frac{n_2 - n_1}{n_{av}} \times \frac{100}{V_1 - V_1} \times 100\%$ per 100 volts

In this problem $(n_2 - n_1)/n_{av} = 0.1\% = 0.1 \times 10^{-2}$ and slope = 3

$\therefore V_2 - V_1 = 0.1 \times 10^{-2} \times 1100 \times 100/3 = 3.3$ volts.

As operating voltage is 1100 volts, hence the voltage fluctuation is 3.3 volts in 1100 volts or 0.3%.

Example 5:

An organic-quenched G.M. tube operates at 1000 volts and has a wire diameter of 0.2 mm. The radius of the cathode is 2 cm, and the tube has a guaranteed life-time of 10^9 counts. What is the maximum radial field and how long will the counter last if it is used on the average, for 60 hours per week at 2000 counts per minute?

Solution:

The maximum radial field or the field along a radius at the wire surface

$$E_{maz} = \frac{V_0}{2.3026\text{x} \log_{10} (b/a)} = \frac{1000}{2.3026 \times 10^{-4} \times \log_{10} 200}$$

$$= 1.8^9 \times 10^6 \text{ Volts/metre.}$$

If the lifetime of the tube is N years the total number of counts recorded will be equal to the guaranteed counts. Hence

$$N \times 50 \times 60 \times 60 \times 2000 = 10^9$$

or $$N = 2.77 \text{ years.}$$

Example 6:

An anthracene crystal and a 10 stage photomultiplier tube are to be used as a scintillation spectrometer. It is desired that a 10 keV β-particle incident on the scintillator produce a 2 mV pulse on the photomultiplier output circuit, which has a capacity of 120 μμf. What average electron multiplication per stage is required in the photomultiplier? Assuming a light collection efficiency unity and a photocathode efficiency 0.1.

Solution:

Anthracene gives the highest yield of photons, about 15 for each 1000 eV of energy dissipated in the crystal. Hence, the number of photons emitted in our problem is 150. Since the light collection efficiency is unity and the photocathode efficiency is 0.1, hence number of photoelectrons emitted is 15. If each dynode produces n secondary electrons, after passing through 10 stages the electrons are multiplied by a factor n^{10}.

∴ No. of electrons in the output = $15 \times n^{10}$,

Hence the output pulse $\Delta V = \dfrac{\Delta q}{C} = \dfrac{15 \times n^{10} \times 1.6 \times 10^{-19}}{120 \times 10^{-12}}$

As it is given to be 2mV, hence

$$n^{10} = \frac{120 \times 10^{-12} \times 2 \times 10^{-3}}{15 \times 1.6 \times 10^{-19}} = 10^5$$

or $$n = 3.16 \text{ (Approx.).}$$

EXERCISES

1. A. 0.1 MeV electrons is absorbed in an anthracene scintillator accompanied by the 10 stage photomultiplier. If it produces a 0.01 V pulse in the photomultiplier output circuit, which has a capacity of 160 μμf. What average electron multiplication per stage is required in the photomultiplier. Assume a photocathode efficiency of 0.07.
2. The plateau of a G.M. counter working at 1 kV has a slope of 2.5% count rate per 100 volts. By how much can the working voltage be allowed to vary if the count rate is to be limited to 0.1%?
3. A silicon detector has an area of 3.0 cm^2 and a depletion layer of 60 microns when polarised with 50 volts. Calculate the time constant of the circuit when used with a 100 megaohm resistance, the time required to collect all of the holes and the amplitude of the voltage pulse produced by the absorption of a 3.78 MeV α-particles.
4. Calculate the resolving time of a pulse counter from the following schcdule of count rates: background 65 cpm, first source in position 8350 cpm, both sources in position 16832 cpm, second source in position 9416 cpm.
5. The paralysis time of a G.M. tube is 400m sec. What is the true count rate for measured count rates of 100, 1,000 and 10,000 counts per min.
6. Scintillator with a background count rate of 260 cpm is used to assay samples with average 900 cpm. What will be the coefficient of variation and the standard deviation of an, "equal-count" assay of 10,000 counts each ? What time will be required for each complete determination.
7. A G.M. tube with a cathode diameter of 4.0 cm, and a wire diameter of 0.016 cm is filled with argon and alcohol to a pressure such that the mean free path is 4.6×10^{-3} cm. Calculate the maximum radius at which secondary ions will be formed when 1200 volts is applied to the tube.
8. A proportional counter has a gas magnification factor of 1000. Estimate the number of mean free paths in which this multiplication takes place and calculate the distance from the

wire within which this multiplication takes place assuming $\lambda = 2 \times 10^{-4}$ cm, for the electrons in the gas.

9. It is required to operate a proportional counter with a maximum radial field of 10 mV/meter. What is the applied voltage required if the radii of the wire and tube are 20 urn and 10 mm respectively ?
10. A sample of uranium, emitting a-particles of energy 4.18 MeV, is placed near an ionization chamber. Assuming that only 10 particles per sec. enter the chamber, calculate the current produced.
11. The output circuit of a ratemeter consists of a resistance of 50,000 ohms shunted by a condenser. The output pulses from the ratemeter have a constant amplitude of 5.0 milliamperes, and a variable width. What pulse width is required to produce a voltage of 5.0 volts across the tank circuit at a count rate of 200 cpm? What pulse length will be required to obtain the same voltage at a count rate of 50,000 cpm? What capacitance will be required to obtain a circuit time constant of 2 sec ?
12. A square wave whose peak to peak value is 1.0 volt extends 0.5 V with respect to earth. The duration of the positive section is 0.1 sec and of the negative section is 0.2 sec. If this waveform is impressed upon a R-C differentiating network of time constant 0.2 sec, what are the steady state maximum and minimum values of the output waveform.

4

Laser Techniques with Photoelectric Effect

INTRODUCTION TO LASER TECHNIQUES

A large percentage of an ensemble of molecules can be brought into an excited state, a way frequently can be found by which large numbers of these excited molecules can be triggered into an almost simultaneous joint transition back to the inactive state, with the resulting emission of a beam of intense coherent radiation. Such a device is now known as laser, an acronym for light amplification by stimulated emission of radiation.

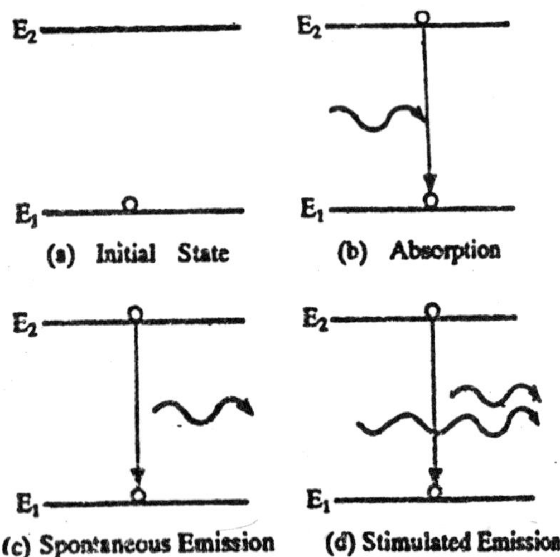

Fig. 1 : Four stage os laser action.

The four steps associated with such a device are illustrated in Fig. 1. In (a) the molecule is inactive, in (b) the molecule gets activated by a photon, in (c) the molecule returns spontaneously to its initial state, emitting radiation, and in (d) the transition to the ground state is triggered by an incoming beam, which is then amplified by the emitted radiation. For effective laser operation the amount of emission form (c) must be small compared with (d). In a typical solid laser like a ruby crystal, the active atoms are those of chromium, which are held in a lattice of aluminium oxide. A flash of intense light raises a large percentage of the chromium atoms to the excited state. Then radiation can trigger the return transition, resulting in the emission of light which is coherent both in space and in time because the stimulating photon on entering the atom causes it to emit a photon which is precisely "in step" with it.

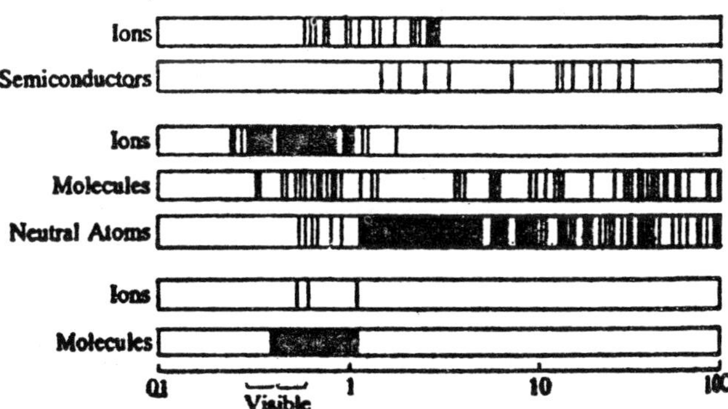

Fig. 2 : Laser and maser radiation sources available.

In this way, monochromatic beams os great intensity can be produced which travel great distances without spreading. There are many important applications possible for such beams both in science and technology. They are valuable in studying Raman scattering and are certainly going to be an asset in communication. As the laser principle applies over the entire range of the spectrum, it is used extensively in the generation of microwaves. The device is then called a *maser*, microwave amplification by simulated emission of radiation. In Fig. 2, some of the wavelengths of various laser and maser sources are depicted.

VALIDITY OF BEER'S LAW

When Beer's law is obeyed by a solution, it means that.

(a) There are no interactions between molecules of the solution.

(b) No new types of molecules will be formed as the concentration is changed.

Deviation from Beer's Law : From Beer's law it follows that if we plot absorbance A against concentration, a straight line passing through the origin should be obtained Fig. 3. But there is usually a deviation from a linear relationship between concentration and absorbance and an apparent failure of Beer's law may ensue. Deviations from the law are reported as *positive* or *negative* according to whether the resultant curve is concave upwards or concave downwards.

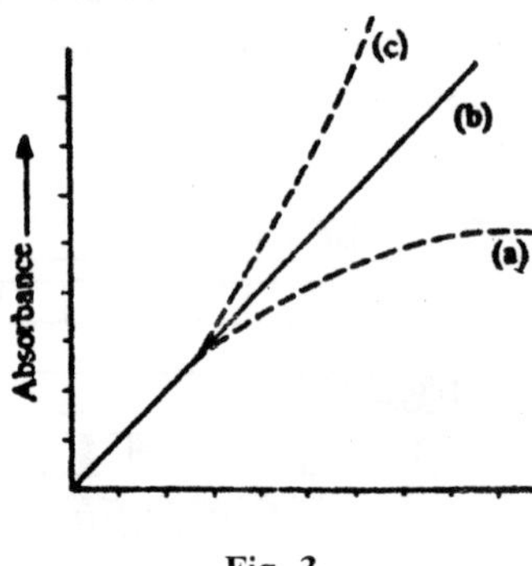

Fig. 3

Deviation from Beer's law can arise due to following factors.

(i) Beer's law will hold over a wide range of concentration provided the structure of the coloured ion or of the coloured non-electrolyte in the dissolved state does not change with concentration. If a coloured solution is having a foreign substance whose ions do not react chemically with the coloured components, its small concentration (foreign substance) does not affect the light absorption whereas its large concentration may affect light absorption and may also alter the value of the extinction coefficient.

(ii) Deviation may also occur if the coloured solute ionises, dissociates or associates in solution.

For example, benzyl alcohol, in chloroform exists in a polymeric equilibrium.

$$4C_6H_5CH_2OH \rightleftharpoons (C_6H_5CH_2OH)_4$$

Dissociation of the polymer increases with dilution. The monomer absorbs at 2.0 to 2.765 μ whereas the polymer absorbs at 3.000 μ. Hence, absorption at 2.050 μ shows negative deviation whereas at 3.000 μ *positive deviation.* Another example of deviation is the change of colour of dichromate ion on dilution. The effect can be represented by the equilibrium shown below.

$$\underset{\text{(Orange)}}{Cr_2O_7^{2-}} + H_2O \rightleftharpoons 2HCrO_4^- \rightleftharpoons 2H^+ + \underset{\text{(Yellow)}}{CrO_4^{2-}}$$

It can be seen from above that a solution of dichromate is converted gradually into chromate. The absorption at 450 μ, is positive and at 350 μ is negative.

(iii) Deviations may also occur due to the presence of impurities that fluoresce or absorb at the absorption wavelength. This interference introduces an error in the measurement of absorption of radiation penetrating the sample.

(iv) Deviations may occur if monochromatic light is not used.

(v) Deviations may occur if the width of slit is not proper and, therefore, it allows undesirable radiations to fall on the detector. These undesirable radiations might be absorbed by impurities present in the sample which would cause an apparent change in the absorbance of the sample. The magnitude of this deviation becomes appreciable at higher concentrations.

(vi) Deviation may occur if the solution species undergoes polymerisation. For example, benzyl alcohol in carbon tetrachloride in high concentration exists in polymeric form $[(C_6H_5CH_2OH)_4]$, association of this polymer increases with dilution. Due to this change, absorbance will change.

(vii) Beer's law cannot be applied to suspensions but the latter can be estimated colorimetrically after preparing a reference curve of known concentrations.

Illustration : From the Beer's law it follows that if the absorption remains constant, then the concentration (c) should be related to the thickness of the absorbing layer (t) in a reciprocal manner. This fact was demonstrated by *Halbon* (1922). He observed that a layer of chlorine 20 cm long at a pressure of 0.1 atmosphere would have the same transmission at 10 cm layer at a pressure of 0.2 atmosphere

Applications of Beer's Law : Beer-Lambert's law can be used for the determination of an unknown concentration by comparison with a solution of known concentration by using colorimeter or spectrophotometer, the principle of which is based upon Beer's law.

LAWS OF PHOTOCHEMISTRY

There are three laws which govern the effect of radiation on chemical reactions. These laws are.

I. Grotthuss–Draper Law : This is sometimes referred to as the first law of photochemistry. On the basis of certain theoretical considerations, Grotthuss in 1818 discovered this law. This law was reaffirmed in 1841 by J.W. Draper as a result of his experimental researches on photochemical reaction between hydrogen and chlorine. It may be stated as follows.

"When light falls on any substance, only the fraction of incident light which is absorbed by the substance can bring about a chemical change', reflected and transmitted light do not produce any such effect."

It is important to remark that all light radiations which are absorbed by reacting system are not effective in producing desired chemical reactions. When conditions are not favourable for the molecules to react, some portion or the whole light absorbed by reacting substances is converted into heat in some cases. While in some other cases, the absorbed light is re-emitted as radiations of the same or another frequency.

Grotthuss-Draper law is purely qualitative. It does not give any relationship between the amount of light absorbed by a system and the number of molecules which have reacted,.

II. Law of Photochemical Equivalence : It is one of the most important laws in photochemistry. According to quantum theory of light, energy is absorbed or emitted only in small packets of quanta of magnitude $\varepsilon = h\nu$ where ε is the energy, ν is the frequency W light and h is the Planck's constant. In 1905, Einstein applied quantum theory to photochemical reactions and enunciated the law of Photochemical Equivalence which states that

"When an atom or molecule absorbs light of a given frequency, it absorbs one quantum only,"

The photochemical importance of the above law emphasized in 1909 by Stark and other. In 1913, Einstein considered the work of Stark and restated his law as.

"Each molecule which takes part in a chemical reaction absorbs one quantum of light which induces the reaction " or briefly "one molecule one quantum."

The term one photon (or one quantum) means energy equal to hv, where *h* is the Planck's constant and v is the frequency of radiation

$$AB + hv \rightarrow AB^*$$

Thus, in the primary process, the number of molecules that are activated is equal to the number of quantas absorbed. Therefore, amount of energy E, for activation of 1 mole will be N hv where N is the Avogadro's number and is equal to 1 mole, *i.e.,*

$$E = Nhv$$

This quantity of energy 'E" *absorbed per mole of the substance is called an Einstein.*

But $N = 6.023 \times 10^{28}$ molecules

$h = 6.624 \times 10^{-27}$ erg sec

$\therefore$ $E = 6.023 \times 10^{28} \times 6.624 \times 10^{-27} \times v$ ergs

$= \underline{6.1023 \times 6.624\times 10^{-27} \times v}$ calories ...(1)

$E = 9.53 \times 11-11 \times v$ calories

[$\because$ 1 cal = 4.186×10^{7} ergs]

But $$v = \frac{c}{\lambda} = \frac{\text{velocity of light in cm/sec}}{\text{wave length in cm}} = \frac{3\times 10^{10}}{\lambda}$$

Substituting the value of v in equation (1), we obtain

$$E = \frac{9.53\times 10^{-11}\times 3\times 10^{10}}{\lambda} \text{ calories}$$

where λ is expressed in cm. But 1 A° = 10^{-8} cm. Therefore,

$$E = \frac{9.53\times 10^{-11}\times 3\times 10^{10}}{\lambda\times 10^{-8}} = \frac{2.859}{\lambda}\times 10^{8} \text{ cal}$$

$$= \frac{2.859}{\lambda}\times\frac{10^{8}}{10^{3}} \text{ K cal} \quad [\because 1 \text{ K cal} = 1000 \text{ cal}]$$

$$= \frac{2.859}{\lambda}\times 10^{5} \text{ K cal}$$

Thus, the value of one Einstein of radiation of given frequency or wavelength is given by the relation (2). It is evident from equation (2) that the energy absorbed per mole decreases with increasing wavelength. In other words, the value of E is inversely proportional to the wavelength of light absorbed, *i.e.,*

$$E \propto \frac{1}{\lambda}$$

From the above facts it is evident that shorter the wavelength of light, the greater the energy. It has been verified that.

1. For the wavelength 4000Å (ultraviolet light), the energy is 71 K cal;
2. For the wavelength 7500Å (red light), the corresponding energy is 38 K, cal.

From the above results, it follows that in ultraviolet and violet portions, the radiations will be more active chemically than those of longer wavelengths.

QUANTUM YIELD OR QUANTUM EFFICIENCY

The photochemical equivalence law applies to the primary photochemical process, *i.e.,* as a result of primary absorption of one quantum, only one molecule undergoes dissociation and the products enter no further reaction. In such cases, there will be 1 : 1 relationship between the number of quantas absorbed and the number of reacting molecules.

In most cases, a molecule activated photochemically initiates a series of thermal reactions called secondary reactions. As a result of such reactions many reactant molecules may undergo chemical change by absorbing one quantum only. Under such conditions, there will be no. 1 : 1 relationship between the number of quanta absorbed and the number of reacting molecules.

In some cases, a molecule activated photochemically undergoes deactivation. Thus, less than one molecule may react per quantum

To describe the deviation from 1 : 1 relationship, the idea of quantum yield or efficiency of a process 'ϕ' was introduced. This can be defined as follows :

"It is the number of molecules which undergo chemical transformation per quantum of absorbed energy."

Mathematically, this quantity is defined as follows.

or $$\phi = \frac{\text{No. of molecules reacting in a given time}}{\text{No. of quanta absorbed in the same time}}$$

or $$\phi = \frac{\text{No. of molecules reacting in a given time}}{\text{No. of einsteions absorbed in the same time}}$$

or $$\phi = \frac{\text{Rate of chemical reaction}}{\text{No. of einsteins absorbed}} \quad ...(1)$$

The number of quanta absorbed (n_a) in a unit time i

$$n_a = \frac{Q}{hv} \quad ...(2)$$

and the number of molecules transformed under the action of light (n_r) is

$$n_r = \frac{Q}{hv} \quad ...(3)$$

On substituting equation (2) in (1), we get

$$\phi = \frac{n_r}{n_a} = \frac{n_r}{Q/hv} \quad ...(4)$$

The rate of a chemical reaction; according to equation (4), is as follows.

$$-\frac{dn}{dt} = \frac{dn_r}{dt} = \phi\frac{dn_a}{dt} = \phi\frac{Q}{hv}$$

The concept of quantum yield or quantum efficiency was first introduced by Einstein. Because of the frequent complexity of photochemical reactions, quantum yields as observed vary from a million to a very small fraction of unity.

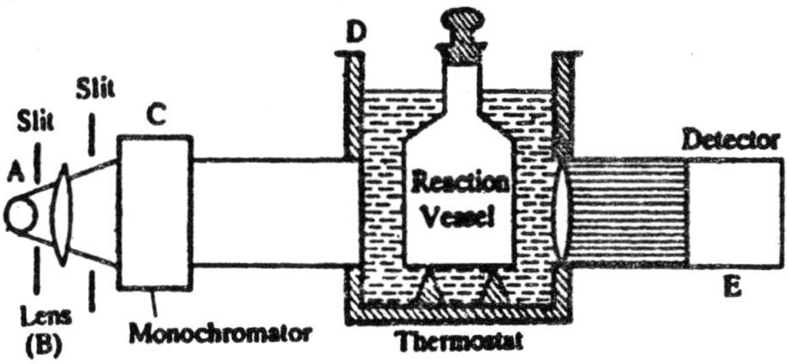

Fig. 4

The concept of quantum yield can be extended to any act, physical or chemical, following light absorption. The quantum yields are dependent on the light intensity. The concept of quantum yield provides a mode of account keeping for partition of absorbed quanta into various pathways. If Stark-Einstein law is correct, then value of ϕ should always be unity. However, later on, it was realised that the law of photochemical equivalence is applicable to only primary processes.

Experimental Determination of Quantum Yields : Knowledge of quantum efficiency of a photochemical reaction offers valuable information about the mechanism of photochemical reactions. In order to determine the value of quantum yield of photochemical reaction, it is essential to know (a) Number of moles reacting, and (b) Number of Einsteins absorbed. For this purpose, an arrangement such as shown in Fig. 5 is required.

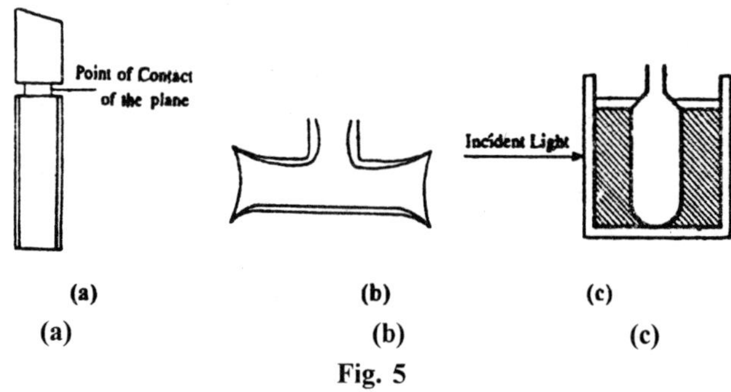

Fig. 5

1. *Light source.* 'A' is the source of light which may be sunlight, arc lamp, mercury vapour lamp, discharge tube, etc. For a quantitative type of work, mercury vapour lamp is the best source of ultra-violet radiations because it emits radiation of suitable intensity in the desired spectral range., For visible or ultraviolet regions, the tungsten lamp is often used in place of mercury vapour lamp.
2. *Monochromator.* C is a monochromator or a filter. When light from source, A is passed through the lens B and then allowed to pass through, C_0 the monochromator (C) will absorbed that these colour filters transmit light within certain range of wavelengths. Monochromators are generally made of gelatin or coloured glass or transparent plates with metal films of suitable thickness which absorb undesired wavelengths by interference.
3. *Reaction cell.* The light from a monochromator enters a reaction cell D which contains the reaction mixture. The reaction cell is generally made of glass or quartz with optical plane windows

for free exit and entrance of light radiation. For visible spectral range, the reaction cell and other parts of the instrument are generally made of glass. For ultraviolet light, all optical parts are made of quartz glass. For solutions, the cell is provided for stirring.

The actual design of the cell depends upon the nature of the reaction, but the front and the back should be plane parallel. The types of cells generally used.

These represent the various types of cells which are generally used. In sealing plane window absorption cell; there is a tubular absorption cell, there is a absorption cell with in blow window.

4. *Detector.* D is the detector which is used for determining the intensity of light coming from a reaction cell. The intensity of light is measured with the reaction cell when empty and then with are action mixture. The difference between the two readings will give the amount of energy absorbed by the reaction system under examination.

The intensity of light can be measured by the following types of detectors :

(a) *Radiomicrometer:* It consists of a coil (thermocouple) which is suspended between the poles of an electromagnet. When the couple gets heated by the light radiations, a current flows in the circuit and it is deflected in the magnetic field. The deflection can be measured by lamp and scale arrangement. This deflection is thus, proportional to the intensity of light radiations.

(b) *Photo-electric cell.* The principle of photo-electric cell is based upon the photo-electric effect. It is a device which converts light energy into electrical energy, A photo-electric cell consists of lithium or sodium metal which is enclosed in a vessel. The metal is generally connected to gold leaves of an electroscope or to an electrometer.

When the light radiations are incident upon lithium metal, it loses electrons and develops a positive charge on itself whereby the gold leaves diverge. This divergence of the gold leaves is proportional to the intensity of light radiations. Photo-electric cells are generally used for measuring very low intensities of light shown in Fig. 6.

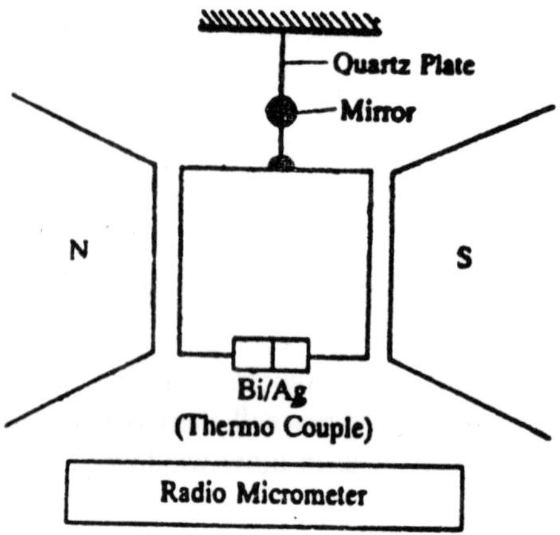

Fig. 6

(c) *Chemical actinometers. When* extreme accuracy is not desired, a chemical actinometer can be used to measure the intensity of light radiation. A chemical actinometer generally consists of gas mixtures or "solutions which are sensitive to light. When radiations Fall upon these substances, a chemical reaction will take place and the extent of which is a direct measure of energy absorbed. Following are the main types of actinometers.

(i) *Bunsen and Roscoe's actinometer.* It is illustrated.

The bulb B′ contains a mixture of hydrogen and chlorine over water. The bulb D′ also contains water. The two bulbs B′ and D′ are connected through a graduated tube C′. When the light is incident on B′, some of the gases in A′ react to form hydrochloric acid. This is absorbed in water of B' and the water moves from C′ to B′. The motion of the water thread will be proportional to the amount of hydrochloric acid formed and hence to the intensity of light absorbed.

(ii) *Eder's actinometer : J. M. Eder (1979)* utilized the following reaction to determine the absorbed radiations

$$2HgCl_2 + (NH_4)_2C_2O_4 \rightarrow 2NH_4Cl + 2CO_2 + Hg_2Cl_2$$

The system was exposed to radiations and the mercurous chloride formed in the reaction was weighed. From the weight of mercurous chloride or CO_2 evolved, the absorbed radiations may be determined quantitatively.

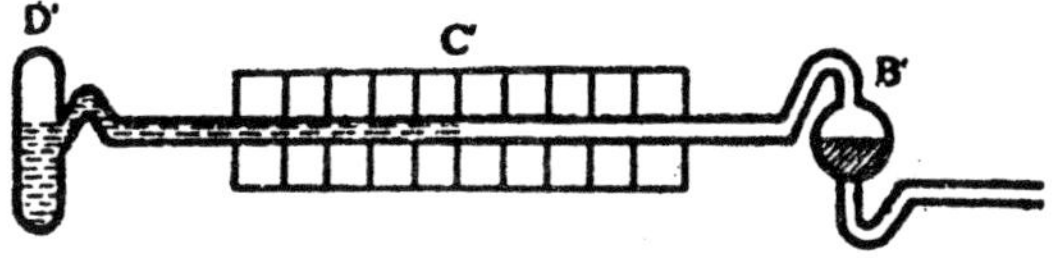

Fig. 7

(iii) *The uranyl oxalate actinometer* : This type of actinometer consists of a solution of 0.05 M oxalic acid and 0.01 M uranyl sulphate in water. When light is incident upon it, oxalic acid undergoes dissociation to form CO, H_2O and CO_2. The extent of the decomposition of oxalic acid is a measure of intensity of light absorbed. The progress of the reaction is determined by titrating it with $KMnO_4$ before and after the exposure.

$$UO_2^{2+} + \begin{matrix} COOH \\ | \\ COCH \end{matrix} \xrightarrow{hv} CO + CO_2 + H_2O + UO_2^{2+}$$

(d) *Thermopile* : It consists of a series of unlike metals which are connected, to a galvanometer. One end of it is coated with lamp black, or platinum black and the other end "is kept as such, The thermopile is kept in, glass or quartz vessel.

When the light from a reaction cell is incident upon the black surface, it gets heated up, resulting a difference in temperatures of the black and the other cold end. Thus, a current flows in the circuit. The current produced is proportional to the intensity of light absorbed. Thermopiles are generally calibrated with standard light sources.

(e) *Geiger Muller counter* : It is a modification of the ionisation chamber and is generally used for measuring very low intensities. The apparatus consists of an ionisation chamber with a thin wire of steel or tungsten. When a charged particle enters the chamber the air present in it gets ionised and discharge is initiated in the gas. A current then .flows between the electrodes which can be detected in the usual manner,.

Process For the Determination of Quantum Yield : Set up the apparatus. The choice of the detector depends upon the nature of the experiment. First keep the reaction vessel empty. Allow the light to pass through it. Determine the light intensity as such. In case of solution, it is filled with the solvent.

Now, fill the reaction vessel with reaction mixture. Allow the light to pass through it for a known time. Then note the intensity of light after the reaction is over. The difference between the two readings will give the amount of energy absorbed by the reaction system under examination. Analyse the contents of reaction mixture. Then, determine the no, of moles reacted in given time. Then, apply the following equation to get the value of quantum yield.

$$\text{Quantum yield} = \frac{\text{No. of moles reacting in a given time}}{\text{No. of einsteins absorbed in the same time}}$$

DEVIATIONS IN THE LAW OF PHOTOCHEMICAL EQUIVALENCE

The photochemical equivalence law applies to the primary photochemical process. As a result of primary absorption of one quanta, only one molecule undergoes dissociation and the products enter no further reaction. In such cases, there will be 1 : 1 relationship between the number of quantas absorbed and the number of reaction molecules. In most cases, a molecule activated photochemically initiates a series of thermal reactions called secondary reactions. As a result of such reactions, many reactant molecules may undergo chemical change by absorbing one quanta only under such conditions.

Class I : High Quantum Yield Reactions : These are the photochemical reactions in which the quantum yield is greater than unity. Examples are :

(i) The quantum yield of the combination of carbon monoxide and chlorine by a light of wavelength ranging from 4000 to 4360 Å is 10^3.

$$CO + C1_2 \rightarrow COC1_2$$

(ii) Another example is the combination of hydrogen and chlorine to form hydrogen chloride by a wavelength less than 4500°. The quantum yield of this reaction has been found to vary from 10^4 to 10^6, *i.e.,* one quantum brings about the combination of 10^4 to 10^6 molecules.

$$H_2 + Cl_2 \rightarrow 2HCl$$

(iii) Another example is the photochemical decomposition of H_2O_2, by a wavelength of about 3100Å". The quantum yield for this reaction is more than 7.

$$2H_2O_2 \rightarrow 2H_2O + O_2$$

Class II. Low Quantum Yield Reactions : This type includes such photochemical reactions in which the quantum yield is less than unity. Examples are.

(i) Combination of hydrogen and bromine to form HBr is an example of this type. The quantum yield of this reaction is about 0.01.

$$H_2 + Br_2 \rightarrow 2HBr$$

(ii) Photolysis of ammonia by a wavelength 2100Å is another example.

$$2NH_3 \rightarrow N_2 + 3H_2,$$

The quantum yield of this is nearly 0.2.

(iii) The quantum yield for the dissociation of acetone to form carbon monoxide and ethane by a light of wavelength 3000Å is 0.1.

$$CH_3COCH_3 \rightarrow CO + C_2H_6$$

(iv) The quantum yield of the concentration of maleic acid into fumaric acid by a light of wave-length ranging from 2000 to 1800 Å is 0.04.

$$\underset{\text{Maleic Acid}}{\begin{array}{c}\text{H—C—COOH}\\ \|\\ \text{H—C—COOH}\end{array}} \longrightarrow \underset{\text{Fumaric Acid}}{\begin{array}{c}\text{H—C—COOH}\\ \|\\ \text{HOOC—C—H}\end{array}}$$

Class III. Small Integer Quantum Yield Reaction : This type includes the photochemical reaction in which the quantum yield is a small integer like 1, 2, 3, etc. Examples are.

(i) The quantum yield for the dissociation of hydriodic acid by a light of wave length 2070 – 2823Å is 2.

$$2HI \longrightarrow H_2 + I_2$$

(ii) The quantum yield of the following reaction is unity.

$$2Fe^{3+} + I_2 \longrightarrow 2Fe^{2+} + 2I^-$$

This reaction is carried out in liquid phase by a wave-length 5790Å.

(iii) The quantum yield of the combination of SO_2 and chlorine to form SO_2C1_2 by a light of wavelength 4200 Å is one.

$$SO_2 + C1_2 \longrightarrow SO_2Cl_2.$$

(iv) The quantum yield of the ozonisation of oxygen by a light of wave-length 1700 – 1900 Å is three.

$$3O_2 \text{ ¾→ } 2O_3$$

REASONS OF HIGH AND LOW QUANTUM YIELD

In order to explain the exceptions to Einstein's Law of photochemical equivalence, *Bodenstein and others* pointed out that photochemical reactions involve two distinct processes I

(a) *Primary process.* It is the process in which each molecule capable of entering into chemical reaction absorbs one quantum of radiation (hv). The absorbed energy may give rise to the formation of excited molecule, *i.e.,*

$$AB + hv \rightarrow AB^*$$

or the molecule which absorbs light may undergo dissociation to yield free atoms (some in excited state) or free radicals.

$$AB + hv \rightarrow A. + B.$$

(b) *Secondary process.* It is the process which involves the excited atoms or molecules or free radicals produced in the primary process. A secondary process can take place in the dark as well. Due to this process, a different number of molecules may undergo reaction, by absorbing one quanta only. As a result, the quantum efficiency of the reaction, as a whole, will differ from unity.

From the above discussion it follows that the value of quantum efficiency or yield will depend upon secondary processes whereas the primary process remains the same for every reaction. We will now discuss the reasons for different quantum yields for different reactions on the basis of secondary process.

Reasons of High Quantum Yield : These are outlined below:

(a) When the excited atoms or molecules or free radicals produced in the primary state undergo secondary processes, each secondary process gives rise to some further excited particles or radicals; this process continues unless it is checked due to one or the other reason. Thus, by absorbing one quanta only, a large number of reactant molecules undergo reaction. Hence, the quantum efficiency of this type of reaction will be greater than unity.

(b) In some reactions, such as the. Combination of hydrogen and chlorine, a chain is set up and quantum yield becomes as high as 10s. These reactions are known as chain reactions. Some evidences in favour of chain reactions are given below.

(i) The chain is broken when a foreign substance is added in the system to remove H_2 or Cl_2.

(ii) **Marshall and Taylor** found that at least one step in the chain reaction can be carried out experimentally.

Reasons of Low Quantum Yield : The low quantum yield is explained on the basis of following reactions.

(a) If the excited particles formed in the primary process are such that they cannot react due to their deactivation by collisions, by fluorescence or by internal arrangements or by coming in contact with inert molecules the quantum yield will be extremely low.

(b) The excited particles produced in the primary process may recombine to form the reactants so as to give low quantum yield.

To illustrate the above mentioned reasons, a few examples will now be taken.

(i) Dissociation of HI : In the primary process, a molecule of hydrogen iodide absorbs light of wavelength less than 4000 Å AND dissociates to form an atom of hydrogen and an atom of iodine. The primary process may be represented as.

$$HI + h\nu \rightarrow H + I$$

The primary process has been established from spectroscopic studies. The possible secondary reactions are.

$$H + HI \rightarrow H_2, + I \quad ...(1)$$

$$I + HI \rightarrow I_2 + H \quad ...(2)$$

$$H + H \rightarrow H_2 \quad ...(3)$$

$$I + I \rightarrow I_2 \quad ...(4)$$

$$I + H \rightarrow HI \quad ...(5)$$

Reaction (2) is endothermic and, thus, it cannot take place at ordinary temperatures. Reactions (3) and (5) are highly exothermic. The heat produced in these reactions is so high that the products formed in these reactions undergo dissociations. Thus, the occurrence of reactions (3) and (5) is highly unlikely. If we eliminate reactions (2), (3) and (5), the reactions (1) and (4) are the only two likely secondary reactions. On this basis, the overall process may be written as.

$$HI + h\nu \rightarrow H + I \quad \text{(Primary)}$$

$$H + HI \rightarrow H_2 + I \quad \text{(Secondary)}$$

$$I + I \rightarrow I_2 \quad \text{(Secondary)}$$

$$\overline{2HI + h\nu \rightarrow H_2 + I_2}$$

Thus, two molecules of hydrogen iodide can undergo dissociation by absorbing one quanta only. Thus, the quantum efficiency of the reaction is 2 which has been found experimentally.

(ii) **Dissociation of HBr :** The mechanism of the reaction is analogous to that proposed above for HI.

(iii) **Combination of Hydrogen and Bromine :** The quantum efficiency of the reaction is about 0.01. The primary process involves the dissociation of bromine molecule into bromine atoms.

$$Br_2 + h\nu \rightarrow 2Br$$

The expected subsequent secondary processes are.

$$Br + H_2 \rightarrow HBr + H \quad ...(1)$$

$$H + Br_2 \rightarrow HBr + Br \quad ...(2)$$

$$H + HBr \rightarrow H_2 + Br \quad ...(3)$$

$$Br + Br \rightarrow Br_2 \quad ...(4)$$

Reaction (1) is highly endothermic and is, therefore, taking place very slowly at ordinary temperature. At this slow speed, most of the bromine atoms recombine to give the bromine molecules. Hence the reactions (2), (3) and (4) which are due to reaction (1) cannot occur. Therefore, the quantum yield will be extremely low.

(iv) Polymerisation of Anthracene : This reaction can be represented as

$$2C_{14}H_{10} \underset{\text{dark}}{\overset{\text{light}}{\rightleftharpoons}} C_{28}H_{20}$$

The quantum yield of this reaction should be 2 but actually it is only 0.5. It is due to the reason that this reaction involves a back thermal reaction. The low quantum yield of isomeric transformation of maleic acid into fumaric acid can be explained on the basis that a state of equilibrium is reached in this reaction which will lower the quantum yield.

FACTORS AFFECTING QUANTUM YIELD

Except those reactions which obey Einstein's law of photochemical equivalence, it is found that the quantum yield depends upon the following factors.

1. **Temperature :** The change of quantum yield with temperature is given by the relation.

$$\frac{d \log \phi}{dT} = \frac{Q}{RT^2}$$

where f = quantum yield.

Q = amount of heat evolved by the formation of one gm mole of the substance.

Kuhn found that the quantum yield of photochemical decomposition of ammonia increases by about 15% for every 100° temperature increase. It reaches seven fold at 500°.

For some photochemical reactions the values of temperature coefficient and activation energy are given in Table 1.

Table 1 : Temperature Coefficient and Activation Energy of Some Photochemical Reactions.

Reactions	Temperature coefficient	ΔE (K cal).
$H_2 + Br_2 \rightarrow 2HBr$	1 – 1.02	40.7
$H_2 + Cl_2 \rightarrow 2HCl$	1 – 1.04	44.9
$H_2 + O_2 \rightarrow H_2O_2$	1.09	47.6
$2O_3 \rightarrow 3O_2$	1.27	80.5
$2C1_4H_{10} \rightarrow C_28H_20$	1.04	44.8

2. **Wave-length :** It has been found experimentally that the power of a quantum or photon of light energy to induce photochemical change is greater, the higher the frequency of incident radiation. In other words, the quantum yield will be lower at the higher-wave lengths.

Bonhoeffer and *Harteck* (1930) suggested that the lower quantum yield at longer wavelength is due to a low efficiency of the primary dissociation process in that region. Generally, the quantum yield is always found to rise over a range of frequencies, as in the case of decomposition of nitrogen' dioxide.

Wavelength	310	365	405	436
Quantum yield	2.97	1.54	0.74	0.009

Henri, Wurmser and Lasareff have obtained a proportionality between the amount of light of different wavelengths absorbed and the photochemical effect produced.

3. **Light intensity :** It was shown by Dhar and his coworkers that decrease in light intensity results in an increase of quantum yield. The light intensity was found to change by varying the iris of the aperture, through which light was allowed to pass into the reacting system. These workers found that.
   (i) The velocity of photochemical reactions is found to be directly proportional to the intensity of light.
   (ii) The quantum yield decreases as the amount of light falling on the reacting substance is increased.
4. **Inert Gases :** It was found by Hartel that in most of the photochemical reactions, the addition of inert gases increases the quantum yield. This may be explained due to the occurrence of an induced pre-dissociation which will increase the number of starting chains as it may retard the diffusion of the atoms to the walls. This latter effect would decrease the rate of chain terminating step thereby increasing the speed of reaction.

LUMINESCENCE

We know that the usual methods of obtaining light are involving the heating of solids, or solid particles to sufficiently high temperatures. For example, by heating a tungsten, platinum wire or a carbon filament electrically or in a flame, bright light can be obtained. These are examples of *incandescence* in which thermal energy is converted into light. If, however, light is produced by processes other than those involving heating, such light is called *luminescence light or cold light.*

The above discussion may be summarised as follows.

"When the emission of visible radiation occurs due to some cause other than temperature, the phenomenon is known as luminescence.

As light is produced at low temperatures, the term luminescence may be regarded as 'light' produced without 'heat' or 'cold light' But in incandescence and luminescence, emission of light occurs due to the return of electrons from excited outer position to lesser excitation position or ground state. Luminescence is of the following types.

1. **Photoluminescence :** Luminescence caused by light is called photoluminescence. Photoluminescence that ceases immediately after the cause of exciation is cut off is called *fluorescence.* Photoluminescence that persists for an appreciable time after the stimulating process is cut off is called *phosphorescence.*
2. **Chemiluminescence :** Luminescence resulting from chemical reactions is called *cathodoluminescence.*
3. **Cathodoluminescence :** Luminescence caused by bombardment of electrons is called *cathodoluminescence.*

   Let us discuss these one by one.
4. **Electroluminescence :** Luminescence resulting from the application of an electric field to matter is called *electro-luminescence.*

FLUORESCENCE AND PHOSPHORESCENCE

Introduction to Fluorescence : *When a beam of light is incident on certain substances, they emit visible light or radiations and they stop emitting light or radiation as soon as the incident light is cut off. This phenomenon is known as fluorescence.*

Such substances which emit radiations during the action of stimulating light are called fluorescent substances.

When a substance absorbs light energy, it will result in the excited state of an atom or molecule. But if the light absorbed is not sufficient energetic to eject an electron, it will cause the electrons to move from inner orbits to outer orbits.

When these excited electrons return to the ground state, the light is emitted which may possess different frequency than the incident light. The successive stages may be put as follows.

$$\underset{\text{Normal state}}{X} + \underset{\text{photon}}{h\nu} \longrightarrow \underset{\text{Excited state}}{X^*}$$

$$X \longrightarrow X^{**} + h\nu'$$

$$X^{**} \longrightarrow X + h\nu''$$

where X** represents an energy state between X and X*. The emitted radiation frequencies v′ and v″ are different from *v, i.e.,* the frequency of incident radiations.

Some characteristics of the phenomenon of fluorescence are as follows

(i) This phenomenon is instantaneous and starts immediately after the absorption of light and stops as soon as the incident light is cut off.

Fluorescence is stimulated by light of the visible or ultraviolet regions of the spectrum. Line, band and continuous spectra of emitted light of fluorescence are observed. The character of the spectrum depends essentially on the state of aggregation of the substance.

(ii) It is a general phenomenon and is exhibited by gases, liquids and solids. No fluorescence will be observed in gases, unless the pressure is low.

(iii) Different substances fluoresce with light of different wavelengths. 'Thus fluorspar fluoresces with blue light, chlorophyll with red light, uranium glass with green light and so on.

(iv) The fluorescent light from solutions is polarised and the degree of polarisation depends in some cases upon the concentration of the solution.

(v) The extent of fluorescence depends upon the nature of the solvent and the presence of certain anions in solution. Thus the thiocyanate, iodide and bromide ions show a marked quenching effect.

(vi) According to *Stoke's law,* during fluorescence light is absorbed at a certain wavelength and should be emitted at a greater wavelength.

The above principle was recognised before the quantum theory was proposed.

(vii) The quantum yield in fluorescence is the ratio of the number of photons of luminescent radiation to the number of photons absorbed from the stimulating light upon a fixed wavelength of the latter. The quantum efficiency of fluorescence increases in proportional to the wavelength l of absorbed radiation. Then, after reaching its maximum value in a certain interval of $\sim \lambda_{max}$ the efficiency drops rapidly to zero upon a further increase in l.

(viii) Fluorescence may be regarded as a secondary effect resulting from the primary process of absorption of a quantum of light by an atom or molecule.

Introduction to Phosphorescence : *When light radiation is incident on certain substances, they emit light continuously even after the incident light is cut off. This type of delayed fluorescence is called phosphorescence and the substances are called phosphorescent substances.*

Some characteristics of the phenomenon of phosphorescence are as follows.

(i) Materials exhibiting fluorescence generally re-emit excess radiation within 10^{-6} to 10^{-4} second of absorption. On the other hand, materials exhibiting phosphorescence re-emit excess radiation within 10^{-4} to 20 seconds or longer. Thus, the life-time of phosphorescence is much longer than fluorescence.

(ii) The phenomenon of phosphorescence is caused chiefly by the ultraviolet and violet parts of the spectrum.

(iii) The phenomenon of phosphorescence is shown mainly by solids.

(iv) The magnetic and dielectric properties of phosphorescent substances are different before and after illumination.

(v) The time for which the light is emitted from phosphorescent substance defends upon the nature of substance and sometimes on the temperature changes.

(vi) Different colours may be obtained by mixing different phosphorescent substances. *Examples of Fluorescent Substances.*

Among the naturally occurring substances, chlorophyll present in green leaves show the phenomenon of fluorescence, *i.e.,* when leaves are strongly illuminated in presence of oxygen, they emit fluorescent light whose intensity changes with time of irradiation. Franck and Herzfeld explained this phenomenon on the basis of complex formation.

Petroleum, vapours of sodium, iodine, acetone and hydrocarbons (paraffins and olefins) have been found to fluoresce in ultraviolet light. For example, acetone absorbs at 2700Å corresponding in $> C = O$ group and emits blue fluorescence. Same thing happens with other aldehydes and ketones.

We know that all molecules are capable of absorption. But fluorescence is not exhibited by a large number of compounds. Fluorescence is generally observed in those organic molecules which have rigid framework and not many loosely coupled substituents through which vibrionic energy can flow out. In analogy with chromophores, following structures are termed as *fluorophores.*

C=C, N=O, —N=N, —C=O, —C=N, —C=S

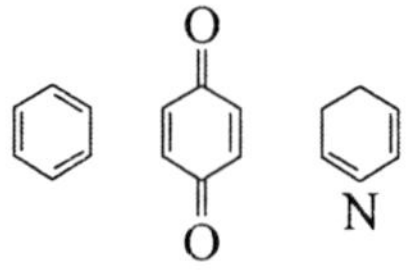

O O N

A large number of substances enhance fluorescence. These are known as fluorochromes in the same analogy as *auxochromes.* Generally electron donors act as auxochromes.

–OH

—NH_2

—CH_2

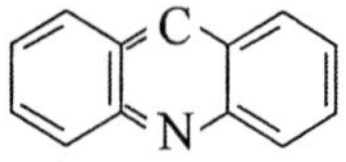

C N

Acridine
(non-fluorescent)

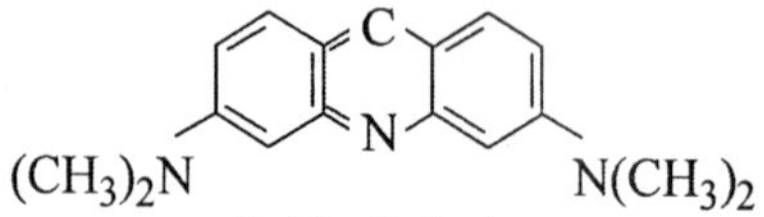

C N $(CH_3)_2N$ $N(CH_3)_2$

Acridine Derivative
(fluorescent)

Electron-withdrawing substituents like – COOH tend to diminish or inhibit fluorescence completely.

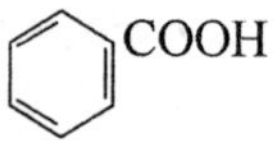

COOH

Benzoic acid
(non-fluorescent)

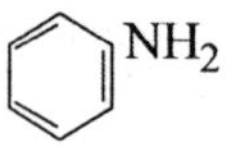

NH_2

Aniline
(Highly fluoresecent)

Some examples of fluorescent substances are as follows:

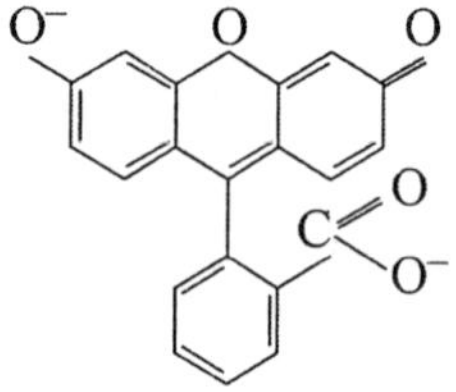

O^- O O C O O^-

Flourescein
(fluorescent)

N=N

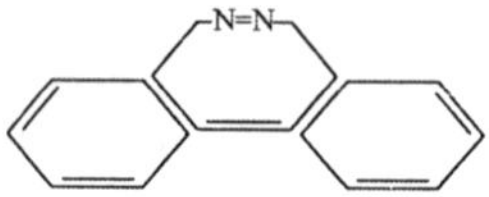

Azophenanthrene
(fluorescent)

Among inorganic substances there are only few examples such as fluorite (the name fluorescence originates from it), uranium compounds and rare earths which exhibit fluorescence.

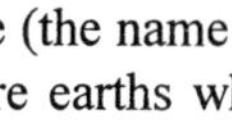

Naphthalene
(fluorescent)

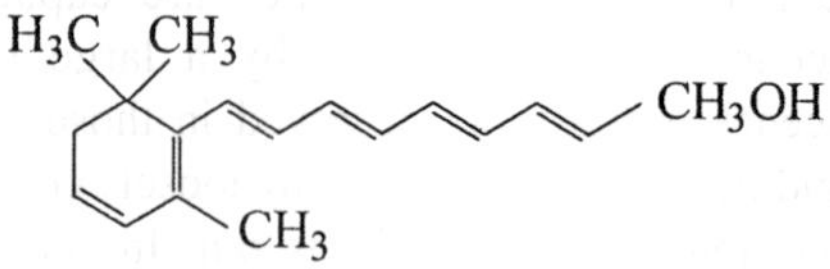

H_3C CH_3 CH_3OH CH_3

Vitamin A
(fluorescence as that of naphthalen)

Nitrogen peroxide in blue light (4920 to 4550A°) shows intense fluorescence which becomes much weaker in violet light (4550 to 3650Å) and vanishes at 3650Å).

EXAMPLES OF PHOSPHORESCENT SUBSTANCES

(i) Many dyes which fluoresce in ordinary light in aqueous solution exhibit phosphorescence when dissolved in fused boric acid or glycerol and cooled to form a glassy solid. The phosphorescence in such a case shows two distinct bands, *i.e.,* one in blue and the other in yellow region.

The former persists for some time at the ordinary temperature and is called α-phosphorescence and is identical with the fluorescence. If the temperature is lowered, α–phosphorescence becomes less marked and disappears altogether at 0°C. The yellow phosphorescence, on the other hand, is practically independent of temperature and has been observed down to – 250°C. The yellow phosphorescence is called β-*phosphorescence.*

(ii) The common substances that exhibit this phosphorescence phenomenon are sulphides of calcium, barium, strontium.

(iii) Many organic substances and certain fungi phosphoresce to emit a faint light This is due to the slow oxidation of organic substances.

(iv) Minerals, *e.g.,* ruby, emarld are the interesting examples of phosphorescent substances.

(v) Many phosphors are prepared by mixing alkaline earth metals with about 2.5 per cent alkali chlorides and a trace of sulphide of some heavy metal.

THEORY OF FLUORESCENCE AND PHOSPHORESCENCE

1. Singlet and Triplet States : In order to understand the theory of fluorescence and phosphorescence, one has to understand the meaning of singlet and triplet states. These terms arise from multiplicity considerations to atomic spectroscopy and simply define the *number of unpaired electrons in the absence of magnetic field.* If there are *n* number of unpaired electrons, it means that (n + 1) -fold degeneracy (equal energy states) will be associated with the electron spin, regardless of the molecular orbital occupied.

Thus, if no unpaired electrons are present (n = 0), where there is only n + 1 or 0 + 1 or 1 spin state. Such a state is a called a *singlet state.* Similarly, systems having 1, 2, 3, 4, ... unpaired electrons refer to doublet, triplet, quartet, etc. respectively.

Most of the molecules in their ground state do not have unpaired electrons *(singlet state).* When such a molecule absorbs ultraviolet or visible radiation of the proper frequency one or more of the paired electrons (generally a re-electron) get raised to an *excited singlet state.* In this excited state, the spin of the electron does not undergo any change and the net spin is still zero.

One more possibility is that one set of electron spins may have undergone unpairing, resulting in two unpaired electrons which make an *excited triplet state.* Fig. 8(a) represents a molecule in the ground singlet state, Fig. 8(b) exhibits a molecule in the excited singlet state. Fig. 8(c) represents a molecule in an excited triplet state.

One should remember that there is very small probability of a direct transition from the singlet ground state to the triplet excited state.

2. Excited-State Processes in Molecules : When molecules are irradi • ed with light of the appropriate frequency, it will be absorbed in about 10^{-15} second. In the process of absorption the molecules may move from the ground to the first excited singlet electronic state.

Although at room temperature molecules may be present in their ground vibrational level, after absorption the excitational molecules can end up in any one of the vibrational levels in the first excited electronic state. From the excited singlet state, one of the following three phenomena will probably occur, depending on the molecules involved and the conditions.

(a) The *first possibility* is that the excited singlet state is relatively *unstable.* In such a situation, the excited molecule will return to the ground state by *collisional deactivation without emitting any radiation.*

(b) The *second possibility* is that the molecule in the excited singlet state may emit an ultraviolet or visible light photon. This process is known as *fluorescence.*

If we compare the absorption and fluorescence spectrum of the same compound, they do not superimpose on each other as expected but they are mirror images of each other with the fluorescence spectrum shifted to longer wavelengths.

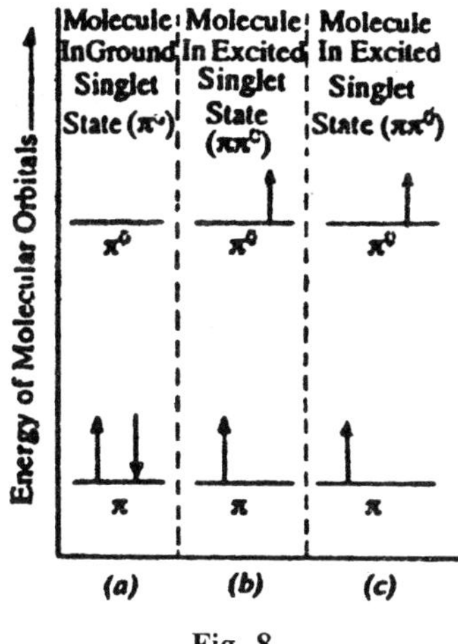

Fig. 8

The reason for this is that as the time required to execute a vibration is about 10^{-13} second which is much shorter than the decay or mean life of 10^{-9} second, most of the excess vibrational energy will be given to the surroundings and the excited molecules will decay in their ground vibrational levels. This is illustrate in Fig. 9.

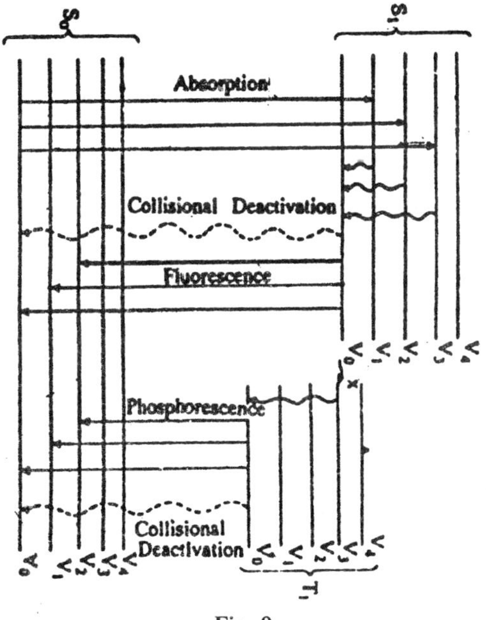

Fig. 9

(c) The third possibility is that the molecule with relatively stable *excited singlet state* may undergo transition to a *metastable triplet state* and some time thereafter returns to the ground states, usually by emission of an ultraviolet or visible light photon. This is known as *phosphorescence emission* Fig. 9, and the process of crossing from a singlet state (no unpaired electron) to a triplet state (two unpaired electrons) is termed as *intersystem crossing.*

The decay from the triplet to the ground state singlet is *forbidden by spin symmetry* and is therefore *slow.* Thus, the life-time of phosphorescence is much longer than flour-essence. The above mechanism of phosphorescence involving singlet-triplet decay scheme has been confirmed by the magnetic susceptibility and ESR measurements.

According to Hund's rule, the triplet level always lies *lower* than the corresponding singlet level and for this reason phosphorescence spectrum is *not the mirror image* of the absorption spectrum and it always occurs at *longer wavelengths* compared with the absorption and flourescence spectrum. Phosphorescence is rarely observed in gases It is almost never observed at room temperature (biacetyl is one exception).

Factors Affecting Fluorescence and Phosphorescence : The various factors are as follows.

(a) All the molecules cannot show the phenomena of fluorescence and phosphorescence. Only such molecules show these phenomena that are able to absorb ultraviolet or visible radiation. In general, the greater the absorbancy of a molecule, the more intense its luminescence. This requirement means that molecules having conjugated double bonds (π bonds) are particularly suitable for this study. On the other hand, aliphatic and saturated cyclic organic compounds, are not suitable.

(b) Substituents often exhibit a marked effect on the fluorescence and phosphorescence of molecules. There are no rigid rules but a few generalities may be useful. These are as follows.

(i) Electron-donating groups like NH_2 and – OH often enhance fluorescence. On the other hand, groups like – SO_2H, – NH_4^+ and alkyl groups do not have much effect on both phosphorescence and fluorescence.

(ii) Electron-withdrawing groups like – COOH, – NO_2. – N = N – and halides decrease or even destroy fluorescence.

(iii) If a high atomic number atom is introduced into a re-electron system, it enhances phosphorescence and decreases fluorescence.

(c) The pH exhibits a marked effect on the fluorescence of compounds. For example, the neutral or alkaline solution of aniline shows fluorescence in the visible region. But if this solution is acidified, the visible fluorescence disappears (aniline shows fluorescence in the ultraviolet regardless of pH).

Relation between Fluorescence Intensity and Concentration : We know that the Beer-Lambert law can be applied to the intensity of radiation transmitted by a substance or a solution. But this cannot be applied to fluorescent radiation directly because it is emitted by a substance. However, the following relation has been developed.

$$F \propto I_D - I$$

or
$$F = K (I_0 - I) \qquad ...(1)$$

where F is the intensity of fluorescent radiation, K is a proportionality constant. I_0 is the intensity of incident radiation and I is the intensity of transmitted radiation. We know that the Beer-Lambert law is

$$I = I_0 \ 10^{-abc}$$

or
$$I_0 - I = I_0 - I_0 \ 10^{-abc} \qquad ...(2)$$

$$= I (1 - 10^{-abc}) \qquad ...(3)$$

On substituting equation (3) in (1), we get

$$F = KI_0 (1 - 10^{-abc}) \qquad ...(4)$$

On rearranging equation (4), we get

$$\log \frac{KI_0}{KI_0 - F} = abc \qquad ...(5)$$

In the above equation, K is the fraction of the incident radiation that is adsorbed, *a* is the absorptivity, *b* is the length of cell path and *c* is the concentration of absorbing substance.

For very dilute solutions, equation (4) simplifies to the following equation

$$F = 2.303 \ KI_0 \ abc$$

or
$$F = K'c$$

For the above equation it follows that the fluorescent intensity is practically proportional to the concentration of the fluorescent substance. Equation (6) holds good for solutions of few parts per million, *i.e.,* dilute solution. For connected solutions, fluorescence concentration curve will bend towards the concentration axis. Fig. 10 shows typical curves at –196°C, and 35°C.

It follows that the curve levels off at high concentration at optimum temperature. Factors such as association, dissociation, or solvation which are responsible for deviations in Beer-Lambert law would be expected to show a similar effect in flourescence.

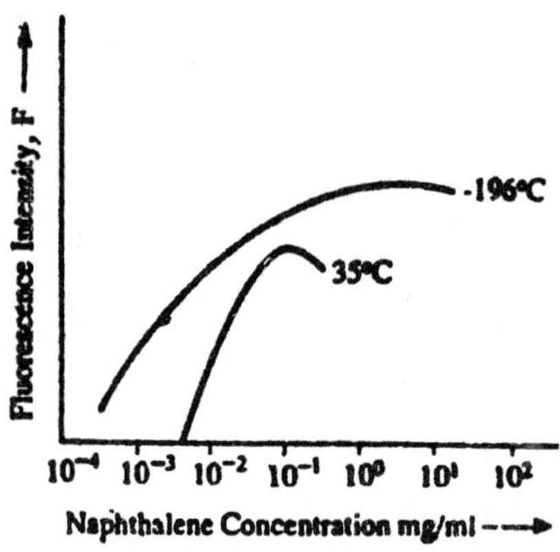

Fig. 10

Another important factor in fluorescence is temperature. In most of the instruments, sources of the very high intensity are used. This causes the heating of the sample or sample chamber. An increasing temperature usually decreases flourescence intensity. In order to minimise temperature effects in fluorescence instruments, shutter mechanisms are provided to minimise exposure time and thermally insulating the sample holder.

Relation between Phosphorescence Intensity and Concentration : Exactly the same considerations as discussed for fluorometry are important in relating phosphorescence intensity P to concentration *c,* and thus the final equation is as follows.

$$\log \frac{K'I_0}{K'I_0 - P} = abc$$

where K′ depends on the instrument used. A typical curve or phosphorimetry is shown in Fig. 11.

Applications of Fluorescene : We know that the phenomenon of fluorescence is a well established analytical tool. A large number of applications are known. But we will describe some of these as follows.

(a) An interesting example is the determination of uranium in salts by fluorescence. This is used extensively in the field of nuclear research.

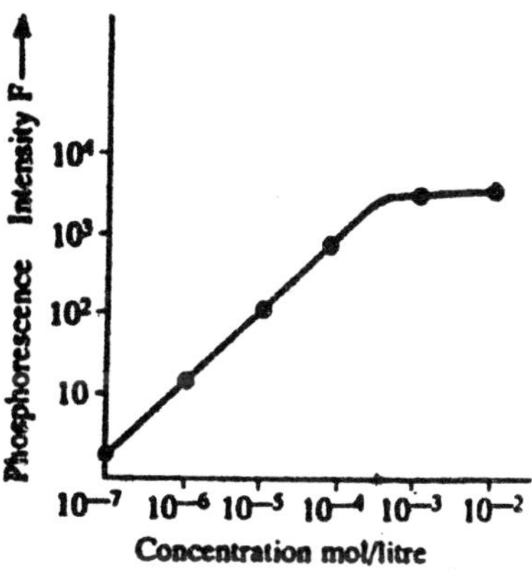

Fig. 11

The uranium sample is evaporated with nitric acid to bring about oxidation. Then the sample is fused with sodium fluoride to a melt having fluorides of sodium and uranium. On cooling, this solidifies to a glass which is examined in a specially designed fluorimeter. By this method, one can determine uranium of the order of 5×10^{-9} g in a 1 g of solid sample.

(b) In general, inorganic ions do not exhibit fluorescence. However, some of these inorganic ions form fluorescent chelates with non-fluorescent organic molecules. This has provided the basis for very sensitive analysis of many elements including most of the transition elements. Some interesting examples are as follows.

(i) An interesting example is the determination of ruthenium ion in the presence of other platinum metals. With 5-methyl-1, 0-phenanthroline ruthenium forms the complex ion which fluoresces strongly at pH 6. By this method, the determination of ruthenium can be carried out in the range of 0.3 to 2.0 μg/ml in the presence of interfering elements of platinum group which may be present to the extent of at least 30μ/ml.

(ii) Another interesting example is the determination of aluminium (III) in alloys. Aluminium (III) forms a complex with dye Pantachrome Blue-black RM at a pH of 4.8 which fluoresces strongly. One can determine aluminium in the range of 0.2 to 25μg in a volume of 50 ml with a sensitivity of 1 part 10^8.

(iii) Another inorganic fluorimetric determination is the estimation of traces of boron in steel by means of the complex formed with benzoin. The boron present in the acid solution of steel is first converted into boric acid which is separated from the other constituents by distillation with methyl alcohol.

The resulting distillate having boric acid is neutralised with sodium hydroxide. Then, the methyl alcohol is distilled off leaving the sodium salt of boric acid in the distillation flask which on treatment with an alcoholic solution of benzoin yields a fluorescent solution. It has been found that the fluorescent power is linear with the concentration upon 100 *ug* of boron in 50 ml volume but drops off at higher concentrations. This method for its determination is superior to old techniques in speed and sensitivity.

(iv) Cadmium can be estimated by precipitating it with 2-(2- hydroxyphenyl)-benzoxazole in the presence of tartrate. The complex on dissolving in glacial acetic acid yields a solution with an orange tint and a bright blue fluorescence in ultraviolet light. The acetic acid solution forms the basis for the determination of cadmium.

(v) Similarly, calcium can be estimated by fluorimetry with calcein solution.

(c) Fluorescent indicators : The intensity and colour of the fluorescence of many substances depend upon the *pH* of the solution, *i.e.,* their colours depend upon the pH range. These are termed *as flourescent indicators.*

These are mainly used in acid-base titrations. These can be employed in the titration of coloured solutions in which the changes in colour of indicators get masked. Some examples of fluorescent indicators are given in Table 2.

Table 2 : Some Fluorescent Indicators.

Name of Indicators	approx. pH	Colour Change
Eosin	3.4–4.0	Colourless to green
Fluorescein	4.0–6.0	Colourless to green
Quinine sulphate	3.0–5.0	Blue to violet
Acridine	5.2–6.6	Green to violet blue
2 – Naphthaquinone	4.4–6.3	Blew to colourless
2 – Hydroxycinnamic acid	7.2–9.0	Colourless to green

(d) **Determination of vitamin B_1 :** Vitamin B_1 (thiamine) is non-fluorescent whereas its oxidation product, thiochrome fluoresces with blue colour. This property is used for the determination of vitamin B_1 in the food samples like meat, cereal, etc.

The food sample is treated with phosphatase which brings about hydrolysis of the phosphate esters of thiamine present in the sample. The solution on filtration removes phosphatase and other insoluble matter. Then, the filtrate is diluted to a known volume. From the filtrate, two equal aliquots are taken, one for analysis and other for a blank.

To both aliquots, equal quantities of sodium hydroxide and isobutyl alcohol are added. To the first oxidising agent like potassium ferricyanide is added. After shaking, the alcoholic solution is separated from the aqueous solution. Then, the alcoholic solution is examined in the fluorimeter. The whole procedure, including a blank, is repeat with a standard thiamine solution.

(e) **Determination of Vitamin B_2 (Riboflavin) :** Determination of vitamin B_2 is done by a fluorescence method because the fluorescent power depends upon the reaction conditions and upon the nature and amount of impurities.

Generally, the method of standard increment is used to become sure about the fact that impurities have the same effect upon the standard and unknown. In the method of standard increment, one measures fluorescence of a portion of the standard in the same solution with the unknown.

The procedure also considers the simple fact that riboflavin on oxidation yields a non-fluorescent substance.

An acid solution of the sample (a food stuff) is treated with different reagents to precipitate various interfering ions. This solution is then oxidised with dilute permanganate, Then, the residual fluorescence is measured as a blank. Now, a slight excess of solid sodium dithionite ($NagSaO_4$) is added.

Fluorescence is again determined. Then, a known volume of standard is added and fluorescence is again measured. Finally, the results are computed by the following scheme.

Table 3

Fluorescence of solution	Designation
10 ml oxidised sample + 1 ml water	F_A
Same + dithionite	F_B
Same + 1 ml standard	F_C

From the above results, the concentration of fluorescing material can be calculated by using the following formula.

$$\frac{F_B - F_A}{F_C - F_A} = \frac{m_x}{m_x + m_s}$$

In the above equation, m_x and m_s are the masses of riboflavin from sample and standard in the cuvette respectively.

(f) Food-Stuffs : The phenomenon of fluorescence is being utilised in examining conditions of food-stuffs.

When ultraviolet light is incident on newly laid eggs, they fluoresce with rosy colour, while bad eggs appear blue. Similarly butter, lard and different kinds of honey can be readily distinguished.

(g) Police Work : The difference in the fluorescence caused by ultraviolet rays in different types of inks enables police to detect forged documents.

(h) Medicine : Ringworm may be detected by the fluorescence caused by ultraviolet radiations.

When fluorescent liquids are injected into an animal body, the internal organs will fluoresce and can be observed by a microscope of special type called fluorescent microscope. In this way the internal organs of human or animal body may be diagnosed by doctors without any difficulty.

When, in fluoroscope. X-rays are allowed to fall on a screen of barium platinocyanide or other materials, there wili be characteristic fluorescence which may be used for diagnosis by doctors. This is the principle of fluoroscope used in X-ray diagnosis.

(i) Analysis : The characteristic fluorescence of various substances when exposed to ultraviolet light offers a good method of quantitative analysis. This has been used in the analysis for the following.

(a) Drugs and dyes

(b) Textile and paper industry.

(c) Medicine and bacteriology,

(d) Fuels and chemicals.

(j) Lamps : Fluorescent lamps which are now widely used for lighting depend upon the fluorescence caused by ultraviolet light on phosphorus coated inside the fluorescent tubes.

(k) Science : Use of the phenomenon may be made in seeing the invisible radiations like ultraviolet rays.

(l) Organic Analysis : Fluorescence has been used to carry out qualitative as well as quantitative analysis for a great many aromatic compounds present in cigarette smoke, air-pollutant concentrates and automobile exhausts. A specific example is the determination of benzopyrene in the nanogram range.

APPLICATIONS OF PHOSPHORESCENCE

Similar to fluorescence, phosphorescence finds applications in biology and medicine. But these applications are not accepted for routine analysis. Some interesting applications are as follows.

(a) One can carry out the determination of aspirin (acetylsalicylic acid) in blood serum with high sensitivity by phosphorimetry at liquid nitrogen temperatures. By this method, 0.02 – 1.00 mg aspirin per ml of serum can be analysed. The metabolic product of aspirin, *i.e.,* salicylic acid, does not exhibit appreciable phosphorescence but is quite strongly fluorescent whereas aspirin is exhibiting strong phosphorescence but not fluorescence. Therefore, it becomes possible to carry out a simple chloroform extraction of the serum and then analysis is done for both the drug (aspirin) and its metabolite (salicylic acid) in the presence of one another by using these two complementary techniques (fluorimetry and phosphorimetry).

(b) Low concentrations of procaine, cocaine, phenobarbital and chloropramazine in blood serum have been determined by phosphorimetry in combination with extraction procedures.

(c) Cocaine and atropine in urine have been determined by employing phosphorimetry in combination with extraction procedures.

(d) Phosphorimetry has been employed in combination with thin layer or paper chromatography. An interesting example is given by Winefordner and Moye. They have separated three tobacco alkaloids (nicotine, nornicotine and anabasine) from crude tobacco by thin layer chromatography on alumina. The R_f values of three alkaloids are 0.80, 0.26 and 0.48 respectively. The separated drugs were scrapped from the plate and determined by phosphorimetry. This method is quite rapid.

COMPARISON OF FLUORESCENCE AND PHOSPHORESCENCE

As phosphorescence is more complicated experimentally than fluorescence, far fewer applications have been realised for phosphorimetry than for fluorimetry. But the potential of phosphorimetry is great and it is hopped that it will find valid applications in chemistry in the near future. However, it is clear that the two techniques are complementary.

Whenever we analyse complex samples having much interfering substances, we prefer phosphorimetry because it is more selective than fluorimetry.

One more advantage of phosphorimetry is that it is more sensitive than fluorimetry due to the following reasons.

(i) Scattering problems do not exist in phosphorimetry whereas these become severe in fluorimetry.

(ii) Quantum efficiencies are much more for phosphorescence than for fluorescence. The reason for this is that phosphorimetric determinations are carried out at – 196°C whereas fluorimetric determinations are carried out at room temperature. One more advantage for the low temperatures in phosphorimetry is that the quenching does not pose a serious problem, than in fluorimetry.

From the above discussion it follows that whenever fluorimetry and phosphorimetry are of equal sensitivity, fluorimetry is preferred because it is experimentally less complicated. However, phosphorimetry is preferred when fluorimetry is not sensitive or selective.

Antistoke's Behaviour : According to G.G. Stoke, the emitted radiation in fluorescence has greater wave-length than the absorbed radiations. There are however, cases where Stoke's law is violated. Two exceptions are.

(a) Resonance Fluorescence : *In certain cases, the fluorescent light has the same frequency as that of incident light. This fluorescent light is called resonance radiation and the phenomenon is known* as resonance fluorescence. This phenomenon is analogous to resonance of sound. For example, when mercury vapour with atoms in normal state ($^1S_0$) is exposed to ultraviolet radiations, it was found to absorb the radiation of wave-length 2537 Å. It then converts to $^3P_1$. If excited electrons return to the ground-state, they emit the radiations of the same wavelength, *i.e.,* 2537 Å.

$$Hg\ (^1S_0) + HV \rightarrow Hg^*\ (^3P_1)$$

$$Hg^*\ (3P_1) \rightarrow Hg\ (^1S_0) + hv.$$

(b) Sensitised Fluorescence : *A substance which is normally non-fluorescent may be made fluorescent in the presence of other fluorescent substances. The phenomenon is known as* Sensitised fluorescence. If the vapour of thallium is added to mercury vapour and then exposed to radiations of wave-length, 2537 Å, the vapour of thallium fluoresces. Thus the thallium which is a non-fluorescent substance can be made fluorescent by mercury vapour. This can be explained by saying that in sensitised fluorescence, mercury vapour gets energy from incident light and thus its atoms get excited to higher energy levels.

$$Hg\ (^1S_0) + hv \rightarrow Hg^*(^3P_1).$$

Then, thallium atoms undergo collision with excited mercury atoms and thus a part of energy of excited mercury atoms is transferred to thallium atoms which are raised to higher energy levels.

$$Hg^* + T_1 \rightarrow T_1^* + Hg$$

These excited atoms of thallium emit their own characteristic fluorescence on reverting to the ground state

$$T_1^* \rightarrow T_1 + h\nu'$$

Quenching of Fluorescence : *When a photochemically excited atom has a chance to undergo collision with another atom or a molecule before it fluoresces, the intensity of the fluorescent radiation may be diminished or stopped. This phenomenon is known as quenching of fluorescence.* Some facts about quenching of fluorescence are as follows.

(i) Quenching of fluorescence depends greatly on the concentration of the fluorescent atoms and of the quenching substance.

(ii) At low pressures, very little quenching will occur. At high pressures, appreciable quenching takes place.

(iii) Quenching of fluorescence takes place appreciably in a liquid medium because collisions are very frequent.

Explanation. The quenching of fluorescence is due to transfer of energy from the photochemically excited atom to the molecule or atom with which it undergoes collisions. As a result of transfer of energy, the following changes may be possible.

(i) A photochemically excited atom may activate the atom with which it undergoes collision.

$$Hg^* + T_1 \rightarrow Hg + T_1^*.$$

This results in sensitized fluorescence.

(ii) The photochemically excited atom may activate a molecule by collision.

$$Cd^* + H_2 \rightarrow Cd + H_2^*$$

(iii) A photochemically excited atom may react chemically with a molecule to form the products. Example is

$$Hg^* + H_2 \rightarrow HgH + H$$

(iv) An excited atom may undergo collision with another molecule and, thus, resulting its dissociation. Example is

$$Hg^* + H_2 \rightarrow Hg + 2H.$$

This phenomenon is known as photosensitisation.

Importance. The photochemical importance of the quenching of fluorescence is that the excited molecules or atoms obtained in the quenching process may undergo further reaction so as to continue the photochemical changes such as (a) photosensitisation (b) sensitised fluorescence, (c) photochemical reactions and (d) dissociation of molecules to form atoms.

PHOTOELECTRIC EFFECT

Photoelectric effect is the phenomenon of ejection of electrons from a metal plate when light of a suitable wave-length falls on it. The electrons emitted are called photo-electrons to indicate their mode of production.

Laws of Photo-Electric Emission : There are two important laws regarding the photoelectric **Emission.**

Law I: *The velocity of the emitted electrons increases with the increase in frequency of the incident light.* This law means that there is a limiting frequency for a metal, below which no electrons are emitted. This is known as the *threshold frequency.*

Law II: *The number of photo electrons ejected is directly proportional to the intensity of the incident light.* Thus, if the intensity of the incident light is increased, the number of electrons emitted is increased, but their velocity remains constant.

Explanation : The laws of photo-electric emission were explained by Einstein in 1905 on the basis of Planck's quantum theory of radiation. According to the quantum theory, the radiation consists of packets of energy called 'quanta' or 'photons' of energy hv, where h is the Planck's constant and v the frequency of radiation. When a photon of energy hv is incident upon a metal surface, its energy is used up in two ways shown in Fig. 12.

(a) A part of its energy is used up in ejecting the electron just out of the surface. The energy depends upon the nature of the metal and is called work-function (ϕ_0).

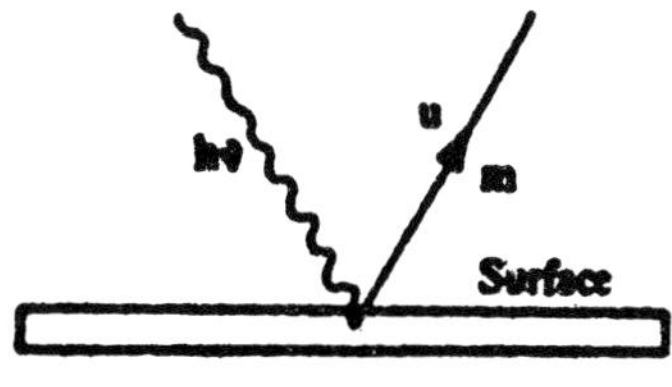

Fig. 12

(b) The rest part of the energy of the striking photon is used up in imparting kinetic energy $\frac{1}{2}$ mu^2 to the ejected electron.

Thus $$h\nu = \phi_0 + \frac{1}{2} mu^2 \qquad ...(1)$$

But $$\phi_0 = h\nu_0$$

where ν_0 is the threshold frequency. Therefore, equation (1) becomes as

$$h\nu - h\nu_0 + \frac{1}{2} mu^2$$

or $$\frac{1}{2} mu^2 = h(\nu - \nu_0) \qquad ...(2)$$

The equation is called Einstein's photo-electric equation.

Explanation of IInd Law : From eq. (2), it follows that the velocity of a photo-electron increases with the increase in frequency. No electron will be emitted when the frequency of the incident radiation ν is less than the threshold frequency ν_0. This explains the II law of photoelectric emission.

Explanation of 1st Law : If the intensity of light is increased keeping the frequency same, more photons are incident on the metal surface but the energy of each photon will remain the same. Hence the number of electrons emitted per unit area per unit second will increase but their velocities remain the same. This is the I law of photoelectric emission. It is clear that the maximum value of velocity u_{max} for a given frequency ν will occur for those electrons for which $\phi = \phi_0$. Hence, equation (2) becomes as

$$1/2 \; mu^2{}_{max} = h\nu - \phi_0 = h\nu - h\nu_0 \qquad ...(3)$$

It is clear from the above equation that the maximum velocity of photo- electrons will increase with increase in frequency ν of incident radiation and no electrons will be emitted when ν is less than ν_0.

Testing. The Einstein's equation explains the observed facts about photo-electric emission. This equation was tested in 1915 by Millikan who observed its results in agreement with experimental results. The successful explanation of photo electric effect by Einstein on the basis of quantum theory offers one of the evidences in favour of the quantum theory.

Importance. Photo-electric effect is used in the construction of photo electric cell which converts light energy into electric energy. The photo-electric cells are widely used in various fields.

PHOTOELECTRIC CELLS

It is a device for converting light energy into electric energy. There are *three* main types of photo-cells.

1. *Photo emission type.* It consists of an evacuated glass, or quartz tube which has its inner surface M coated with sodium, potassium or caesium, and this coating is connected to – ve end of a battery. A window is left open through which light can enter the tube, and a wire loop P, or a cylindrical wire in the centre is connected to the +ve end of the battery and it collects the electrons, Fig. 13, 14 and 15. The two parts are insulated from each other and are brought out to two legs at the bottom of the photo cell. On admitting light, which is of a frequency above the threshold frequency, electrons from the photometal are emitted. They travel towards the loop wire and a current is produced in the circuit.

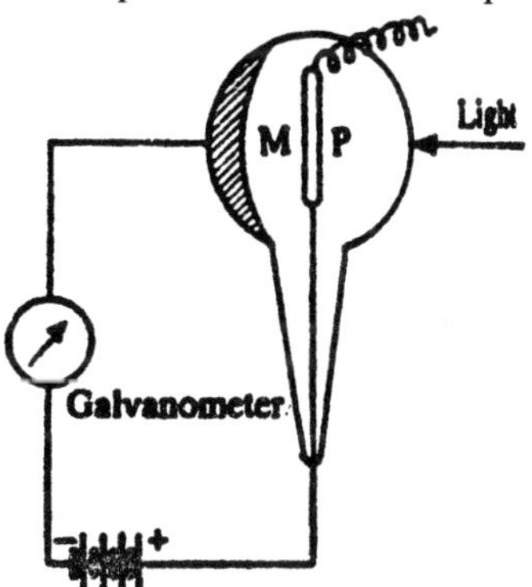

Fig. 13

Since a photo cell responds to the light which falls on it, as does the human eye, it is therefore, called an *electric eye.* There are two types of photo emission cells, (1) high vacuum type, and (2) gas filled type.

In vacuum cells, the photoelectric current is very small but these cells keep a strict proportionality between the current and intensity of light. The chief advantages of vacuum type are.

(i) The sensitivity of these cells remain unaltered for a very long time provided the cathode is properly selected.

(ii) There is no time lag between the incident light and the photoelectrons and the photoelectric current is proportional to the intensity of illumination. Thus, they are extremely accurate in response.

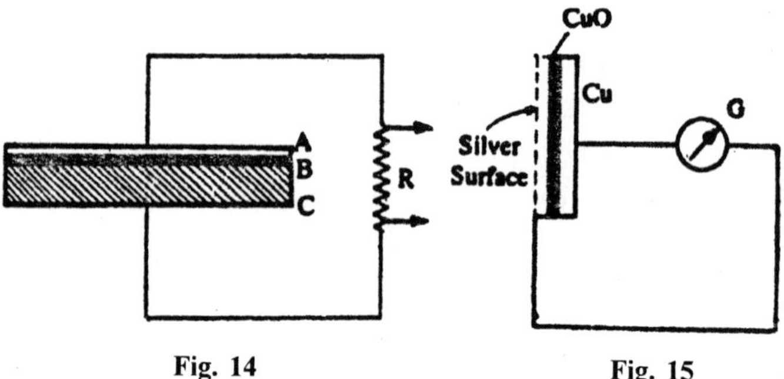

Fig. 14 Fig. 15

Due to the above mentioned advantages, these cells are being used in television and photometry.

In gas filled cells, the response to light is not so quick but since an inert gas is being used in the cell, ionisation of the gas is produced by the emitted electrons which therefore increase the photo current. Photoelectric current in this case is not proportional to the intensity of illumination and there is some time lag. Such cells are used in cinematography both for recording and reproduction of sound.

2. *Photo-conductive cell.* These cells are based on the property that the resistance of selenium and certain other metals decreases with increase of illumination. No photo-electrons are emitted in this case. A photo-conductive cell consists of a thin film of a semiconductor. *e.g.,* selenium, B, placed below a thin semi-transparent metal film A. The combination is placed on a block of iron C in contact. Ordinary selenium is a semi-conductor. When light is incident on A, it is absorbed by the selenium layer, electrons are emitted and if a small external e.m.f, is applied, a current flows in the resistance and can operate a relay when the intensity of incident light is high. These cells are generally connected directly to a micrometer or a low resistance relay and are preferably used in a Wheatstone bridge arrangement. Their response to light is not as quick as in photo-emission cell, but they are more sensitive to red light.

3. *Photo voltaic cell.* This type of photo-cell consists of a thin-layer of cuprous oxide coated on a disc of copper. The oxide film has sputtered silver or gold film on its upper surface. When light falls on the oxide,

electrons are emitted from it not into the surrounding air but into the copper. The oxide layer thus becomes + vely charged and copper negatively. An e.m.f. is thus set up. If contacts are made to the sputtered film and to the copper, the photo voltaic e.m.f. Produces a current. Photo-voltaic cells are used as exposure meters or light intensity meters.

Applications of Photoelectric cells. Main applications are:

(1) Reproduction of sound in films,

(2) In televisions

(3) In meteorology as day-light recorders,

(4) For measuring the complexion of persons,

(5) In automatic light switching for switching on and switching off the street light.

(6) In determination of temperature of stars.

(7) Inburglar alarms to detect thieves and in fire alarms to indicate the outbreaks of fire,

(8) In traffic signals.

CHEMILUMINESCENCE

It is the phenomenon of emission of visible light as a result of a chemical change at a temperature at which a black body normally does not emit visible light.

From the above definition, it is evident that chemiluminescence is the reverse of photochemical reaction in which light absorption causes chemical action. Some examples are.

(i) A classical example of chemiluminescence is the phosphorus which produces a greenish-yellow glow. It is probably due to the oxidation of phosphorus vapours by atmospheric oxygen. Phosphorus trioxide also produces glow. This is probably due to the oxidation of P_2O_3 to P_2O_5.

(ii) When a solution of strontium chloride is added to dilute H_2SO_4 in dark, a feeble glow is produced along with the precipitate of $SrSO_4$.

$SrCl_2 + H_2SO_4 \rightarrow SrSO_4 + 2HCl + h\nu.$

(iii) It was observed by Evans that a solution of p-bromophenyl magnesium bromide (Grignard reagent) in ether produces greenish- blue glow consisting of a single broad band.

(iv) When alkali metal vapours react with halogens or with mercuric halides, luminescence occurs which consists of the spectrum of the alkali metal. The mechanism of this is based on the fact that sodium metal contains monatomic and diatomic molecules. The mechanism is.

$$Na + Cl_2 \rightarrow NaCl + Cl$$
$$Na + Cl \rightarrow NaCl$$
$$Cl + Na_2 \rightarrow Na + NaCl^*$$
$$NaCl^* + Na \rightarrow NaCl + Na^*$$
$$Na^* \rightarrow Na^* + h\nu \text{ (yellow spectrum)}$$

(v) When the surface of mercury is allowed to come in contact with a stream of atomic hydrogen, a blue fluorescence appears. The spectrum of this blue glow consists of the resonance line 2537 Å and a band of spectral lines from 4520 to 3250 Å due to mercuric hydride. The possible mechanism is due to

$$Hg + H \rightarrow HgH$$
$$H + H \rightarrow H_2 + \text{Energy}$$
$$HgH + \text{Energy} \rightarrow HgH^*$$
$$HgH^* + HgH \rightarrow Hg + Hg^* + H_2$$
$$Hg^* \rightarrow Hg + h\nu$$

Theoretical Explanation of Chemiluminescence : The phenomenon of chemiluminescence can be explained on the basis of quantum theory.- According to this theory, certain chemical reactions result in the formation of products which are in electronically excited states, When these excited products return to their ground levels, they do so by emitting the extra energy in the form of visible light radiations. Thus, by this process, the chemical energy is converted into light energy.

Sensitised Chemiluminescence : In 1925 Kautsky reported that oxidation of unsaturated silicon hydride by $KMnO_4$ is accompanied by a luminescent glow. If the same oxidation is carried out in the presence of contain dyestuffs such as rhodamine-B, a strong red fluorescence characteristics of the dye is observed. This type of change is known as *sensitised chemiluminescence.*

Explanation. During the oxidation of silicon hydride, the energy is given out and is subsequently transferred to the dyestuff, raising it to an excited state. When it returns to normal state, a fluorescence is observed.

PHOTOCHEMICAL KINETICS

The rate laws which photochemical reactions follows are generally more complex than those for thermal reactions because more variables are involved.

Like the kinetics of reactions as discussed in chemical kinetics we will apply the steady state treatment. We will now apply steady state-treatment to reactions which do not involve chains. Examples are :

(1) Dissociation of HI : The dissociation of hydriodic acid was investigated by *Warburg* (1911) in ultraviolet light of wavelength 2070, 2530 and 2820 Å.

$$2HI \xrightarrow{hv} H_2 + I_2$$

The experimental quantum efficiency of this reaction was 2.0. In order to account for this,. Warburg postulated the following mechanism.

(a) HI + hv ¾→ H + I Rate = I_{abs}

(b) $H + HI \xrightarrow{k_2} H_2 + I$ Rate = k_2 [H] [HI]

(c) $I + \xrightarrow{k_3} I_2$ Rate = k_3 $[I]^2$

In the above mechanism, hydrogen iodide is consumed insteps (a) and (b). Thus, its rate of dissociation is given by

$$\frac{-d[HI]}{dt} = I_{abs} + k_2 [H] [HI] \qquad ...(1)$$

Also, the net rate of formation of hydrogen atoms is given by

$$\frac{d[H]}{dt} = I_{abs} - k_3 [H] [HI] \qquad ...(2)$$

As [H] atom is short lived, the steady treatment, therefore, can be applied to it. Thus, equation (2) becomes

$$\frac{d[H]}{dt} = 0\ I_{abs} - k_2 [H] [HI] \text{ or } k_2 [HI [HI] = Iabs$$

Substituting this value in equation (1), we get

$$\frac{-d[HI]}{dt} = 2I_{abs} \qquad ...(3)$$

Quantum yield. By definition, the quantum yield is given by

$$\phi = \frac{\frac{-d[HI]}{dt}}{I_{abs}} = \frac{2I_{abs}}{I_{abs}} = 2 \qquad \text{[From (3)]}$$

This value of quantum efficiency is quite in agreement with experimental value Table 4.

Table 4 : Photochemical Decomposition of HI

Wavelength	Moles per K cal	Quantum yield
2070 Å	1.44×10^{-2}	1.98
2530 Å	1.85×10^{-2}	2.08
2830 Å	2.09×10^{-2}	2.10

(ii) Dissociation of HBr. The quantum yield for the dissociation of HBr is also two. Spectroscopic studies revealed that the mechanism followed by this reaction is the same as represented above for the dissociation of hydrogen iodide.

We will now apply steady-state to following reaction which involves chain mechanism.

1. Hydrogen and chlorine reaction and
2. Hydrogen and bromine reaction.

For details of chain reactions place see the chapter on 'Chemical Kinetics"

(i) Hydrogen and Chlorine Reaction : It is a classical example of photochemistry which involves chains. The reaction is

$$H_2 + Cl_2 \xrightarrow{hv} 2HCl$$

This reaction was first of all studied by *W. Cruickshank* and later by *J. W. Draper, R Bunsen* and so many other workers.

Facts : Special facts about this reaction are.

(i) The reaction is provoked by the light absorbed by chlorine in the region of 4785 Å.

(ii) An induction period is often observed. This is a period after the start of illumination during which the reaction is either very slow or does not take place at all.

The induction period is due to the presence of impurities like ammonia, NO_2, etc. That is why with pure hydrogen and chlorine no induction period is observed.

(iii) It was found by Draper (1841) that hydrogen-chlorine combination was followed by an increase in volume at constant pressure. This is known as Draper's effect.

(iv) The effect of oxygen is to influence the reaction after it has started. It has been found that the rate of reaction is inversely proportional to the concentration of oxygen.

(v) The quantum yield varies from 10^4 to 10^6 in the absence of oxygen. The value of quantum yield depend upon the impurities and oxygen concentration in the reaction vessel.

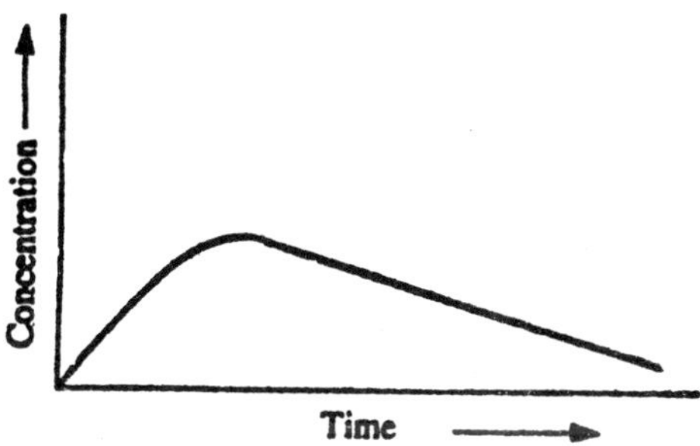

Fig. 16 : Combination of H_2 and Cl_2.

On the suggestion of Bodenstein, Nernst (1918) explained the reaction by chain mechanism which is universally accepted.

We will now consider the chain mechanism in two ways to explain.
(a) Combination of hydrogen and chlorine in the absence of oxygen and
(b) Combination of hydrogen and chlorine in the presence of oxygen.

In the Absence of Oxygen : (a) When a mixture of hydrogen and chlorine is exposed to light in the region of chlorine spectrum, chlorine molecule dissociates into atoms.

Initiation.

(i) $Cl_2 + h\nu \xrightarrow{k_1} 2Cl$] Rate = $k_1 I_{abs}$

The initiation reaction is followed by the following reactions.
Chain Propagation.

(ii) $Cl + H_2 \xrightarrow{k_2} HCl + H$ Rate = k_2 [Cl] [H_2]

(iii) $H + Cl_2 \xrightarrow{k_3} HCl + Cl$ Rate = k_3 [H] [Cl_2]

Chain Termination.

(iv) $Cl + \text{wall} \xrightarrow{k_4} \frac{1}{2}Cl_2$ Rate = k_4 [Cl]

Derivation of Rate Law : The total rate of formation of hydrogen chloride is given by reactions (ii) and (iii). Therefore, its rate is given by

$$\frac{d[HCl]}{dt} = k_2 [Cl] [H_2] + k_3 [H] \{Cl_2] \qquad ...(1)$$

The rate of formation of chlorine atoms is given by step (i) and (ii) and the rate of removal of chlorine atoms is given by steps (ii) and (iv). Therefore, the net rate of formation of [Cl] atoms is given by

$$\frac{d[Cl]}{dt} = k_1 I_{abs} + k_3 [H] [Cl_2] - k_2 [Cl] [H_2] - k_4 [Cl] \qquad ...(2)$$

Similarly, for hydrogen atoms we can write as

$$\frac{d[H]}{dt} = k_2 [C_1] [H_2] - k_3 [H] [Cl_2] \qquad ...(3)$$

When the stationary state is reached, it follows that

$$\frac{d[H]}{dt} = \frac{d[Cl]}{dt} = 0.$$

$$k_1 I_{abs} + k_3 [H] [Cl_2] - k_2 [Cl] [H_2] - k_4 [Cl] = 0 \qquad ...(4)$$

and $\quad k_2 [Cl] [H_2] - k_3 [H] [Cl_2] = 0 \qquad ...(5)$

By adding Eqs. (4) and (5), we get

$$k_1 \text{ Iabs} - k_4 [Cl] = 0 \text{ or } [Cl] = \frac{k_1}{k_4} I_{abs}. \qquad ...(5a)$$

Also, Eq. (5a) gives as

$$k_2 [Cl] [H_2] = k_3 [H] [Cl_2] \qquad ...(6)$$

Substituting Eqs. (6) in (1) we get

$$\frac{d[HCl]}{dt} = 2k_2 [H_2] \frac{k_1}{k_4} I_{abs}$$

$$= 2\left(\frac{k_1 k_2}{k_4}\right) I_{abs} [H_2] \qquad ...(7)$$

Expression (11) is in perfect agreement with experimental data. In the presence of high chlorine content, step (iv) is replaced by following complicated steps.

(v) $Cl + Cl_2 \rightarrow Cl_3$

(vi) $2Cl_3 \rightarrow 3Cl_2$

If we calculate the rate expression by using steps (i), (ii), (iii), (v) and (vi), we get an expression which is very complex.

(b) In the Presence of Oxygen : When the reaction is carried out in the presence of oxygen, the following mechanism was proposed in 1921 by Gohring.

(i) $Cl_2 + h\nu \xrightarrow{k_1} 2Cl$ Rate = $k_1\ I_{abs}$

(ii) $Cl + H_3 \xrightarrow{k_2} HCl + H$ Rate = $k_2\ [Cl]\ [H_2]$

(iii) $H + Cl_2 \xrightarrow{k_3} HCl + Cl$ Rate = $k_3\ [H]\ [Cl_2]$

(iv) $H + O_2 \xrightarrow{k_4} HO$ Rate = $k_4\ [H]\ [O_2]$

(v) $Cl + O_2 \xrightarrow{k_5} ClO_2$ Rate = $k_5\ [Cl]\ [O_2]$

(vi) $Cl + X \xrightarrow{k_6} ClX$ Rate = $k_6\ [Cl]\ [X]$

In step (iv), X is any substance which removes chlorine atom. Applying the steady-state treatment to chlorine atoms, we get

$$\frac{d[Cl]}{dt} = k_1\ I_{abs} - k_2\ [Cl]\ [H_2] + k_3\ [H]\ [Cl_2]$$
$$- k_5\ [Cl]\ O_2] - k_6\ [Cl]\ [X] = 0$$

This equation simplifies to

$$[Cl = \frac{k_1\ I_{abs} + k_3\,[H]\,[Cl_2]}{k_2\,[H_2] + k_5\,[O_2] + k_6\,[X]} \qquad ...(8)$$

The rate of formation of hydrogen atoms is given by

$$\frac{d[H]}{dt} = k_2\ [Cl\ [II_2] - k_3\ [II]\ [Cl_2]$$
$$- k_4\ [H]\ [O_2] = 0 \qquad ...(9)$$

This equation on simplification yields.

$$[Cl] = \frac{k_3\,[H]\,[Cl_2] + k_4\,[H]\,[O_2]}{k_2\,[H_2]} \qquad ...(10)$$

By equating Eqs. (8) and (10), we get

$$k_2\ [H_2]\ (k_1\ I_{abs} + k_3\ [H\ [Cl_2]) = (k_2\ [H_2] + k_5\ [O_2]$$
$$+ k_6\ [X]) + (k_3\ [H]\ [Cl_2] + k_4\ [H]\ [O_2])$$

By omitting small term $k_4 k_5\ [H]\ [O_2]^2$, the above equation is simplified to give the value of [H], *i.e.*,

$$[H] = \frac{k_1 I_{abs} k_2 [H_2]}{k_3k_6 [Cl_2][X] + [O_2](k_1k_4 (H_2] + k_3k_5 [Cl_2] + k_4k_6 [X])} \quad ...(11)$$

The rate of formation of hydrogen chloride is given by steps (ii) and (iii). Thus,

$$\frac{d[HCl]}{dt} = k_2 [Cl] [H_2] + k_3 [H] [Cl_2] \quad ...(12)$$

Reaction (iii) is much faster than (ii), consequently the rate of formation is given by step (iii) only. Therefore, Eq. (12) becomes as

$$\frac{d[HCl]}{dt} = k_3 [H] [Cl_2] \quad ....(13)$$

Introduction of expression (15) for [H] into this equation, we obtain.

$$\frac{d[HCl]}{dt} = \frac{k_1k_2k_3 [H_2][Cl_2] I_{abs}}{k_3k_6 [Cl_2][X] + [O_2](k_2k_4 [H_2] + k_3k_5 [Cl_2] + k_4k_5 [X])} \quad ...(14)$$

or
$$\frac{d[HCl]}{dt} = \frac{\left(\frac{k_1k_3}{k_4}\right) I_{abs} [H_2][Cl_2]}{\frac{k_3k_6}{k_2k_4}[Cl_2][X] + [O_2]\left([H_2] + \frac{k_3k_5}{k_2k_4}[Cl_2] + \frac{k_6}{k_2}[X]\right)}$$

$$= \frac{k I_{abs} [H_2][Cl_2]}{m[Cl_2] + [O_2]\left([H_2] + \frac{[Cl_2]}{10}\right)} \quad ...(15)$$

where $k = \frac{k_1k_3}{k_4}$, $m = \frac{[X][k_3k_6]}{k_2k_4}$

and $\frac{1}{10} = \frac{k_3k_5}{k_2k_4}$ ...(16)

Eq. (15) is in full agreement with the experimental rate equation as given by Bodenstein and Unger.

(2) Reaction Between Hydrogen and Bromine : In 1924, Bodenstein and Lurke-Meyer found that the photochemical reaction proceeds according to the empirical equation

$$\frac{d[HBr]}{dt} = \frac{k'[H_2] I_{abs}^{1/2}}{1 + \frac{[HBr]}{m'[Br_2]}} \quad ...(17)$$

where k′ and m′ are constants, and I_{abs}, is the intensity of the light absorbed.

Mechanism : The photochemical combination between hydrogen and bromine to form hydrogen bromide is an example of a chain reaction. Its mechanism is similar to the thermal reaction with the difference that the initiation is brought about by the absorption of a photon by a bromine molecule.

Chain initiation

(i) $Br_2 \xrightarrow[k_1]{hv} 2Br$ Rat $= k_1\ I_{abs}$

The rest part of the mechanism is similar to the thermal reaction

Chain Propagation. Rate $= k_2\ [Br]\ [H_2]$

(ii) $Br + H_2 \xrightarrow{k_2} HBr + H$ Rate $= k_2\ [Br]\ [H_2]$

(iii) $H + Br_2 \xrightarrow{k_3} HBr + Br$ Rate $= k_3\ [H]\ [Br_2]$

Chain inhibition :

(iv) $H + HBr \xrightarrow{k_4} H_2 + Br$ Rate $= k_4\ [H]\ [HBr]$

Chain breaking : Rate $= k_5\ [Br]^2$

(v) $Br + Br \xrightarrow{k_5} Br_2$

Here k_1, k_2, k_3, k_4, and k_5 represent the specific rates of all the five reactions respectively.

Derivation of Rate Law : Since HBr is produced by reactions (ii) and (iii) and removed by the reaction (iv), the net rate of formation of HBr is given by

$$\frac{d[HBr]}{dt} = k_2\ Br]\ [H_2] + k_2\ [H]\ [Br_2] - k_4\ [H]\ [HBr] \qquad ...(18)$$

As. Eq. (18) involves concentrations of hydrogen and bromine atoms which are too small quantities to be measured directly, it is required that their concentration must be expressed in measurable quantities. This is done by writing down and solving steady-state equations for [H] and [Br].

The Br atoms are formed by (i), (ii) and (iv) and are removed by (ii) and (v), so the net rate of formation is given by

$$\frac{d[Br]}{dt} = k_1\ I_{abs} + k_3\ [H]\ [Br_2] + k_4\ [H]\ [HBr]$$
$$- k_2\ [H_2]\ [Br] - k_5\ [Br]^2 \qquad ...(19)$$

As [H] atoms are formed by reaction (ii) and removed by (iii) and (iv), (v), so the net rate of formation is given by

$$\frac{d[H]}{dt} = k_2\ [H_2]\ [Br] - k_3\ [H]\ [Br_2] - k_4\ [H]\ [HBr] \quad ...(20)$$

When the stationary state is reached, it follows that

$$\frac{d[H]}{dt} = 0 \text{ and } \frac{d[Br]}{dt} = 0$$

Therefore, Eqs. (19) and (20) become as

$$k_1\ I_{abs} + k_3\ [H]\ [Br_2] + k_4\ [H]\ [HBr] - k_2\ [H_2]\ [Br] - k_5\ [Br]^2 = 0 \quad ...(21)$$

$$k_2\ [H_2]\ [Br] - k_3\ [H]\ [Br_2] - k_4\ [H]\ [HBr] = 0 \quad ...(22)$$

By adding Eqs. (21) and (22), we get

$$k_1 I_{abs} - k_5\ [r]^2 = 0$$

or

$$[Br] = \sqrt{\left(\frac{k_1}{k_5} I_{abs}\right)} \quad ...(23)$$

Putting the value of Br in Eq. (23). we get

$$k_2\ [H_2]\ \sqrt{\left(\frac{k_1\ I_{abs}}{k_5}\right)} - k_3\ [H]\ [Br_2] - k_4\ [H]\ [HBr] = 0$$

or $k_3\ [H]\ [Br_2] + k_4\ [H]\ [HBr] = k_2\ [H_2]\ \sqrt{\left(\frac{k_1\ I_{abs}}{k_5}\right)}$

or $[H]\ (k_3\ [Br_2] + k_4\ [HBr]) = k_2\ [H_2]) = k_2\ [H_2]\ \sqrt{\left(\frac{k_1\ I_{abs}}{k_5}\right)}$

$$[H] = \frac{k_2\ [H_2] \sqrt{\left(\frac{k_1\ I_{abs}}{k_5}\right)}}{k_3\ [Br_2] + k_4\ [HBr]} \quad ...(24)$$

On substituting the values of [Br] and [H] from Eqs. (23) and (24) in Eq. (22), we obtain

$$\frac{d[HBr]}{dt} = k_2\ [H_2] \sqrt{\left(\frac{k_1\ I_{abs}}{k_5}\right)} + \frac{k_3\ [Br_2]\ k_2\ [H_2] \sqrt{\left(\frac{k_1\ I_{abs}}{k_5}\right)}}{k_3\ [Br_2] + k_4\ [HBr]}$$

$$- \frac{k_4 [HBr] k_2 [H_2] \sqrt{\left(\frac{k_1 I_{abs}}{k_5}\right)}}{k_3 [Br_2] + k_4 [HBr]}$$

$$\frac{d[HBr]}{dt} = k_2 [H_2] \sqrt{\left(\frac{k_1 I_{abs}}{k_5}\right)} \left(1 + \frac{k_3 [Br_2]}{k_3 [Br_2] + k_4 [HBr]} - \frac{k_4 [HBr]}{k_3 [Br_2] + k_4 [HBr]}\right)$$

or
$$\frac{d[HBr]}{dt} = \frac{k_2 [H_2] \sqrt{\left(\frac{k_1 I_{abs}}{k_5}\right)}}{k_3 [Br_2] + k_4 [HBr]} (k_3 [Br_2] + k_4 [HBr] + k_3 [Br_2] - k_4 [HBr])$$

$$= \frac{2k_2 [H_2] k_3 [Br_2] \sqrt{\left(\frac{k_1 I_{abs}}{k_5}\right)}}{k_3 [Br_2] + k_4 [HBr]} \quad ...(25)$$

Dividing the numerator and denominator by k_3 $[Br_2]$, we get

$$\frac{d[HBr]}{dt} = \frac{2k_2 [H_2] \sqrt{\left(\frac{k_1 I_{abs}}{k_5}\right)}}{1 + \frac{k_4 [HBr]}{k_3 [Br_2]}}$$

$$= \frac{k' [H_2] \sqrt{\left(\frac{k_1 I_{abs}}{k_5}\right)}}{1 + \frac{HBr}{m'[Br_2]}} \quad ...(26)$$

where $k' = 2k_2$ and $m' = k_3/k_4$ are two constants. Equating (26) in similar to obtained by Bodenstein on the basis of this experimental data. Inspite of the chain mechanism similar to hydrogen and chlorine reaction, the quantum yield of this photochemical reaction is very low, being about 0.01 at ordinary temperatures. This is due to following reasons.

1. Reaction (ii) is highly endothermic and therefore, takes place so slowly at ordinary temperatures that most of the bromine atoms recombine to form bromine molecules. Therefore, the reactions

(iii), (iv) and (v) which are due to reaction (i) cannot occur. Hence, the quantum yield is extremely slow.

2. With increasing time, the reversal reaction becomes predominant and hence, the rate of formation of HBr decreases.

(3) Photolysis of Acetaldehyde : The photolysis of acetaldehyde has attracted much attention, in recent years. While discussing this reaction, the following facts must be kept in mind.

(a) The quantum yield of carbon monoxide formation increases with decreasing wave-length.

(b) When the temperature is raised, the quantum yield increases.

(c) The quantum yield of carbon monoxide formation at room temperature increases with decreasing pressure at 3130 A°.

(d) The photolysis of acetaldehyde in the light of 2500=3100 A° wavelength yields methane, carbon monoxide, ethane and a number of other products.

$$CH_3CHO + hv \rightarrow CH_4 + C_2H_6 + CO + \text{other products}$$

To account for the above facts, the following mechanism has been postulated.

(a) $CH_3CHO + hv \longrightarrow CH_3 + CHO$ Rate = Iabs

(b) $CH_3CHO + CH_3 \xrightarrow{k_2} CH_4 + CH_3CO$

Rate $= k_2\,[CH_3CHO]\,[CH_3]$

(c) $CH_3CO \xrightarrow{k_3} CO + CH_3$ Rate $= k_3$ [CH3Co]

(d) $CH_3CH_3 \xrightarrow{k_4} C_2H_6$ Rate $= k_4\,[CH_3]^2$

In the above mechanism, carbon monoxide is formed only in step (c), hence the rate of formation of this substance must be

$$\frac{d\,[CO]}{dt} = k_3\,[CH_3CO] \qquad ...(27)$$

The CH_3CO radical is formed by step (b) and is used up in the step (c). Therefore, the net rate of its formation is given by

$$\frac{d\,[CH_3CO]}{dt} = k_2\,[CH_3CHO]\,[CH_3] - k_3\,[CH_3CO] \qquad ...(28)$$

Again, $[CH_3]$ radical is formed by steps (a) and (c) and is consumed in steps (b) and (d). Hence its net rate of formation is given by

$$\frac{d[CH_3]}{dt} = I_{abs} + k_3\ [CH_3CO] - k_2\ [CH_3CHO]\ [CH_3] - k_4\ [CH_3]^3 \quad ...(29)$$

Applying the steady-state treatment to $[CH_3CO]$ radical, we obtain

$$\frac{d[CH_3CO]}{dt} = k_2\ [CH_3CHO]\ [CH_3] - k_3\ [CH_3CO] = 0$$

or $k_2\ [CH_3CHO]\ [CH_3] - k_3\ [CH_3CO] = 0$ ...(30)

Similarly, the steady-state treatment on applying to $[CH_3]$ radical yields

$$\frac{d[CH_3]}{dt} = I_{abs} - k_2\ [CH_3CHO]\ [CH_3] + k_3\ [CH_3CO] - k_4\ [CH_3]^2 = 0$$

$$I_{abs} - k_2\ [CH_3CHO]\ [CH_3] + k_3\ [CH_3CO] - k_4\ [CH_3]^2 = 0 \quad ...(31)$$

By adding equations (30) and (31), we obtain

$$I_{abs} - k_4\ [CH_3]^2 = 0$$

or $$k_4\ [CH_3]^2 = I_{abs}$$

or $$[CH_3] = \left[\frac{I_{abs}}{k_4}\right]^{1/2}$$

On substituting the value of $[CH_3]$ in equation (31) , we get

$$k_2\ [CH_3CHO]\left(\frac{I_{abs}}{k_4}\right)^{1/2} - k_3\ [CH_3CO] = 0$$

or $$k_3\ [CH_3CO] = k_2\ [CH_3CHO]\left(\frac{I_{abs}}{k_4}\right)^{1/2} \quad ...(32)$$

Substituting this value of $k_3\ [CH_3CO]$ in equation (32), we obtain

$$\frac{d[CO]}{dt} = k_2\ [CH_3CHO]\left(\frac{I_{abs}}{k_4}\right)^{1/2}$$

$$= \frac{k_2}{(k_4)^{1/2}}\ [CH_3CHO]\ I_{abs})^{1/2}$$

$$= I_{abs}^{\ 1/2}\ [CH_3CHO] \quad ...(33)$$

where $$k = \frac{k_2}{(k_4)^{1/2}}$$

Quantum Yield : If the rate of formation of carbon monoxide is divided by the intensity of absorbed radiation, the value of quantum yield is obtained. Thus,

$$\text{Quantum yield} = \frac{d[CO]/dt}{I_{abs}}$$

$$= \frac{k\, I_{abs}^{1/2}\, [CH_3CHO]}{I_{abs}} \qquad \text{[From eq. (33)]}$$

$$= \frac{k\, [CH_3CHO]}{I_{abs}} \qquad ...(34)$$

General Discussion

1. From equation (34), it follows that the quantum yield of carbon monoxide formation increases as the intensity of light is decreased.
2. The activation energy of chain carrying process (b) has an appreciable energy, *e.g.*, about 10 K cal.

(b) $CH_3CHO + CH_3 \rightarrow CH_4 + CH_3CO$

This high activation energy accounts for the fact that quantum yield decreases as the temperature is lowered.

3. The effect of pressure on quantum yield may be explained by assuming that primary process

(a) $CH_3CHO + h\nu \rightarrow CH_3 + CHO$

is the main one at 3130 Å. At this wavelength, the CHO radical does not possess enough energy to decompose spontaneously. At high pressure, the CHO radicals may combine to form glyoxal. At low pressure, the CHO radicals may diffuse to the walls to form carbon monoxide.

PHOTOCHEMICAL REACTIONS

Photochemistry is the branch of chemistry which is mainly concerned with rates and mechanisms of reactions resulting from the exposure of reactants to light radiations.

This board definition of photochemistry would include reactions produced by all radiations of wave lengths ranging from those of radio waves to chose γ rays. But for practical purposes, the light radiations of the visible and ultraviolet regions lying between 2000 to 8000 Å are mainly concerned in bringing about such reactions which are termed as *Photochemical Reactions*. Therefore, the definition of photochemistry may be summarised as follows.

"It is the study of chemical effects produced by light radiations ranging from 2000 to 8000 Å wave-length"

The study of photochemistry helps us to know the changes when a molecule absorbs radiations.

TYPES OF CHEMICAL REACTIONS

A chemical reactions is one in which, *the identity of molecules is changed due to the rupture and formation of chemical bonds.* Chemical reactions are of two types.

Dark of Thermal Reactions : *These are the ordinary chemical reactions which are influenced or induced by temperature, concentration of reactants, presence of a catalyst, etc., except light radiations.*

(i) $N_2 + 3H_2 \rightleftharpoons 2NH_3$

(ii) $H_2 + I_2 \rightleftharpoons 2HI$

(iii) $PCl_5 \rightleftharpoons PCl_3 + Cl_2$

(b) Photochemical Reactions : *A photochemical reaction may be defined as any reaction which is induced or influenced by the action of light on the system.*

Some of the photochemical reactions are accompanied by increase in free energy unlike dark reactions. Examples of some photochemical reactions are.

(a) *Dissociation.*

$$2HBr \rightarrow H_2 + Br_2$$

(b) *Rearrangement.* Fumaric acid → Maleic acid

(c) *Addition Reaction.*

$$Br_2 + (C_6H_5)_2C = C(C_6H_5) \rightarrow (C_6H_5)_2\ CBr - CBr(C_6H_5)_2$$

(d) *Polymerisation.*

$$nCH_2 \rightarrow (C_2H_5)Nx$$

(e) Photo-catalytic *Reaction*

$$CO_2 + H_2O + \text{chlorophyll} \xrightarrow{\text{light}} \frac{1}{n}(H_2CO)_n + O_2 + \text{Chlorophyll}$$

(f) *Combination.*

$$H_2 + Cl_2 \rightarrow 2HCl$$

(g) *Decomposition.*

$$2O_3 \rightleftharpoons 3O_2$$

(h) *Double decomposition.*

$$C_6H_{12} + Br_2 \rightarrow C_6H_{11}Br + HBr$$

DIFFERENCES BETWEEN DARK AND PHOTOCHEMICAL REACTIONS

Photochemical reactions differ from ordinary dark or thermal reactions in certain aspects which are as follows.

(i) In ordinary thermal reactions the energy of activation is provided by the collisions. On the other hand in photochemical reactions the energy required is gained through the absorption of quanta of visible or ultraviolet light.

(ii) All ordinary chemical reactions are always accompanied by a decrease in free energy. On the other hand, the free energy of the light or photo-chemical reactions increases as some of the light energy is convered into free chemical energy of the products. Some examples of such photochemical reactions are ozonisation of oxygen, polymerisation of anthracene, and photosynthesis occurring in plants.

In these photochemical reactions, the energy of absorbed radiation gets transformed into free chemical energy of the products. However, when the source of radiations is removed, the system tends to return to its original state, though at a very slow rate.

A feature of photochemical activation is its selectivity. The absorbed quanta of light excite and thus activate a separate atom or group of atoms in a given molecule. This is a great advantage of activating molecules with light in comparison with thermal activation.

(iii) The difference between photo excitation and thermal excitation is that while in the former a few molecules succeeding in absorbing photons are strongly excited, in the case of latter a significant rise in temperature increases the average energy of all the molecules by a small amount. Thus, light absorption can bring about reactions at room temperature which might otherwise require a raising of temperature by several tens or hundreds of degrees centigrade.

While the rate of thermal reactions depends upon the temperature, the photochemical reaction rate is independent of temperature. But the reaction rate does depend upon the intensity of the radiation used. In certain photochemical reactions, the rate is seen to vary with the temperature, but this is due to the temperature dependence of thermal reactions which follow the light absorption step.

PHOTOSENSITISATION

Certain reactions are known which are not sensitive to light. These reactions can be made sensitive by adding a small amount of foreign material which can absorb light and stimulate the reaction without itself taking part in the reaction. Such an added material is known as photo-senistiser and the phenomenon as photosensitisation.

Formerly, it was thought that photosensitisation might be akin to catalysis. It has now been confirmed that photosensitisation is different in nature from ordinary catalysis. Photosensitized reactions are spontaneous involving an increase in free energy of the system.

Role Played by a Photosensitiser : The formation of a photosensitiser is to absorb light, become excited and then pass on this energy to one of the reactants and thereby activate them for reaction, without itself taking part in the reaction. Thus, a photosensitiser acts as a carrier of energy. Some examples are.

REACTIONS SENSITISED BY MERCURY ATOMS

(i) When a mixture of mercury vapour and hydrogen gas is illuminated by light of wavelength 2537 Å, the dissociation of molecular hydrogen into atomic hydrogen takes place.

$$Hg + H_2 \rightarrow 2H + Hg$$

The *mechanism* which is now suggested is the excitation of mercury atom and then to transfer the energy to hydrogen molecule which will dissociated to form atoms.

$$Hg + hv \rightarrow Hg^*$$

$$Hg^* + H_2^* \rightarrow H_2^* + Hg$$
$$H_2^* \rightarrow 2H$$

(ii) When the reaction is carried out between hydrogen and oxygen in the presence of mercury vapour under the influence of light radiations, the reaction leads to the formation of H_2O_2 and H_2O.

$$3H_2 + 2O_2 \xrightarrow[\text{Light}]{\text{Hg}} 2H_2O + H_2O_2$$

The accepted *mechanism* of the above reaction is that the first stage is the formation of hydrogen atoms by collision between excited mercury atom and hydrogen molecule.

$$Hg + h\nu \rightarrow Hg^*$$
$$Hg^* + H_2 \rightarrow 2H + Hg$$

The above reaction may be followed by reactions such as

(1) $H + O_2 + X \rightarrow HO_2 + X$

(2) $HO_2 + HO_2 \rightarrow H_2O + O_2$

(3) $HO_2 + H_2 \rightarrow H_2O_2 + H$

Now H_2O_2 may either be isolated as such or further decomposed to form H_2O and O_2

$$H_2O_2 \rightarrow H_2O + \frac{1}{2}O_2$$

(iii) The reaction between hydrogen and carbon monoxide is photosensitised by mercury atoms. The main products are formaldehyde and glyoxal in similar amounts. The quantum yield for this reaction in nearly 2. The accepted *mechanism* is that the primary process is the formation of hydrogen atom which further carries out chain reactions as given below.

$H + CO + X \rightarrow HCO + X$ (where X is the foreign body)

Either $2HCO \rightarrow HCHO + CO$ or $2HCO \rightarrow (CHO)_2$

(iv) Mercury photosensitises the decomposition of ammonia in the presence of light of wavelength 2537 Å. Quantum efficiency of this reaction is 7 and the rate of this photosensitised reaction is, about 200 times as great as the photolysis of ammonia (in the absence of mercury vapour), if both the reactions are carried out at the same wavelength.

The possible mechanism is

$$NH_3 + Hg^* \rightarrow NH_3^* + Hg$$
$$NH_3^* \rightarrow NH_2 + H$$
$$H + H \rightarrow H_2$$
$$NH_2 + NH_2 \rightarrow N_2H_4$$
$$N_2H_4 + H \rightarrow NH_3 + NH_2$$
$$NH_2 + NH_3 \rightarrow N_2 + 2H_2$$

Other example which are photosensitised by mercury atoms are. (a) Decomposition of phosphine, (b) Ozonolysis of oxygen, (c) Decomposition of acetone, (d) Decomposition of water vapour, (e) Decomposition of ethyl alcohol.

(b) Chlorine as A photosensitiser : In many reactions, chlorine acts as a photosensitiser. Let us consider the decomposition of ozone in the presence of chlorine (as a photosensitiser) under the influence of ultra-violet light.

$$2O_3 \xrightarrow[hv]{Cl_2} 2O_2 .$$

The rate of above reaction is independent of concentration, but is proportional to the intensity of absorbed light. Many mechanisms have been proposed from time to time. The most satisfactory mechanism is given below.

1. $Cl_2 + hv \rightarrow 2Cl$
2. $Cl + O_3 + X \rightarrow ClO_3^* + X.$

   The excited ClO3* radical may be absorbed on the walls of the containing vessel to form a mixture of Cl_2O_6, Cl_2 and O_2.
3. $ClO_3^* + ClO_3^* + M \rightarrow Cl_2 + 3O_2 + M.$

   In addition to the above reactions, $ClO_3^* + M \rightarrow Cl_2 + 3O_2 + M$

   In addition to the above reactions, ClO_3^* in the gas phase may react to form ClO_3 and O_2.
4. $ClO_3^* + O_2 \rightarrow ClO_2 + 2O_2$
5. $ClO_2 + O_3 \rightarrow ClO_3 + O_2$

(b) *Bromine as a photosensitiser.* Bromine acts as photosensitiser in the conversion of maleic acid into fumaric acid.

```
H—C—COOH                      H—C—COOH
  ||          hv                ||
  ||       ———————>             ||
H—C—COOH      Br2          HOOC—C—H
Maleic acid                Fumaric acid
```

(c) *Cadmium vapour as a photosensitiser.* Cadmium vapour acts as a photosensitiser for the polymerisation of ethylene, for the decomposition of ethane and propane to hydrogen, methane and higher hydrocarbons.

$$nC_2H_4 \xrightarrow[hv]{Cd} (C_2H_4)_n$$

Photosensitisation in Solid Phases : When silver halides in photographic plates are exposed to red (7500 Å) and yellow (5900 Å) light, they are not appreciably affected. But addition of certain photo-sensitisers like red dye to the emulsion makes the photographic plate to respond not only to red but even to far infrared radiations.

$$AgBr + hv \xrightarrow{\text{Red dye}} Ag + Br$$

Photosensitisation in Solution : When the decomposition of oxalic acid is carried out by light of shorter wavelength in presence of uranyl ions, the quantum yield for this reactions is 0.5 or more. The mechanism of this reaction is not clear.

$$UO_2^{2+}\ hv \rightarrow [UO_2{}^{+2}]$$

$$[UO^{2+}]^{2*} + \begin{matrix} COOH \\ | \\ COOH \end{matrix} \longrightarrow CO_2 + CO + H_2O + UO_2^{+2}$$

Uranyl ions also act as a photosenitizer in the photolysis of formic acid.

Chlorophyll as a photosensitiser : Chlorophyll acts as a photosenitiser in the photosynthesis of carbohydrates from CO_2 and H_2O.

$$\text{Chlorophyll} + hv \rightarrow [\text{Chlorophyll}]^*$$

$$6CO_2 + 6H_2O + [\text{Chlorophyll}]^* \xrightarrow[\text{sunlight}]{} C_6H_{12}O_6 + 6O_2 + [\text{chlorophyll}]$$

ABSORPTION OF LIGHT

Introduction : *When light* (monochromatic or heterogeneous) *is incident upon a homogeneous medium, a part of the incident light is reflected, a part is absorbed by the medium and the remainder is allowed to transmit as such.* If I_0 denotes the incident light, I, the reflected light, I_r the absorbed light and I_a, the transmitted light, then one can write

$$I_0 = I_a + I_t + I_r \qquad ...(1)$$

It a comparison cell is used, the value of I_r which is very small (about 4 per cent), can be *eliminated for air-glass interfaces.* Under the condition, equation (1) becomes as

$$I_0 = I_a + I_t \qquad ...(2)$$

Bouguer actually investigated the range of absorption of light with the thickness of medium. But the credit was enjoyed by *Lambert* who simply extended the concepts developed by *Bouguer*. Beer later applied Lambert's concept to solution of different concentrations and reported his results just prior to those of *Bernard*. However, the two separate laws governing, absorption are generally known as *Lambert's* Law and *Beer's Law*. We will now discuss these one by one.

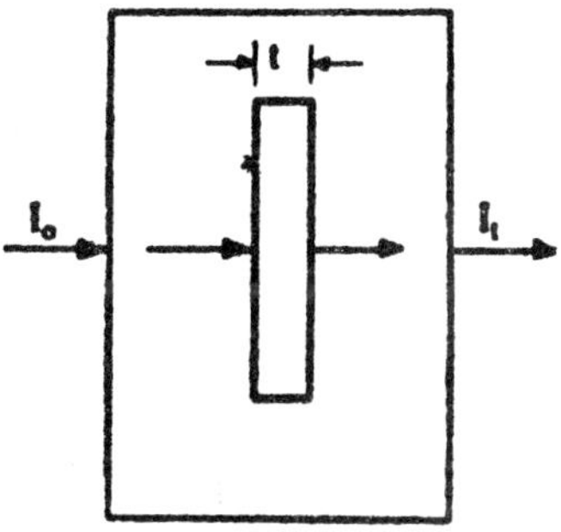

Fig. 17

Lambert's Law : This law can be stated as follows. *"When a beam of light is allowed to pass through a transparent medium, the rate of decrease of intensity with the thickness of medium is directly proportional to the intensity of the light."*

Mathematically the Lambert's law may be stated as follows.

$$-\frac{dI}{dt} \propto I \text{ or } -\frac{dI}{dt} = kI \qquad ...(3)$$

where I denotes the intensity of incident light of wavelength λ, t denotes the thickness of the medium and k denotes the proportionality factor. On integrating equation (3) and putting $I = I_0$, when $t = 0$, we get

$$\ln \frac{I_0}{I_t} = kt \text{ or } I_t = I_0 \, e^{-kt} \qquad ...(4)$$

where I_0 denotes the intensity of the incident light, I_t denotes the intensity of the transmitted light and k is a constant which depends upon the wavelength and absorbing medium used. On changing equation (4) from natural to common logarithms, we get

$$\ln \frac{I_0}{I_t} = kt \text{ or } I_t = I_0 \, e^{-Kt} \qquad ...(5)$$

where $$K = k/2.3026 \qquad ...(6)$$

In equation (6) K is the *absorption coefficient* which is defined as.

"*It is the reciprocal of the thickness which is required to reduce the light to 1/10 of its intensity.*"

The above definition follows from equation (5),

$$It/I_0 = 0.1 = 10^{-Kt}$$

or $$Kt = 1 \quad \text{or} \quad K = K \propto \frac{1}{t}$$

The ratio I_t/I_0 is termed as the *transmittance,* T and the ratio log I_0/I_t is termed as the *absorbance* A, of the medium. Formerly absorbance was termed as *optical density* D or *extinction coefficient* ≤. The ratio $I_0/I_{t,}$ is termed as *opacity,*

$$A = \log \frac{I_0}{I_t} \qquad ...(7)$$

Lambert's law is very rigid and like Faraday's laws of electrolysis, it has no exception.

Beer's Law : Lambert's law shows that there exists a logarithmic relationship between the transmittance and the length of the optical path through the sample. Beer observed that a similar relationship holds between transmittance and the concentration of a solution, *i.e., the intensity of a beam of monochromatic light decreases exponentially with the increase in concentration of the absorbing substance arithmetically.*

Thus, equation (4) becomes as

$$I_0 = I_0\, e^{-k'0} \qquad ...(8)$$

$$= I_0 .\, 10^{-0.4348k'c} = I_0\, 10^{-K'2} \qquad ...(9)$$

where k′ and K′ are constants and *c* is the concentration of the absorbing substance. On combining equations (5) and (9), we get

$$I_t = I_0 . 10^{-act}$$

or $$\log (I_0/I_t) = act \qquad ..(10)$$

where a is the new constant.

Equation (10) is termed as mathematical statement of *Beer-Lambert law*. This is also the fundamental equation of colorimetry and spectrophotometry..

In equation (10), the value of a depends upon the units of concentration. If c is expressed in mole dm^{-3} and *t* in centimeters, then *a* is replaced by the symbol ε and is termed as the *molar absorption coefficient* or *molar absorptivity* (formerly the molar extinction coefficient). It is important to remark here that there exists a relationship between the absorbance A, the transmittance T and the molar absorption coefficient ε, *i.e.*,

$$A = \varepsilon\, ct = \log \frac{I_0}{I_t} = \log \frac{1}{T} = -\log T \qquad ...(11)$$

In spectrophotometers, the scales are calibrated to read directly absorbances. In colorimeters, I_0 is considered to be the light transmitted by the pure solvent whereas I_t is considered to be the light transmitted by the solution.

Equation (11) may be put as

$$\varepsilon = A/ct \qquad ...(12)$$

If c = 1 mole dm^{-3} and T = 1 cm, equation (12) becomes as

$$\varepsilon = A \qquad ...(13)$$

From equation (13) it follows that the molar absorption coefficient is the specific absorption coefficient for a concentration of 1 mole dm^{-3} and a path length of 1 cm. When a system contains several absorbing substances, each of these contributes to the rate of absorption of light.

$$A = \varepsilon_1 c_1 t + \varepsilon_2 c_2 t + ... = \sum_i \varepsilon_t \; c_i \, t$$

Nature of Molar Absorptivity and Absorbance : Absorbance is an *extensive property* of a substance whereas absorptivity is its *intensive property.*

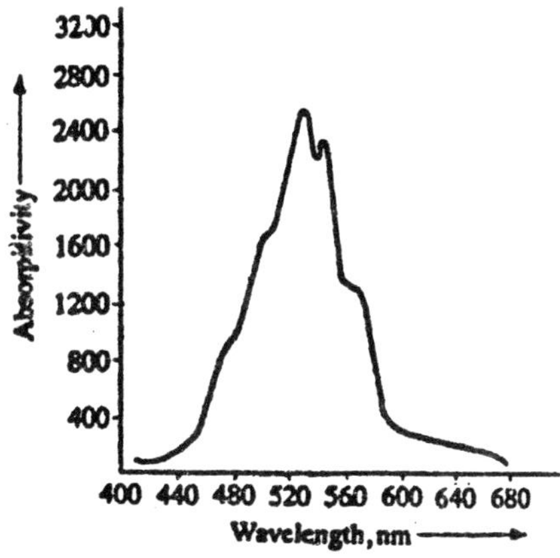

Fig. 18 : Molar absorption at different wave lengths.

If there is a change in concentration and in the thickness of the container, the value of molar absorptivity will remain constant within the Beer's law range but the absorbance will change significantly. The value of molar absorptivity will be different at different wavelengths. It is alto evident why a monochromatic radiation is used in spectrophotometry and colorimetry. Similarly absorbance also varies with wavelength.

SOLVED EXAMPLES

Example 1:

The ionisation potential of an atom is 14.2 eV. Calculate the series limit in its absorption spectrum. Dates as in last problem.

Solution:

The series limit in the absorption spectrum of an atom corresponds to the energy which when absorbed by a ground-state atom, ionises it. Thus, if V be the ionisation potential of an atom, the wavelength at the series limit is given by

$$\lambda = \frac{hc}{eV}$$

$$= \frac{(6.62 \times 10^{-34}\ \text{J s}) \times (3 \times 10^{8}\ \text{m s}^{-1})}{(14.2 \times 1.6 \times 10^{-19}\ \text{J})}$$

$$= 0.874 \times 10^{-7}\ \text{m}$$

$$= 874\ \text{Å}$$

Example 2:

With Franck-Hertz type of experiment on sodium, the first spectral line to appear is the D-line, $\lambda = 5.89 \times 10^{-7}$ m. What is the first excitation potential of sodium ? Given : h = 6.63×10^{-34} J, s, c 3×108 m/s and 1 eV = 1.6×10^{-19} J.

Solution:

Let V volt be the excitation potential. Then the (excitation) energy imparted to the electron will be eV joule, where e coulomb is the charge on the electron. This energy is re-emitted as photon (radiation) when the electron returns to the normal state. If v be frequency of the emitted radiation, the photon energy will be hv. Thus

$$eV = hv$$

or $$V = \frac{hv}{e}.$$

If l be the wavelength of the emitted radiation, then $v = c/\lambda$.

$$\therefore \quad V = \frac{hv}{e\lambda}$$

$$= \frac{(6.63\times10^{-34}\,\text{J s})(3\times10^{8}\,\text{m s}^{-1})}{(1.6\times10^{-19}\,\text{C})(5.89\times10^{-7}\,\text{m})}$$

$$= 2.1\ (\text{J/C}) = 2.1\ \text{V}.$$

Example 3:

The first two excitation potentials of atomic hydrogen in Franck-Hertz experiment are 10.2 and 12.09 volts. Draw an energy level diagram and show all possible transitions for emission and absorption along their wavelengths.

($h = 6.63 \times 10^{-34}$ J s, $c = 3.0 \times 108$ m/s, 1 eV $= 1.6 \times 10^{-19}$ J.)

Solution:

The energy level diagram, and the *three* possible transitions (a), (b) and (c) for emission are shown in the figure.

The frequency of radiation resulting from the transitions (a) is given by

$$v_a = \frac{E_2 - E_1}{h},$$

and the corresponding wavelength is

$$\lambda_a = \frac{hc}{E_2 - E_1}. \qquad [\because c = v\lambda]$$

Now, from the Fig., $E_2 - E_1 = 10.2$ eV $= 10.2 \times (1.6 \times 10^{-19})$ J. Therefore

$$\lambda_a = \frac{(6.63\times10^{-34}\,\text{J s})\times(3.0\times10^{8}\,\text{m s}^{-1})}{(10.2\times1.6\times10^{-19}\,\text{J})}$$

$$= 1.216 \times 10^{-7}\ \text{m} = 1216 \times 10^{-10}\ \text{m} = 1216\ \text{Å}.$$

Similarly, $$\lambda_b = \frac{hc}{(E_3 - E_1)}$$

$$= \frac{(6.63 \times 10^{-34}) \times (3.0 \times 10^{8})}{(12.09 \times 1.6 \times 10^{-19})}$$

$$= 1.026 \times 10^{-7} \text{ m} = 1026 \times 10^{-10} \text{ m} = 1026 \text{ Å}.$$

Also, $\lambda_c = \frac{hc}{(E_3 - E_2)}$

$$= \frac{hc}{(E_3 - E_1) - (E_2 - E_1)}$$

$$= \frac{hc}{(12.09 - 10.2)\,\text{eV}} = \frac{hc}{1.89\,\text{eV}}$$

$$= \frac{(6.63 \times 10^{-34}) \times (3.0 \times 10^{8})}{(1.89 \times 1.6 \times 10^{-19})}$$

$$= 6.567 \times 10^{-7} \text{ m} = 6567 \times 10^{-10} \text{ m} = 6567 \text{ Å}.$$

In absorption, only the transitions starting from n = 1 shall be observed which correspond to 1216 Å and 1026 Å.

Example 4:

(a) Compare the assumptions made by Planck in discussing cavity radiation, those by Einstein in connection with the photoelectric effect and those by Bohr in connection with the hydrogen spectrum.

(b) Explain the implication of the fact that- the energies of the hydrogen atom orbits are negative.

(c) Can a hydrogen atom absorb a photon of energy greater than the binding energy of the atom ?

Solution:

(a) *Assumptions of Planck, Einstein and Bohr :* Planck had assumed that the atoms of a hot body behave as oscillators and have discrete (quantised) energies. They do not emit radiant energy continuously, but only in 'jumps' or 'quanta'. Planck, however, still maintained that radiation propagates continuously through space as electromagnetic waves.

Einstein, in order to explain the photoelectric effect, went a step ahead. He proposed that the radiation not only is emitted as quantum at a time but also propagates as individual quanta (photons). He thus treated the propagation of radiation as particle propagation rather than wave propagation.

Bohr's in order to explain the hydrogen spectrum, adopted the Planck's quantum hypothesis that the radiation is emitted discontinuously from the atom. He started with the quantisation of the angular momentum of the electron in the orbit which ultimately results in the quantisation of energy of the atom.

(b) *Negative Energy of Hydrogen Orbits :* An electron revolving in an hydrogen orbit has a negative potential energy by virtue of its attraction towards, the nucleus, and also kinetic energy (which is positive) by virtue of its motion. The potential energy is greater in magnitude .than the kinetic energy so that the net energy is negative. The negative energy signifies that the electron cannot escape from The atom. A positive energy for a nucleus electron combination would mean that electron is not bound to the nucleus. Such a combination cannot constitute an atom.

Example 5:

A μ^- meson (charge – e, mass = 207 m, where m is mass of electron) can be captured by a proton to form a hydrogen- like "mesic" atom. Calculate the radius of the first Bohr orbit, the binding energy, and the wavelength of the first line in the Layman series for such an atom. The mass of the proton is 1836 times the mass of electron. The radius of first Bohr orbit and the binding energy of hydrogen are 0.53 Å and 13.6 eV respectively. $R_{\yen} = 109737\ cm^{-1}$.

Solution:

The reduced mass of the system is

$$m = \frac{(207\,m)(1836\,m)}{207\,m + 1836\,m} = 186\ m.$$

From Bohr theory, the radius of the first orbit (n = 1) of a hydrogen -like atom for Z = 1, is given by (taking finite mass of nucleus in consideration).

$$r_1 = 4\pi\varepsilon_0 \frac{h^2}{4\pi^2\mu e^2} = 4\ \pi\varepsilon_0 \frac{h^2}{4\pi^2(186m)e^2}$$

$$= \frac{1}{186}\left(4\pi\varepsilon_0 \frac{h^2}{4\pi^2 me^2}\right).$$

The quantity in the bracket is the first Bohr orbit of hydrogen atom which is 0.53 Å. Therefore

$$r_1 = \frac{1}{186} \times 0.54\ \text{Å} = 2.85 \times 10^{-3}\ \text{Å}.$$

Again, from Bohr's theory, the ground-state energy for a hydrogen-like atom with Z = 1 is given by

$$E_1 = -\frac{\mu e^4}{8\varepsilon_0^2 h^2} = -186\ \frac{me^4}{8\varepsilon_0^2 h^2}$$

$$= -186 \times 13.6 = -2530\ \text{eV}.$$

Hence the binding energy is 2530 eV.

The wavelength of the Lyman lines are given by

$$\frac{1}{\lambda} = R_M\left(\frac{1}{1^2} - \frac{1}{n^2}\right),\ n = 2, 3, 4,...$$

where R_M is the Rydberg constant for the music atom. For the first line, n = 2 so that

$$\lambda = \frac{4}{3R_M}.$$

Now, $R_M = \dfrac{\mu e^4}{8\varepsilon_0^2 ch^3}$ and $R_¥ = \dfrac{me^4}{8\varepsilon_0^2 ch^3}$, so that

$$R_M = \frac{\mu}{m} R_¥ = 186\ R_¥ = 186 \times 109737\ \text{cm}^{-1}.$$

Here $\lambda = \dfrac{4}{3R_M} = \dfrac{4}{3 \times 186 \times 109737\ \text{cm}^{-1}}$

Example 6:

A positronium atom is a system consisting of a position and an electron. Calculate the reduced mass, the Rydberg constant and the wavelength of the first Balmer line for positronium. (Give m = 9.1 × 10^{-31} kg, R_H = 1.09737 × 10^{-3} Å^{-1} and H_α = 6563 Å).

Solution:

The positron has the same mass m as the electron and has equal but positive charge. The reduced mass of the electron-positron atom is therefore

$$m = \frac{(m)(m)}{m+m} = \frac{1}{2}\ m = 4.55 \times 10^{-31}\ \text{kg},$$

while the reduced mass of electron in hydrogen is very nearly m.

The Rydberg constant $\left(\dfrac{\mu e^4}{8\varepsilon_0^2 ch^3}\right)$ for positronium is therefore half that for hydrogen (with infinitely heavy nucleus). Thus

$$R_P = \frac{1}{2} R_H = 0.54868 \times 10^{-3} \text{ Å}^{-1}.$$

The wavelength of first Balmer line (H_α) for hydrogen atom is given

$$\frac{1}{\lambda_H} = R_H \left(\frac{1}{2^2} - \frac{1}{3^2}\right),$$

while that for positronium atom is $\dfrac{1}{\lambda_p} = R_P \left(\dfrac{1}{2^2} - \dfrac{1}{3^2}\right)$.

Thus $$\frac{\lambda_p}{\lambda_H} = \frac{R_H}{R_P} = 2.$$

$$\therefore \quad \lambda_P = 2\lambda_H = 2 \times 6563 = 13126 \text{ Å}.$$

Example 7:

Find the recoil speed of hydrogen atom after it emits a photon in going from n = 3 to n = 1 state. Electron mass is 9.11×10^{-31} kg and $h = 6.626 \times 10^{-34}$ J s.

Solution:

The energy of the hydrogen atom in the n th state is given by

$$E_n = -\frac{Rhc}{n^2},$$

where R is Rydberg constant. From this, we get

$$E_1 - E_3 = -\frac{Rhc}{1^2} + \frac{Rhc}{3^2} = -\frac{8}{9} R h c.$$

The energy of the emitted photon is

$$\Delta E = E_1 \sim E_3 = \frac{8}{9} R h c.$$

The momentum of the photon is

$$p = \frac{\Delta E}{c} = \frac{8}{9} R h.$$

By conservation of momentum, the recoil momentum of the hydrogen atom will be equal (and opposite) to the momentum of the emitted photon. The recoil speed of the atom is

$$v = \frac{\text{momentum}}{\text{mass}} = \frac{8}{9}\frac{Rh}{m_H}.$$

But m_H = 1836 m, where m is electron mass.

$$\therefore \quad v = \frac{8}{9}\frac{Rh}{(1836\,m)}.$$

Putting the given values of h, m and using R=1.097 × 10^7 m^{-1}, we get

$$v = \frac{8}{9}\frac{(1.097\times10^{7}\ m^{-1})(6.626\times10^{-34}\ Js)}{1836\,(9.11\times10^{-31}\ kg)}$$

$$= 3.86\ m/s.$$

Example 8:

The ionisation potential of hydrogen tom is 13.6 volt. Find the wavelength of the Lyman series limit.

Solution:

When an atom absorbs energy so that its two stationary states becomes existed simultaneously, two superimposed sets of radiations are produced to give a 'beat' variation. If v¢ and v¢¢ represent the frequencies of two superimposed vibrations, the frequency v of the emitted beat is

$$v = v' - v''.$$

The wave mechanical vibrations are related to the energies of the corresponding state by the usual quantum expressions E′ = hv′ and E′ – E″ = hv″ so that

$$E' - E'' = h\,(v' - v'')$$

$$\Delta E = 13.61\left(\frac{1}{(1)^2} - \frac{1}{\infty}\right) = 13.61\ eV.$$

Therefore, ionisations potential = 13.61 eV.

Example 9:

A beam of monochromatic photons of energy 9 eV is incident on hydrogen gas all of whose atoms are in the ground state. It is found that the beam is fully transmitted without absorption. Why ? The ground state energy of an electron in the hydrogen atom is E_1 = – 13.6 eV.

Solution:

The minimum energy that can be absorbed by ground-state hydrogen atom is $E_1 \sim E_2$, which would excite it to the next state (n = 2). Now,

$$E_1 \sim E_2 = E_1 \sim \frac{E_1}{4} \qquad \left[\because E_n = \frac{E_1}{n^2}\right]$$

$$= \frac{3}{4} E_1$$

$$= \frac{3}{4} \times 13.6 = 10.2 \text{ eV.}$$

Hence photons of energy 9 eV cannot be absorbed by hydrogen atoms.

Example 10:

Calculate the ionisation potential of hydrogen from the following date.

Solution:

According to Bohr's theory, the energy of an electron in the nth orbit of hydrogen atom is

$$E = -\frac{2\pi^2 e^4 m}{n^2 h^2}$$

When the atom is in the normal state, the only electron in the hydrogen atom in the first orbit, *i.e.*, n = 1.

$$\therefore \quad E = -\frac{2\pi^2 me^4}{h^2}$$

or $$E = \frac{2 \times (3.14)^2 \times 9 \times 10^{-28} \times (4.8 \times 10)^4}{(6.6 \times 20^{27})^2} = 2.165 \times 10^{-11} \text{ eg.}$$

To remove the electron from first orbit to infinity 2.165×10^{-11} erg of energy must be supplied. The amount of energy is called the ionisation potential of hydrogen atom.

Ionisation potential of hydrogen atom

$$= 2.165 \times 10^{-11} \text{ erg}$$

$$= \frac{2.165 \times 10^{-11}}{1.6 \times 10^{-12}} \text{ electron volts} = 13.53 \text{ eV.}$$

Example 11:

Energy in a Bohr's orbit is given to be equal to – B/n² *where* B = 2.179 × 10^{-11} *erg. Calculate the frequency of radiation and also the wave number when the electron jumps from the third orbit to the second* (h = 6.62 × 10^{-27} *erg sec) orbit.*

Solution:

Energy of a Bohr's orbit is given by

$$E = \frac{B}{n^2} = \frac{2.179 \times 10^{-11}}{n^2} \text{ erg.}$$

For the third orbit, n = 3

$$E_2 = \frac{2.179 \times 10^{-11}}{9} = -\ 0.2421 \times 10^{-11} \text{ erg.}$$

For the second orbit, n = 2

$$E_2 = \frac{2.179 \times 10^{-11}}{4} = -\ 0.54475 \times 10^{-11} \text{ erg.}$$

$$E_3 - E_2 = (0.54475 \times 10^{-11}) - (0.2421 \times 10^{-11})$$

$$= 0.30265 \times 10^{-11} \text{ erg.}$$

When electron jumps from the third to the second orbit, a photon is emitted, the energy of which is given by

$$hv = E_3 - E_2 \text{ where v is the frequency}$$

$$v = \frac{E_3 - E_2}{h} = \frac{0.30265 \times 10^{-11}}{6.62 \times 10^{-27}} = 4.572 \times 10^{14} \text{ sec}^{-1}.$$

The corresponding wave number is given by

$$\bar{v} = \frac{1}{\lambda} = \frac{v}{c} = \frac{4.572 \times 10^{14}}{3 \times 10^{-10}} = 15240 \text{ cm}^{-1}$$

Example 12:

Give using spectral notation for the following states of the atom.

(i) n = 4, L = 2, S = 0

(ii) n = 4, L = 1, S = 1, J = 0 and

(iii) n = 3, L = 2, multiplicity 2.

Solution:

(i) Multiplicity (2S + 1) = 1, J = L + S = 2

$\therefore$ State will be 4 $^1D_2$.

(ii) Multiplicity (2S + 1) = 3 $\therefore$ State will be 4 $^3P_0$.

(iii) Multiplicity will be (2S + 1) = 2 or S = 1/2

As L = 2, J = 5/2 or 3/2.

The two states which are positive are 3 $^2D_{5/2}$ or $^2D_{3/2}$. *Give using spectral notation for the following states of the atom.*

(i) n = 4, L = 2, S = 0

(ii) n = 4, L = 1, S = 1, J = 0 and

(iii) n = 3, L = 2, multiplicity 2.

Solution:

(i) Multiplicity (2S + 1) = 1, J = L + S = 2

$\therefore$ State will be 4 $^1D_2$.

(ii) Multiplicity (2S + 1) = 3 $\therefore$ State will be 4 $^3P_0$.

(iii) Multiplicity will be (2S + 1) = 2 or S = 1/2

As L = 2, J = 5/2 or 3/2.

The two states which are positive are 3 $^2D_{5/2}$ or $^2D_{3/2}$.

Example 13:

Calculate the energy in calories per mole or per Einstein for radiations of wavelength 100 Å..

Solution:

Energy per quantum = hv

But h = 6.62 × 10^{-27} erg-sec.

$$\therefore \quad V = \frac{c}{\lambda} = \frac{3 \times 10^{10}}{1000\,\text{Å}} = \frac{3 \times 10^{10}\,\text{cm.sec}^{-1}}{1000 \times 10^{-8}\,\text{cm}} = 3 \times 10^{15}\ \text{sec}^{-1}$$

Energy per quantum = hv

$$= (6.62 \times 10^{-27}\ \text{erg. sec})\ (3 \times 10^{15}\ \text{sec}^{-1})$$

$= 19.86 \times 10^{-12}$ ergs.

But $N = 6.02 \times 10^{28}$ molecules/mole

and $h\nu = 19.86 \times 10^{-12}$ erg.

Therefore, the energy per mole or per Einstein

$= N h\nu$

$= (6.02 \times 10^{23} \text{ molecules mole}^{-1})\ 19.86 \times 10^{-12}$ ergs/mole

$= 11.94 \times 10^{12}$ ergs/mole

$= \frac{11.94 \times 10^{12}}{10^7}$ Jule/mole [$\because$ 1 Joule = 10^7 ergs]

$= \frac{11.94 \times 10^{12}}{4.184 \times 10^7}$ cal/mole [$\because$ 1 cal = 4.184 Joules]

$= \frac{2.8590 \times 10^5}{10^3}$ K cal/mole [$\because$ 1 K cal = 10^3 cal]

$= 285.90$ K cal/mole.

$= \frac{285.90}{23.06}$ electron volt [$\because$ 1 eV=23.06 K cal/mole]

$= 12.390$ electron-volts.

Example 14:

$$B \rightarrow C$$

1.0×10^{-5} mole of B was formed on absorption of 6.62×10^7 ergs at 3600 Å. Calculate the quantum yield or efficiency.

Solution:

No. of moles reacting

$= 1.0 \times 10^{-5} \times 6.02 \times 10^{23}$ molecules

$= 6.02 \times 10^{18}$ molecules

No. of quanta absorbed

$$= \frac{\text{Total energy absorbed}}{\text{Energy of one quantum}} = \frac{6.62 \times 10^7 \text{ ergs}}{h\nu}$$

$$= \frac{6.62 \times 10^7}{hc/\lambda}$$

$$= \frac{6.62 \times 10^{7} \times \lambda}{hc} \qquad \left[\because v = \frac{c}{\lambda}\right]$$

But $\quad l = 3600$ Å $= 3600 \times 10^{-8}$ cm.

$c = 3 \times 10^{10}$ cm/sec, $h = 6.62 \times 10^{-27}$ erg/sec.

No. of quanta absorbed $= \dfrac{6.62 \times 10^{7} \times 3600 \times 10^{-8}}{6.62 \times 10^{-27} \times 3 \times 10^{10}} = 1.2 \times 10^{19}$

Therefore, equation yield $= \dfrac{\text{No. of molecules reacting}}{\text{NO. of quanta absorbed}}$

$$= \frac{6.02 \times 10^{18}}{1.2 \times 10^{19}} = 0.506.$$

Example 15:

The bond energy in a molecule is 142.95 cal/mole. What is the longest wave-length of light capable of dissociating this molecule ?

Solution:

Energy per mole = N hv

where N= 6.02×10^{23} molecules, $h = 6.62 \times 10^{-27}$ erg-sec

$$v = \text{frequency} = \frac{c}{\lambda} = \frac{3.0 \times 10^{10}}{\lambda}$$

$\therefore$ Energy per mole $= \dfrac{6.02 \times 10^{23} \times 6.62 \times 10^{-27} \times 3 \times 10^{10}}{\lambda}$ ...(A)

The energy per mole = 142.95 cal/mole

$= 142.95 \times 10^{3}$ cal/mole

$= 142.95 \times 10^{3} \times 4.184 \times 10^{7}$ ergs/mole ...(B)

Equating eqs. (A) and (B), we get

$$\frac{6.02 \times 10^{23} \times 6.62 \times 10^{-27} \times 3 \times 10^{10}}{142.95 \times 10^{3} \times 4.184 \times 10^{7}}$$

$= 142.95 \times 103 \times 4.184 \times 107$

or $\quad \lambda = \dfrac{6.02 \times 10^{23} \times 6.62 \times 10^{-27} \times 3 \times 10^{10}}{142.95 \times 10^{3} \times 4.184 \times 10^{7}}$ cm

$= 2000 \times 10^{-8}$ cm $= 2000$ Å.

Example 16:

The dissociation energy of hydrogen is 102900 cal/mole. If H_2 is dissociated by illumination with radiation of wave-length 2537 Å. What fraction of the radiant energy will be converted in to kinetic energy ?

Solution:

Dissociation energy = 102900 cal/mole

$$= \frac{102900}{6.023 \times 10^{23}} = 1.708 \times 10^{-19} \text{ cal/molecule.}$$

Also, one quantum of light is able to dissociate one molecule of hydrogen. Therefore, Energy of one quantum = hν = $\frac{hc}{\lambda}$.

But $h = 6.625 \times 10^{-27}$ erg sec,

$c = 3 \times 10^{10}$ cm/sec

$\lambda = 2537$ Å $= 2537 \times 10^{-8}$ cm

$$\therefore \text{ Energy of one quantum } = \frac{6.625 \times 10^{-27} \times 3 \times 10^{10}}{2537 \times 10^{-8}} \text{ ergs}$$

$$= \frac{3 \times 6.625 \times 10^{-9}}{2537} \text{ ergs}$$

$$= \frac{3 \times 6.625 \times 10^{-9}}{2537 \times 4.18 \times 10^{7}} \text{ cal}$$

[∵ 1 cal 4.18 × 10^7 ergs]

But the dissociation energy per mole = 1.708×10^{-19} cal

∴ Energy converted into kinetic energy

$$= (1.874 \times 10^{-19} - 1.708 \times 10^{-19}) \text{ cal}$$

$$= 0.166 \times 10^{-19} \text{ cal}$$

Hence, fraction converted into K.E. = $\frac{0.166 \times 10^{-19}}{1.874 \times 10^{-19}}$.

Example 17:

Calculate the time taken by the electron to traverse the first orbit in the hydrogen atom. Electron mass and charge are 9.1×10^{-31} kg and 1.6×10^{-19} C and $h = 6.63 \times 10^{-34}$ J-s.

Solution:

The radius of the n the Bohr orbit is

$r_n = 4\pi\varepsilon_0 \dfrac{n^2h^2}{4\pi^2 me^2}$ and the velocity of electron in the n th orbit is

$$v_n = \frac{1}{4\pi\varepsilon_0}\frac{2\pi e^2}{nh}.$$

The time taken by the electron to traverse the n the orbit is therefore

$$T_n = \frac{2\pi r_n}{v_n}$$

$$= 2\pi\,(4\pi\varepsilon_0)^2 \frac{n^2h^2}{4\pi^2 me^2}\frac{nh}{2\pi e^2}$$

$$= \frac{4\varepsilon_0^2 h^3 n^3}{me^4}.$$

For the first orbit, n = substituting the given values of h, m, e and ε_0 = 8.85 × 10^{-12} $C^2/N\text{-}m^2$, we get

$$T_1 = \frac{4.(8.85\times10^{-12})^2\,(6.63\times10^{-34})^3}{(9.1\times10^{-31})\,(1.6\times10^{-19})^4} = 1.5\times10^{-16}\text{ s.}$$

Example 18:

Find an expression for the radius of the electron orbit in the hydrogen atom in its n th state. What will be the approximate quantum number n for an electron in an orbit of radius 1 mm ? (Take required values from problem 1).

Solution:

The basic equation in the Bohr's theory of hydrogen atom are.

$$\frac{mv^2}{r} = \frac{1}{4\pi\varepsilon_0}\frac{e^2}{r^2} \quad ...(1)$$

and

$$mvr = \frac{nh}{2\pi}. \quad ...(2)$$

Squaring (2) and dividing by (1), we get

$$r = 4\pi\varepsilon_0 \frac{n^2h^2}{4\pi^2 me^2}.$$

Substituting the given values, we get

$$r = \frac{1}{9\times10^{9}\ \text{Nm}^2/\text{C}^2}\times$$

$$\frac{n^2\,(6.63\times10^{-34}\ \text{Js})^2}{4\times(3.14)^2\times(9.1\times10^{-31}\ \text{kg})\times(1.6\times10^{-19}\ \text{C})^2}$$

$$= 0.53 \times 10^{-10}\ n^2 \text{ meter}$$

$$= 0.53\ n^2\ \text{Å}.$$

For r = 1 mm = 10^7 Å, we have

$$10^7 = 0.53\ n^2.$$

$$\therefore \qquad n^2 = \frac{10^7}{0.53} = 18.87 \times 10^6 \quad \text{or n; } 4350$$

Example 19:

The average life-time of an electron in an excited state of hydrogen atom is about 10^{-8} s. How many revolutions does an electron in the n = 2 state make before dropping to the n=1 state? (R = 1.097 × 107 m^{-1}).

Solution:

Let v be the velocity of electron (mass m, charge e) in an orbit of radius r. This basic equations are.

$$\frac{m v^2}{r} = \frac{1}{4\pi\varepsilon_0}\frac{e^2}{r^2}$$

and

$$mvr = n\,\frac{h}{2\pi}.$$

These equations give

$$v = \frac{nh}{2\pi mr},\ r = \frac{n^2h^2\varepsilon_0}{\pi me^2}.$$

The number of revolutions of the electron in the orbit per second is

$$f = \frac{v}{2\pi r} = \frac{2Re}{n^3} \qquad \text{(as proved in Q. 8)}$$

For n = 2 state, we have

$$f = \frac{2\times(1.097\times10^7\ \text{m}^{-1})\times(3\times10^8\ \text{ms}^{-1})}{8}$$

$$= 8.2 \times 10^{14}\ \text{s}^{-1}.$$

Hence the number of revolutions of the electron in its life-time of 10^{-8} second is $(8.2 \times 10^{14}) \times 10^{-8} = 8.2 \times 10^{6}$.

Example 20:

An orange photon of wavelength 600 nm is emitted from an atom. Find the difference in energy in the two atomic states involved. Find the same result for the red 6563 Å line of hydrogen. ($h = 6.63 \times 10^{-34}$ *j s*, $c = 3 \times 10^{8}$ *m* s^{-1}, *1 eV* $= 1.6 \times 10^{-19}$ *J*).

Solution:

By Bohr's postulate, the emitted frequency is given by

$$v = \frac{\Delta E}{h},$$

where ΔE is the difference in energy. But $v = c/\lambda$. Therefore

$$\frac{c}{\lambda} = \frac{\Delta E}{h} \text{ or } \Delta E = \frac{hc}{\lambda}.$$

Here $\lambda = 600 \text{ nm} = 600 \times 10^{-9}$ m.

$$\therefore \quad \Delta E = \frac{(6.63 \times 10^{-34}\,\text{Js}) \times (3.0 \times 10^{8}\,\text{ms}^{-1})}{600 \times 10^{-9}\,\text{m}}$$

$$= 3.31 \times 10^{-19} \text{ J}$$

$$= \frac{3.31 \times 10^{-19}}{1.6 \times 10^{-19}} = 2.07 \text{ eV.}$$

For $\lambda = 6563$ Å $\times 10^{-10}$ m, we can show that

$$\Delta E = 3.03 \times 10^{-19} \text{ J} = 1.9 \text{ eV.}$$